LIMESTONE-BUILDING ALGAE AND ALGAL LIMESTONES

LIMESTONE-BUILDING ALGAE AND ALGAL LIMESTONES

by

J. HARLAN JOHNSON

Published through funds provided by the Colorado School of Mines Foundation Inc., by the Department of Publications: William R. Peters, Director of Publications, Hendrietta Jenson, Assistant.

Printed and bound by the Johnson Publishing Company
Boulder, Colorado

Engravings by the Colorado Engraving Company
Denver, Colorado

TABLE OF CONTENTS

LIST OF TABLES

PLATES

PREFACE

The present publication is an outgrowth of an earlier booklet published in 1954, entitled "An Introduction to the Study of Rock Building Algae and Algal Limestones." That booklet was prepared originally as a text to be used in connection with a graduate course having the same title, which was given by the author at the Colorado School of Mines. It also seemed to meet a need felt by the petroleum geologists for an introductory book on fossil algae, and the demand for the publication gradually grew until it went out of print during the summer of 1959. During the interval since 1954, the author was retired from teaching.

This book is a new approach. While it does include much material that appeared in the earlier publication and while a number of the old plates have been used, it has been entirely rewritten, and it is intended to give a fair coverage of the entire field of fossil algae and algal limestones. Inasmuch as it is prepared primarily to meet the needs of graduate students and petroleum geologists, the author has attempted to present in concise non-technical language the basic information that they would require. It has been impossible to avoid the use of some technical language and expressions, but a distinct effort has been made to keep these at a minimum. A glossary of the technical terms is appended at the end of the book. A short list of pertinent references is included at the end of each chapter. Most of these publications contain some reference data, but the last reference list in the book is restricted to bibliographies or publications which contain much bibliographical material.

A number of the more important genera are described and illustrated, but not all of the algal genera are included as can be seen from the lists given in the Appendix. Persons desiring information on additional genera or greater detail on the genera that are mentioned are referred to the more detailed papers by the author and to the references given in the lists accompanying the various chapters.

INTRODUCTION

In 1913 Dr. E. J. Garwood delivered a presidential address on rock-building algae to the geological section of the British Association for the Advancement of Science, and shortly thereafter, he published a somewhat amplified version of the speech in the *Geological Magazine*. These two events were essentially the beginning of the study by geologists of the geological work of the lime-secreting algae and may also be said to have started the study of fossil algae as a branch of paleobotany.

Prior to 1913, a few notes on fossil algae had appeared in print, but these publications were brief and widely scattered and gave very little real information. A few were by geologists or paleontologists, but most of them were written by botanists and gave scarcely more than a general mention of the occurrence of such a fossil. Several groups of fossils which we now know to represent algae had been described before 1913. Both Dasycladaceae and Solenoporaceae had been described a generation or so earlier, but they were considered first as problematical organisms and then were attributed to several different phyla of animals or plants before finally being identified as algae.

Thus, Brown (1894) showed that *Solenopora* possessed structures definitely akin to those of the coralline red algae and paved the way for them to be considered as fossil algae. Rothpletz, particularly in his papers of 1891, 1908, and 1913, described a variety of fossil algae and showed them to possess structures so analogous with those of present day algae that they too were accepted as algae. Madame Lemoine, in her classic work of 1911 and her short paper on algae as builders of limestone (1911), called attention to the fossil coralline algae and showed that they could be carefully studied and classified on the basis of structures that were to be found both in fossil and Recent forms.

Following the publication of Garwood's paper, a number of geologists, paleontologists, and paleobotanists became actively interested both in the study of fossil algae and in the geological work of algae. Pia in Vienna and the Morellets in Paris began their studies of fossil Dasycladaceae. Lemoine in 1916 began a series of studies on fossil coralline algae. Many field geologists began to look for fossil algae, and from time to time, paleontologists described some forms that were algal or supposed to be algal. Walcott, in 1914, published a paper on the Precambrian algal flora, and his studies not only demonstrated the presence of fossil algae in ancient rocks but considerably stimulated the interest of American geologists in such objects.

So, an interest was created and gradually developed. In 1931 Garwood, then President of the Geological Society of London, again chose the subject of fossil algae for a presidential address to that Society and was able to point out substantial progress and a great increase in the knowledge of algae and their geologic work in the period between 1913 and 1931. Publication of this address again had a very stimulating effect upon both European and American geologists. A gradually increasing interest in the subject and a steady increase in the number of publications continued until about 1950.

At that time, the petroleum geologists of the world became actively interested in reef limestones as possible reservoir rocks for the accumulation of petroleum. This greatly stimulated the study of both fossil and Recent reefs and reef limestones and of the organisms that contribute to them. The study of reef builders showed that in many areas algae played an important part in the building of reefs and bioherms and that they had done so throughout much of the geologic past. This greatly accelerated the study of fossil algae and the work being done by various types of algae, both at the present and in the past, in the building of limestones.

Recently, it has been shown that fossil algae are of real value to the geologists and paleontologists because these fossils can be used to give a great deal of environmental information and also because certain types, particularly the Dasycladaceae, some of the Codiaceae, and the fossil coralline algae, have considerable possibilities for use in dating the rocks.

Because of the above facts and because of the interest being shown by the petroleum industry in a study of carbonate rocks, there is now a demand for information on rock building algae, and a number of geologists and graduate students are beginning to enter this most interesting field. The writer hopes that this book may be of real service to them and that it may stimulate them to continue the work and to add to our information on the subject.

The literature on fossil algae is quite extensive, but it is widely scattered in journals published in many countries and in a number of languages. Most of the literature published prior to 1955 was collected by Johnson and was included in two bibliographic lists published in 1943 and 1957.*

*These appeared as two Colorado School of Mines *Quarterlies* vol. 38, No. 1, 102 p., 23 figs., 1943, and vol. 52 no. 2, 92 p., 1957.

ACKNOWLEDGMENTS

In the preparation of the present book, the author has drawn freely on the literature and on his previous publications. He is most indebted to the late Julius Pia and to Madame Lemoine who assisted him greatly when he was beginning the study of fossil algae. Their advice, publications, and many suggestions have been invaluable.

During the period from 1947 to 1959, the author had the good fortune to work with the Military Geology Branch of the U. S. Geological Survey in the study of the geology of a number of the mandated islands in the Pacific. This gave him the opportunity to study many Recent and fossil reefs, to collect large quantities of material, and to work with large collections of fossil algae and algal limestones of Cenozoic age.

The Colorado School of Mines has assisted most generously in supplying laboratory space, equipment, photographic supplies, and secretarial help. The Colorado School of Mines Foundation, Inc., has assisted the author, since his retirement in 1957, with research funds and facilities which have enabled him to continue his studies on an almost full-time basis. This publication has been prepared under a Colorado School of Mines Foundation, Inc., research grant which was administered by the Colorado School of Mines Research Foundation.

The Colorado School of Mines Library has been used extensively, and many thanks are due Madeleine Gibbons, Virginia Wilcox, Helen Johnston, and other members of the Library staff.

The members of the Colorado School of Mines Department of Publications have assisted greatly in the editing and processing of this book. Thanks are due Mrs. Mary Baldwin who edited and typed the manuscript. My wife, Merle K. Johnson, has been very helpful with criticisms, suggestions, and encouragement.

The illustrations used have come from many sources. About 40 percent of them were taken from our original publication "An Introduction to the Study of Rock Building Algae and Algal Limestones" and about 20 percent were taken from the various Colorado School of Mines *Quarterlies* on fossil algae, published in 1956, 1958, 1959, and 1960. The U. S. Geological Survey kindly permitted me to use photographs I took in connection with the studies on Cenozoic algae from the tropical Pacific.

LIMESTONE-BUILDING ALGAE

Classification

The algae are a group of plants characterized by having most of the functions of life carried on by almost all parts of each plant. This is in contrast to higher types of plants which have their tissues differentiated into roots, stems, leaves, and other parts, each of which has a separate and distinct function. The algae comprise a large, highly diversified group and include everything from microscopic unicellular plants to enormous plants of rather complicated structure. The great majority of the algae live in the water, although some of the smaller types may live in moist soils. The greatest number and variety of algae live in the oceans, but there are also many brackish- and fresh-water forms.

The algae have been divided into a number of groups or phyla on the basis of the color of the living plant. These major groups are sub-divided into progressively smaller divisions. There is still considerable difference of opinion among present-day students of the subject as to detailed classification of the algae. However, most of the authorites agree on the major divisions and on the genera. The chief differences of opinion are largely in the sub-divisions between the phyla and the genera. The classification that will be followed in the present work is the one proposed by Papenfuss in 1955. It is given below.

Phylum RHODOPHYCOPHYTA Papenfuss 1946
  Syn.: Rhodophyta Wettstein 1901
    Class Rhodophyceae Ruprecht 1851
        Subclass Florideophycidae (Lamouroux) Engler orth. mut. L. M. Newton 1953
            Order Nemalionales Schmitz, *in* Engler 1892
                Family Helminthocladiaceae (J. Agardh) Harvey orth. mut. Hauck 1883
                Family Chaetangiaceae Kützing orth. mut. Hauck 1883
            Order Cryptonemiales Schmitz, *in* Engler 1892
                Family Squamaraceae (J. Agardh) Hauck 1883
                Family Solenoporaceae Pia 1927
                Family Corallinaceae (Lamouroux) Harvey 1849
                Family Gymnocodiaceae Elliott 1955

Phylum CHLOROPHYCOPHYTA Papenfuss 1946
  Syn.: Chlorophyta Pascher 1914
    Class Chlorophyceae Kützing 1843
            Order Siphonales Wille, *in* Warming (1884) orth. mut. Blackman et Tansley 1902

Family Codiaceae (Trevisan) Zanardini 1843
Order Dasycladales Pascher 1931
Family Dasycladaceae Kützing orth. mut. Stizenberger 1860

Phylum CHAROPHYCOPHYTA Papenfuss 1946
Syn.: Charophyta Migula 1890
Class Charophyceae G. M. Smith 1938
Order Charales Lindley 1836
Syn.: Sycidiales Mädler 1952, Trochiliscales Mädler 1952
Family Palaeocharaceae Pia 1927
Family Characeae Richard ex C. Agardh 1824
Family Clavatoraceae Pia 1927?
Family Trochiliscaceae Karpinsky orth. mut. Peck 1934
Family Sycidiaceae Karpinsky orth. mut. Peck 1934

Phylum SCHIZOPHYTA (Falkenberg) Engler 1892
Class Schizophyceae Cohn 1880
"Section" Porostromata Pia 1927
"Section" Spongiostromata Pia 1927

Each phylum contains a wide variety of forms ranging in structural complexity from simple unicellular to very complex multicellular types, and there is a great deal of parallelism in the development within each group. In the course of their long history, a few members of several of the groups of algae have developed the ability to secrete lime. These are the forms commonly known as the calcareous algae, and they are almost the only algae that have any chance of being preserved as fossils. They are also the only ones that are of any importance in the building of limestones.

The way in which the lime is deposited and the amount of calcification may vary from group to group and even from genus to genus among the calcareous forms. Among the coralline red algae, calcium carbonate is secreted within and between the cell walls. Among many of the green calcareous algae, calcification involves most of the tissue and starts at the outer portions and works inward. In a few of the calcified forms, we find that calcification starts just within the outer walls and involves only a thin zone of tissue around the outer portion of the plant.

Calcareous algae comprise the oldest fossils known, and their remains occur in greater or lesser abundance in rocks of all ages from far back in the Precambian to the Recent.

The classification of fossil algae presents something of a problem. Most of the fossil algae found in rocks of Cenozoic age seem to belong to genera and families that still exist, and for these algae, the classification used for Recent algae is applicable. However, although the fossils from more ancient rocks may show structures somewhat similar to those found in modern algae, they definitely do not belong to Recent genera, and the more ancient the fossils are, the less are they like present-day forms. Consequently, the older the fossil, the greater the amount of uncertainty that comes into the classification and the more probable it is that the classification used is artificial. Any classification, however, is merely a method of giving the student a number of pigeon holes into which he can sort his specimens, and one does not have to worry too much about the system used as long as it will work for most of the specimens studied.

For the red and green algae, the classification used is essentially that used for Recent forms with, of course, the addition of a number of species and even a few genera and families. Among the lower groups of algae, however, there are many types that are known only in the fossil state, and the classification that has been devised for them is undoubtedly very artificial in the sense that plants put in a single group may actually belong to several quite unrelated groups.

The more important families of fossil rock-building algae are shown in Table I. A more detailed classification is considered in the discussion of each family.

TABLE I
IMPORTANT FAMILIES OF FOSSIL ALGAE

Rhodophycophyta (red algae)	Corallinaceae	Closely packed layers or threads of cells, rectangular in section. Sporangia or conceptacles develop within or partly immersed in the tissue.
	Solenoporaceae	Rows of closely packed cells with polygonal cross section, commonly rectangular in longitudinal section. Sporangia seldom seen, probably external in most cases.
	Gymnocodiaceae	Thallus segmented or not. Tissue of loosely packed threads. Calcification variable, rather weak, limited commonly to an irregular outer zone. Sporangia internal.
Phaeophycophyta (brown algae)	Laminariales and others (?)	Corded strands of parallel threads (as in framework of *Archimedes* Hall). Fronded types.
Chlorophycophyta (green algae)	Codiaceae	Composed of freely branched tubular filaments packed or interwoven to form a plant body of definite and consistent external form. Tubes commonly round in cross section.
	Dasycladaceae	A central stem surrounded by tufts or whorls of branches. Calcification occurs around the central stem and branches. Fossils commonly a mold with central stem and branches appearing as canals or pores.
Charophycophyta	Characeae	Highly developed small bushy plants. Fossils usually consist of calcified, heavily ribbed, spherical oogonia and the whorled branches which bear them.
Schizophyta (blue-green algae)	Porostromata	Small tubes so loosely arranged as not to compress each other. No cross partitions visible.
	Spongiostromata	Cellular structure seldom preserved. The $CaCO_3$ is deposited as crusts on the outside of colony or cell or between the tissues, not in the cell wall. Classified on the basis of growth habit and form of colony.

REFERENCE

Papenfuss, G. F., 1955, Classification of the algae: Century of Progress of Natural Science (Calif. Acad. Sci.), p. 155-224.

CHEMICAL COMPOSITION

General

An enormous number of chemical analyses of marine algae have been made and the results published. As might be expected, however, most of these analyses have been of algae that have economic possibilities as food, fertilizer, or raw materials for the chemical industry. Only a relatively small number of studies have covered the lime-secreting, rock-building algae which are of major interest in this treatise.

Most of the data presented in this chapter have been taken from the magnificent memoir by A. P. Vinogradov, "The Elementary Chemical Composition of Marine Organisms." This publication and the earlier "Inorganic Constituents of Marine Invertebrates" by Clarke and Wheeler should be on the reference list of all geologists.

The Calcareous Algae

The calcareous algae are so named because they contain abnormally high concentrations of calcium carbonate. Before considering in detail the chemical composition of the calcareous algae, we should mention the chemical composition of algae in general, particularly with regard to calcium and magnesium content.

The amount of calcium and magnesium in the ash of marine algae varies widely with the genus, the species, and the time of year. In general, calcium predominates over magnesium, but there is no apparent relation or ratio between the two elements. A survey of the published analyses of particular species of algae, covering species belonging to all of the major phyla, shows that the content of calcium oxide in the ash will range from about 1 percent to a high of over 64 percent, while the magnesium oxide content ranges from about .3 percent to a high of about 15 percent. The magnesium content seldom goes above 10 percent of the ash or .5 percent of the living material and has a much smaller range of variation than the calcium content.

Analyses of Brown Algae

Among the brown algae, the amount of magnesium present is no greater than the average for all algae and, in some cases, is considerably less. A calcium content larger than the average is very seldom seen among the brown algae. Only one species of brown algae, *Padina pavonia,* is known that is sufficiently active in the deposition of calcareous material to take part in the formation of calcareous sediments. The mineralogical character of the calcium carbonate among the brown algae is not known. Judging by the composition of the ash, it probably is in an amorphous form or is aragonite, although some magnesium and calcium protein compounds and calcium pectinate have been observed.

In general, it may be said that among the brown algae, their metabolism of calcium and magnesium is not as important as this function in other phyla of algae. It cannot be compared in any way with the potassium metabolism which is a most spectacular geochemical function of the majority of the brown algae.

Analyses of Red Algae

The most obvious chemical characteristic of many red algae is the calcium function, which is more strongly developed than in the brown algae. Potassium metabolism occurs, as in all algae, but it is much less important than among the brown algae. Sulphur, iodine, and some other elements occur in about the same proportions as in all algae. Among the Rhodophycophyta, however, the most significant element is calcium and, for certain species, magnesium also. In some red algae, the calcium function is specifically different than it is in others. The red algae may be divided into three groups with respect to the calcium content.

Group 1 includes the greatest number of the Rhodophycophyta. Among members of this group, the calcium content is similar to the average for all algae. Typical examples may be found in the genera *Chondrus, Delesseria, Polysiphonia,* and *Nitophyllum.*

Group 2 includes forms in which there is a concentration and precipitation of calcium carbonate, probably as aragonite. The amount of calcification is greater than the average for all algae, but it is usually appreciably less than in group 3. Genera characteristic of group 2 are *Galaxaura, Laurencia,* and *Liagora.*

Group 3 includes algae in which the calcium carbonate or, more precisely, the calcium-magnesium carbonate is concentrated and deposited usually in the form of calcite. This group is exemplified by and is largely restricted to the family Corallinaceae.

The red algae belonging to the first group are, for all purposes, normal algae and are of little or no interest to us in the study of rock-building algae. The second group contains some strongly calcified forms, which are capable of being preserved as fossils, but only rarely have members of this group occurred in sufficient numbers to be of any importance as rock builders. Apparently, the only known family of fossil algae belonging to group 2 that was at all important as a rock builder was the family Gymnocodiaceae, which during Permian times in some areas, for example in northern Italy and a few places along the Adriatic coast, appears to have been a local rock builder. The third group is the one that is of real interest to us in our consideration of rock-building forms.

Table II, taken directly from the Sears Foundation Monograph (1953), shows the composition of a number of species of coralline algae. These compositions are in terms of percentage of the dried material. From these tables, we can see that the Corallinaceae are organisms with large amounts of magnesium and calcium carbonate in the ash, with the ash residue ranging up to about 50 percent of the living weight of the plant.

Among the different species of the Corallinaceae, the amount of calcium carbonate and magnesium carbonate varies comparatively little. Attempts have been made to correlate the amount of magnesium carbonate present with the temperatue of the water in which the algae lived, and while the general rule can be made that species living close to the equator contain larger amounts of magnesium than do those living in cold water, this is not universally true, and there are quite a number of rather surprising exceptions. For example, some species of *Lithothamnium* from Java contain only up to 3.76 percent magnesium carbonate, while specimens of *Lithothamnium glaciale* from the Arctic have been observed to contain as much as 13.19 percent magnesium carbonate. In the great majority of cases, however, the rule does apply that the magnesium carbonate content appears to depend upon the temperature of the water.

To judge from a comparatively small number of analyses, some variation in the amount of calcium carbonate and magnesium carbonate appears to occur in plants of different ages within the same species. There may also be some fluctuation in the amount of these carbonates with the time of year.

Numerous observations and experiments have shown that among the Corallinaceae the carbonate is present in the form of calcite. The coralline algae contain higher pecentages of magnesium carbonate than have been found in the skeletons of any other animals or plants. Of the various genera of coralline algae that have been analyzed, the genus *Goniolithon* is the richest in magnesium carbonate.

In summary, the Corallinaceae typically are calcium-magnesium organisms containing calcite with the maximum amount of magnesium carbonate known in any organism. This chemical characteristic of being particularly rich in magnesium carbonate is very distinctive for this family of algae and distinguishes them from all other lime-encrusting and lime-secreting algae.

Table III shows the amount of $MgCO_3$ and $CaCO_3$ present in a number of genera and species of coralline algae from widely distributed localities (Sears Foundation Monograph, 1953, p. 56-58).

Analyses of Green Algae

The Chlorophycophyta, probably more than any other group of algae, have spread from marine waters into brackish and fresh waters. The marine forms, although occurring in all seas, have their greatest development in warm and tropical marine waters. As a result of their diversity in geographical and environmental distribution, the green algae show a very great variety of chemical characteristics.

The green algae can be divided into two groups on the basis of the calcium content. These groups resemble the first and second groups of the red algae. Members of group 1 contain calcium carbonate in amounts comparable to the average of all algae, while members of group 2 are rich

TABLE II

COMPOSITION OF CORALLINE ALGAE

(% of dried material)

ALGAE	Na_2O	K_2O	MgO	CaO	SO_3
Lithothamnium ramulosum	—	—	3.06	45.88	—
Lithothamnium sp.	—	—	1.90	48.09	—
Lithothamnium nodosum	—	—	2.66	47.14	—
Lithothamnium glaciale	—	—	4.78	45.41	0.14
Lithothamnium kaiseri	1.10	0.24	7.69	40.76	0.78†
Lithothamnium erubescens	—	—	7.52	42.96	0.61
Lithothamnium crassum	—	0.96	1.90	40.60	6.89†
Lithothamnium calcareum	—	—	2.13	48.35	0.96†
Lithophyllum sp.	0.55	0.27	5.89	43.32	0.95
Lithophyllum pallescens	—	—	6.42	40.39	0.71
Lithophyllum craspedium f. *mayorii*	2.06	0.25	8.0	42.22	0.81†
Lithophyllum craspedium	—	—	8.64	41.56	0.15
Lithophyllum daedaleum	—	—	8.14	40.48	0.44
Lithophyllum antillarum	—	—	7.24	43.34	0.57
Lithophyllum oncodes	—	—	8.07	42.57	0.27
" "	1.42	0.45	6.65	43.78	0.87†
Lithophyllum intermedium	—	—	7.25	42.55	0.02
Lithophyllum tarniense	—	—	8.81	41.07	0.69
Lithophyllum pachydermum	—	—	6.47	42.60	0.59
Lithophyllum pachydermum (young)	—	—	10.69	37.66	0.79
Lithophyllum pachydermum (old)	—	—	6.54	44.88	0.71
Lithophyllum kaiseri	—	—	7.09	45.92	0.00
" "	2.83	0.49	7.84	40.28	0.94†
" "	1.91	0.87	7.28	41.76	0.90†
Lithophyllum proboscideum	2.78	0.34	3.88	40.31	0.97†
Archaeolithothamnium episporum	—	—	5.93	44.80	0.53
Phymatolithon compactum	—	—	4.02	38.19	0.52
Goniolithon acropetum	—	—	8.27	40.60	0.62
Goniolithon strictum	—	—	10.39	38.54	0.60†
" "	—	—	10.93	38.03	0.61†
Goniolithon strictum (young)	—	—	9.64	37.64	0.63
Goniolithon strictum (old)	—	—	10.26	38.25	0.67
Goniolithon frutescens	—	—	6.29	46.16	0.00
" "	1.57	0.38	7.28	41.76	0.91†
Goniolithon orthoblastum	—	—	5.71	42.39	0.00
Amphiroa tribulus	0.89	0.39	8.83	39.74	1.05
Amphiroa fragilissima	—	—	6.71	34.82	0.56
Amphiroa foliacea	—	—	7.53	39.02	1.08†
Melobesia sp.	1.75	0.65	6.41	32.76	1.25

* Fe_2O_3 without Al_2O_3.
† SO_4.
Fe_2O_3 CO_2+

TABLE II (Continued

$(+Al_2O_3)$	SiO_2	Organic matter	Locality	Author
0.41*	1.91	44.98	Gulf of Naples	Schwager (see Walther, 1885)
0.28*	1.59	44.93	" " "	" " " "
2.96		42.62	——	Gümbel, 1871
0.23	0.41	48.0	Newfoundland	Clarke and Wheeler, 1922
—	0.02	49.09	Tutuila, Samoa	Lipman and Shelley, 1924
0.12	0.19	47.89	Timor (Indian Archipelago)	Clarke and Wheeler, 1922
1.93	—	10.39	Normandy, France	Vincent, 1924
0.9	—	0.45	Concarneau, France	" "
0.08	—	44.17	Mediterranean	Damour, 1851
0.22	1.01	49.94	California, U.S.A.	Clarke and Wheeler, 1922
—	0.28	46.95	Rose Atoll	Lipman and Shelley, 1924
0.09	0.03	48.37	Palmyra I. (Pacific)	Clarke and Wheeler, 1922
0.08	0.18	49.99	Puerto Rico	" " " "
0.10	0.04	48.11	" "	" " " "
0.12	0.09	47.96	Coetivy I. (Indian O.)	" " " "
—	0.06	47.29	Tutuila, Samoa	Lipman and Shelley, 1924
0.21	0.27	48.63	Jamaica, B.W.I.	Clarke and Wheeler, 1922
0.16	0.12	48.77	New Guinea	" " " "
0.08	0.04	49.55	Bahama Is.	" " " "
0.11	0.04	50.64	" "	Kamm (see Clarke and Wheeler, 1922)
0.14	0.09	47.51	" "	" " " "
0.28		45.72	Cocos-Keeling I. (Bahamas)	Chambers (see Vaughan, 1918)
—	0.18	49.79	Tutuila, Samoa	Lipman and Shelley, 1924
—	0.16	48.71	" "	" " " "
0.64*	8.38	42.31	California, U.S.A.	" " " "
0.89	1.47	45.69	Panama	Clarke and Wheeler, 1922
0.23	0.10	55.42	Newfoundland	" " " "
0.04	0.06	49.56	Puerto Rico	" " " "
0.01	0.02	49.91	Bahama Is.	" " " "
0.05		49.91	Florida	" " " "
0.32	0.02	51.14	Bahama Is.	Kamm (see Clarke and Wheeler, 1922)
0.57	0.08	48.76	" "	" " " "
0.07		46.7	Cocos-Keeling I. (Bahamas)	Chambers (see Vaughan, 1918)
—	0.22	48.85	Tutuila, Samoa	Lipman and Shelley, 1924
0.11		50.97	Murray Isle, Australia	Chambers (see Vaughan, 1918)
—	—	45.73	Antilles (W. Indies)	Damour, 1851
0.31	2.82	48.78	Puerto Rico	Clarke and Wheeler, 1922
0.64*	0.32	50.87	Tutuila, Samoa	Lipman and Shelley, 1924
0.20	—	49.88	Algiers	Damour, 1851

TABLE III

$MgCO_3$ and $CaCO_3$ IN CORALLINACEAE FROM DIFFERENT LOCALITIES (IN % of ASH)

ALGAE	$MgCO_3$	$CaCO_3$	Lat.†	Long.	Locality	Author
Lithothamnium						
sp.	8.67	84.83	78° N	(20° E)	Spitzbergen	Högbom, 1894*
soriferum	9.56	80.90	75	—	Arctic O.	" "
glaciale	13.19	83.10	75	—	" "	" "
sp.	9.94	74.26	72	—	Bering I.	" "
fornicatum	10.09	88.61	—	—	Norway	Charf (see Lemoine, 1910)
polymorphum	9.10	74.22	57	12	Kattegat	Högbom, 1894
"	15.15	83.4	57	12	"	Vesterberg, 1900-01
Corallina officinalis	12.06	86.68	49°33′	1°12′W	Saint-Wast, Normandy	Payen, 1843
Lithothamnium calcareum	12.04	84.60	48°5′	4°W	Jlenan Isle (Finisterre)	Charf (see Lemoine, 1910)
"	12.4	85.1	48°5′	—	English Channel	" " "
"	12.52	87.48	48	—	Roscoff (Finisterre)	Chalon, 1900
Lithophyllum incrustans	11.14	87.10	48	3	English Channel (Gatteville)	Charf (see Lemoine 1910)
Lithothamnium calcareum	12.70	87.30	48	3	Concarneau, France	Vincent, 1924
crassum	5.30	94.70	—	3	?	" "
glaciale	10.93	88.11	48	53	Topsail, Newfoundland	Clarke and Wheeler, 1922
Phymatolithon compactum	10.93	87.21	48	55	Torbay, Newfoundland	" " "
Lithophyllum tortuosum	9.83	77.58	44	9°E	Genoa, Italy	Charf (see Lemoine 1910)
expansum	13.11	75.99	44	7	Villefranche	Novácek, 1930
tortuosum	14.12	73.91	43	3	Banyuls-sur-Mer, France	" "
Lithothamnium polymorphum	13.76	74.15	43	3	" "	" "
Corallina officinalis	11.13	72.12	43	3	" "	" "
"	10.09	68.36	—	—	" "	" "
Lithophyllum incrustans	15.90	84.10	43	3	" "	" "
Lithothamnium ramulosum	9.46	63.00	41	14	Naples, Italy	Högbom, 1894
racemus	11.33	77.39	41	14	" "	Schwager (see Walther, 1885)
ramulosum	6.81	86.90	41	14	" "	
sp.	4.19	90.30	41	14	" "	
tortuosum	12.35	87.65	41	14	" "	Chalon, 1900
(*calcareum*)	12.76	86.36	40	—	Mediterranean	Damour, 1851
Lithophyllum proboscideum	8.15	72.0	37	122°W	Monterey, Calif., U.S.A.	Lipman and Shelley, 1924

TABLE III (Continued)

ALGAE	$MgCO_3$	$CaCO_3$	Lat.†	Long.	Locality	Author
Melobesia sp. . .	14.44	84.36	36	3°E	Algiers . . .	Damour, 1851
Lithophyllum incrustans . .	10.73	85.08	33	7°W	Mazagan, Morocco . .	Charf (see Lemoine, 1910)
Lithothamnium sp. . . .	12.73	82.44	32°N	65°W	Bermuda . . .	Högbom, 1894
racemus . .	5.35	—	25	75	Bahamas (B.W.I.)	Nichols, 1906
Goniolithon strictum . .	24.00	74.85	25	75	" "	Clarke and Wheeler, 1922
*strictum*α . .	22.98	75.42	25	75	" "	Kamm, 1922 (see Clarke and Wheeler, 1922)
*strictum*β . .	23.74	74.29	25	75	" "	
*Lithophyllum pachydermum*α	24.95	73.65	25	75	" "	" " "
*pachydermum*β	15.43	83.06	25	75	" "	" " "
pachydermum .	15.08	83.68	25	75	" "	Clarke and Wheeler, 1922
Goniolithon strictum . .	25.17	73.63	24	81	Soldiers' Key, Fla., U.S.A. .	" " "
Lithophyllum pallescens . .	15.46	81.48	24	110	Bay La Paz, Calif., U.S.A. . .	" " "
Lithothamnium sp. . . .	25.32	73.23	23	80	South Florida, U.S.A. .	Phillips (see Clarke and Wheeler, 1922)
Amphiroa tribulus . .	19.29	79.81	20	—	Antilles (W.Indies)	Damour, 1851
Lithothamnium sp. . . .	9.39	84.01	20	161	Honolulu, T.H.	Högbom, 1894
Goniolithon acropectum .	19.24	79.05	18	68	Culebra I., Puerto Rico . .	Clarke and Wheeler, 1922
Lithophyllum daedaleum . .	19.03	79.85	18	68	Salinas Bay, Puerto Rico . .	" " "
antilarum . .	16.35	82.46	18	68	Culebra I., Puerto Rico . .	" " "
intermedium .	16.59	82.85	18	72	Kingston, Jamaica	" " "
Amphiroa fragilissima .	17.47	76.23	18	68	Lemon Bay Puerto Rico . .	" " "
Lithophyllum craspedium . .	16.82	72.77	18	150	Tahiti	Charf (see Lemoine, 1910)
Lithothamnium kaiseri . . .	16.15	72.80	15°S	172	Tutuila, Samoa .	Lipman and Shelley, 1924
Goniolithon frutescens . .	15.29	74.57	15	172	" "	" " "
Porolithon craspedium . .	16.80	75.39	15	172	Rose Atoll . .	" " "
oncodes . .	13.75	78.20	15	172	Tutuila, Samoa .	" " "
Lithophyllum kaiseri . . .	16.46	71.93	15	172	" "	" " "
Amphiroa foliacea . .	15.80	69.64	15	172	" "	" " "

TABLE III (Continued)

ALGAE	$MgCO_3$	$CaCO_3$	Lat.†	Long.	Locality	Author
Lithothamnium						
sp. . . .	36.36	—	13	175	Samoa	Phillips (see Clarke and
" . . .	19.47	74.4	13	175	" . . .	Wheeler, 1922)
Goniolithon . .					Cocos Is. . . .	Chambers (see
frutescens	13.80	86.13	12°5′	96°5′E	(Bahamas)	Vaughan, 1918)
Lithophyllum						
kaiseri . . .	15.33	84.38	12°5′	96°5′	" "	" "
Goniolithon . .					Murray Isle . .	Chambers (see
orthoblastum	13.66	86.22	10	145	(Australia)	Vaughan, 1918)
Lithothamnium .					Timor (Indian .	Clarke and Wheeler,
erubescens	16.96	81.59	10	123	Archipelago)	1922
Archaeolithothamnium						
episporum . .	13.09	83.47	10	80	Panama . . .	" " "
Lithothamnium					Funafuti . . .	Skeats (see Royal Soci-
philippi var.						ety of London, 1904)
funafutiensis	5.85	—	8	178		
Lithophyllum						
oncodes . .	18.17	80.93	7	56	Madagascar . .	Clarke and Wheeler,
craspedium . .	19.60	72.92	5°5′N	172°W	Palmyra I. . . .	1922
tarniense . .	20.02	78.43	5°S	145°E	Nw Guinea . .	" " "
Lithothamnium						
sp.	3.76	72.03	5	—	Java Sea . . .	Högbom, 1894
"	6.53	83.60	0	90°W	Galapagos Is. . .	" "
"	12.17	83.75	0	90		Vesterberg, 1900-01
nodosum . .	6.06	91.04	—	—		Gümbel, 1871

* Analyses of Mauzelius, Sahlbom and Guinchard.
† For analyses in which only the geographical name was given, we have calculated the approximate latitude and longitude.
α Young. β Old.

TABLE No. IV

MEAN $MgCO_3$ IN DIFFERENT GENERA OF CORALLINACEAE (IN % OF ASH)

ALGAE (Genus)	No. of analyses	Minimum $MgCO_3$	Maximum $MgCO_3$	Mean $MgCO_3$
Lithothamnium	29	3.76	36.36	12.30
Lithophyllum	24	5.30	24.95	14.81
Melobesia	1	—	—	14.44
Archaeolithothamnium	1	—	—	13.09
Porolithon	2	13.75	16.80	15.28
Goniolithon	8	13.66	29.98	20.61
Amphiroa	3	15.80	19.29	17.52
Corallina	3	11.13	12.06	11.09

in calcium. This analogy with the groups of the Rhodophycophyta can be carried further, inasmuch as group 2 of the marine green algae also contains calcium carbonate in the form of aragonite.

Among the green algae, the first group consists mainly of fresh-water forms which apparently do not need, or at least use, calcium. This group also includes a considerable number of marine types.

TABLE V

COMPOSITION OF CHLOROPHYCEAE RICH IN CALCIUM (IN % OF DRY MATTER)

ALGAE	MgO	CaO	SO_3	P_2O_5	Fe_2O_3	SiO_2	CO_2 and org. matter	Locality	Author
Halimeda simulans	0.19	50.20	0.23	trace	0.18	0.44	48.38	Puerto Rico	Clarke and Wheeler, 1922
Halimeda tridens	0.44	45.32	0.49	trace	0.58	0.88	51.94	" "	
Halimeda monile	0.45	48.21	0.02	trace	1.10	1.90	47.56	" "	
Halimeda opuntia	0.01	50.32	0.07	trace	0.20	0.37	48.20	Florida, U.S.A	
*Halimeda opuntia**	0.29	48.47	0.33	?	—	—	46.49	Red Sea	Damour, 1851

*Na_2O = 1.13; K_2O = 0.54; Cl = 0.84.

TABLE VI

$MgCO_3$ AND $CaCO_3$ IN CHLOROPHYCEAE (IN % OF ASH RESIDUE)

ALGAE	$MgCO_3$	$CaCO_3$	Locality	Author
Halimeda opuntia	0.64	98.73	Red Sea	Damour, 1851
" "	0.02	99.21	Florida, 24° N 82° W . .	Clarke and Wheeler, 1922
" "	0.60	86.50*	Funafuti Atoll, Ellice Is., 8° S 178° E	Skeats, 1904 (see Royal Society of London)
" "	5.50	90.16*	Martinique	Payen, 1843
Halimeda simulans	0.44	98.45	Puerto Rico, 18° N 66° W	Clarke and Wheeler, 1922
Halimeda tridens	1.09	96.21	" " " "	" " " "
Halimeda monile	1.04	95.58	" " " "	" " " "
Halimeda sp.	1.39	98.20	Funafuti Atoll, Ellice Is., 8° S 178° E	Cullis, 1904 (see Royal Society of London)

*Possibly in percent of dry matter.

The second group consists mainly of warm-water marine forms. Most of them belong to two families, the Codiaceae and the Dasycladaceae. Among the Recent Codiaceae, probably the best known genera are *Halimeda, Udotea,* and *Penicillus,* while among the Dasycladaceae, *Cymopolia, Acetabularia, Neomeris,* and *Bornetella* are probably the most widely distributed and best known Recent forms.

Unfortunately, relatively little chemical work has been done on the green algae. Practically all the analyses that have been made are limited to the genus *Halimeda.* The two tables given above show the compositions obtained (Table V and Table VI).

Some fresh-water types of green algae assist in the deposition of calcium carbonate. In most of these fresh-water carbonate deposits, the magnesium content is low, but in a few cases, an appreciable amount of magnesium carbonate has been observed.

Among the marine forms of group 2, some of the carbonate deposited by members of the families Codiaceae and Dasycladaceae contains appreciable amounts of magnesium carbonate. The amount present, however, is usually far less than is found among the coralline algae.

Recent work has shown that the calcium carbonate is deposited in the form of aragonite in the genera *Halimeda, Cymopolia,* and *Acetabularia.*

The mineralogical character of the carbonate deposited by various fresh-water forms is not well known. In some cases, the sediment produced by the life activities of *Cladophora* and *Vaucheria* consists of calcite and that deposited by the charophytes is also in the form of calcite. There is a strong suggestion that the marine green algae deposit carbonate in the form of aragonite, which is usually very poor in magnesium, while the fresh-water forms may deposit carbonate chiefly in the form of calcite.

There are a few fresh-water green algae belonging to the family Desmediaceae (de Bary) that appear to deposit gypsum.

Among the great variety of green algae, there are several small groups that do destructive geologic work. These groups include boring algae which perforate shells, limestones, and other calcareous matter, leading to the disintegration and destruction of the rocks.

Analyses of Blue-Green Algae

The Schizophyta are a most interesting group of organisms whose members live under exceptionally diverse conditions and have a wide geographical distribution. From a chemical point of view, they are probably the most interesting of all algae because of the large number of geochemical processes in which they participate. Unfortunately, only a relatively small number of chemical studies have been made of this group.

Blue-green algae take part in the formation of lake muds, the muds of estuaries, the peculiar organic sediments known in Japan as the *tengu,* some chalky sediments, combustible schists, sapropelite and boghead coals, and some oils. Strangely enough, relatively little study has been made of their geological work, although the various activities mentioned suggest that they are of considerable geological importance.

The available evidence suggests that probably all pre-Devonian sapropelites, oils, and similar bitumen were formed with the participation of algae, particularly blue-green algae. It is not until the Devonian period that humus coals with origin obviously depending on the higher land plants begin to appear in the rocks. Coals formed by the Schizophyceae do not produce phenols upon sublimation, and such coals can be easily distinguished chemically from the humus coals formed by the higher land plants.

Commonly, the nitrogen content of the Schizophyceae is higher than it is in other algae. In many cases, a higher than average phosphorus content is also observed among the blue-green algae.

Certain genera and species of blue-green algae are clearly associated with the deposition of calcium carbonate, but how they accomplish the work is not well known. Algae of this type participate in the deposition of certain types of calcareous tufas and travertines, but unfortunately, this aspect of their work has not received the attention that it deserves. Most of the chemical studies that have been made so far have involved fresh-water forms. The information available shows that the cell walls of some species of Schizophyceae become impregnated with calcium carbonate, rarely with silica. Large numbers of marine blue-green algae, as well as forms living in slightly saline and fresh waters, take part in the precipita-

TABLE VII

COMPOSITION OF CYANOPHYCEAE, *MICROCYSTIS AERUGINOSA* (IN % OF DRY MATTER), FROM VINOGRADOV, 1939

Ash	Ca	Mg	Si	P	Fe	Locality
42.18	0.90	2.06	0.09	0.69	0.05	Sea of Azov
43.37	0.95	1.82	0.12	0.70	0.05	" " "
39.31	0.89	1.70	0.20	0.75	—	" " "
21.37	1.00	0.50	4.61	1.61	—	" " "
43.78	0.25	1.91	0.11	0.80	—	" " "
12.88	1.61	0.45	1.50	1.03	—	" " "
6.84	0.30	0.23	—	—	0.10	(freshwater*)

*There was 2.5 X 10^{-2}% chlorine in the dry matter of *Microcystis*.

tion of calcium carbonate. Examples of lime-depositing blue-green algae are *Neulandia, Rivularia,* and *Oscillatoria.* To date, probably more chemical work has been done on *Rivularia* and *Oscillatoria* than on any other genera.

Most of the Precambrian algal limestones have been attributed to the activity of blue-green algae.

At the present time, many fresh-water types of Schizophyceae precipitate calcium carbonate on their surfaces and aid in the formation of travertines and tufas. This is particularly true of the genera *Phormidium, Chroococcus,* and *Gloeocapsa.*

The calcium-magnesium ratio among the lime-secreting blue-green algae is not well known, but in most cases, the percentage of magnesium is relatively low as most of the algal limestones attributed to these organisms are poor in magnesium. In this respect, they chemically resemble many other planktonic organisms, such as the foraminifera, whose skeletons are also poor in magnesium. Apparently, the calcium carbonate in most of the blue-green algae is deposited as calcite.

Destructive geologic work is done by certain of the Schizophyceae. A number of species of perforating algae bore or drill into limestones, shells, and other calcareous deposits and aid in the ultimate destruction of these materials.

REFERENCES

Clarke, F. W., and Wheeler, W. C., 1917, revised 1922, Inorganic constituents of marine invertebrates: U. S. Geological Survey Professional Paper 102, 56 p.; revised edition, U. S. Geological Survey Professional Paper 124, 62 p.

Vinogradov, A. P., 1953, The elementary chemical composition of marine organisms: (translated by J. Efron and J. K. Setlow) Sears Foundation for Marine Research, Yale University.

ECOLOGY

GENERAL

The ecological distribution of algae represents their adaptation to a considerable number of factors. Most important of these factors are light, depth of the water, character of the bottom, salinity of the water, circulation of the water, clarity of the water, and temperature of the water. It is very difficult to separate the influence of one factor from that of another. For example, we may say that certain algae develop best at a given depth, but the intensity of light decreases with depth, and, in most places, the temperature also changes with depth. So, when we say that a particular depth is the most desirable, or optimum, depth for a given species, does that mean the favorable condition is a result of the depth, or the result of the intensity of light, or the result of the temperature, or, what is more probable, is it not the combined result of all factors? Each of the factors mentioned above will be considered in our discussion of the coralline algae, but for the other groups of algae, the discussion of the various ecological factors will be more generalized.

Light

Algae, being plants, depend upon the energy of light to carry on their basic metabolism. Consequently, light is probably one of the most important factors in their distribution. Algae do not grow at depths greater than light penetrates. Indeed, very few types of algae extend down to the extreme limit of light. Most of them live in strong light at, or very close to, low-tide level, and at least half of these species of algae are restricted to depths of less than 10 to 12 fathoms, that is 60 to 75 feet.

Character of the Bottom

The character of the bottom strongly influences the distribution of different types of algae. Most of the crustose coralline algae develop best growing firmly attached to rocks and other solid objects. Certain types of green algae prefer a sandy bottom, and many other types of algae will grow only on muds.

Salinity of Water

Some algae have adapted themselves to waters of almost every degree of salinity, and algae of one sort or another can be found in waters ranging from practically pure fresh water to the very highly saline waters of salt lakes. We find, however, that most of the blue-green algae live in fresh waters. Only a few occur in normally marine waters. On the other hand, almost all red algae are purely marine, very few of them extending into even brackish water. The green algae are much more widely distributed with respect to salinity of the water, but on the generic level, there is quite an adaptation to degree of salinity even among the greens, with many genera being purely marine, others living solely in fresh waters, and a few living only in brackish waters.

Circulation of the Water

In general, at least a fair circulation of water is desirable in order to obtain a luxuriant growth of algae. Areas where the bottom water is stagnant or has but slight circulation usually are lacking in or very poorly populated with algae. The most luxuriant developments occur at shallow depths where wave action is not strong but where there is good circulation.

The strength of the circulation may have a definite bearing on the growth form of the algae. Thus, among the crustose coralline algae, we commonly find that the growth forms with rounded heads of tightly packed branches are largely restricted to areas of strong circulation, frequently on the edges of the reefs that are exposed to the violent action of the waves. Forms of algae that are composed of long slender widely separated branches generally grow slightly attached or loose on the bottom, in water sufficiently deep for wave action to be slight. Completely crustose forms are limited, in most cases, to relatively quiet waters. They develop behind the zone of strong wave action or in the deeper waters.

Clarity of the Water

This factor depends on the presence or absence of fine silt and other suspended matter in the water. The presence or absence of suspended matter influences the intensity of light at a given depth and also influences the character of the bottom, and, consequently, affects the growth and distribution of various types of algae.

Temperature of the Water

This is a very important factor as can readily be seen if one examines the distribution of existing genera of marine algae. A large number of species occur only in tropical waters; others are found only in temperate waters; while a few forms are restricted to cold waters. Temperature not only influences the geographic distribution but also affects the depth distribution of algae.

Ecological Distribution of the Crustose Coralline Algae

Littoral or Inter-tidal Zone

The crustose corallines may be observed in the inter-tidal zone. Commonly, they are encrusting forms which show conspicuously because of their white, pink, or purplish color. They occur in rock pools or on bare rocks exposed at low tide. At first glance, these exposed occurrences may seem surprising because the algae will not stand much drying in the hot sun. However, several factors are at work. Usually those places on the exposed rocks where such algae grow are either kept moist by spray from the breakers or are well shaded. In other places, the algae are completely out of the water but are more or less covered by other algae, especially brown algae, which protect them from the sun and minimize dessication. Crustose forms also occur associated with barnacles which seem to assist in keeping the algae moist.

In areas of exceptional tides, such as the Bay of Fundy and the English Channel, algal crusts of this type have been observed to grow as much as 3 meters above low tide level in situations where they are shaded from the sun or are bathed in spray. Similarly, in small sea caves and grottos, where there is high humidity and much spray, encrusting forms have been observed to grow above the permanent high-tide level. Such occurrences, however, appear to be rather unusual.

The species found in the inter-tidal zone are primarily encrusting forms (Plate 6, figure 1), although there may be some other types. The crusts may be smooth, warty, or mammillated. In some areas, loose or nearly loose foliated and branching forms occur, as along the coasts of Normandy and Brittany. Rounded, nodular forms seem to be rare along the coasts of France and Great Britain. The author has observed them growing in great numbers, however, in highly agitated, slightly deeper waters at a few localities along the reef at Bikini in the Marshall Islands. There on the reef flat, just inside the surf zone, large areas covered by nodular growths were uncovered at abnormally low tide, but usually were immersed in several inches of agitated water at normal low tides. Mme. Weber van Bosse (1904, p. 4-5) mentions areas of the reef flat at Haingsisi, an island southwest of Timor, covered by rounded masses of the branching *Lithothamnion erubescens,* which were exposed at very low tide.

A characteristic feature of the inter-tidal zone is a narrow belt or rim of deposits around the headlands, or rocky islands, or across the mouths of shallow bays. These rims are formed by an intergrowth of calcareous algae associated with Bryozoa, *Serpula,* corals, or vermiform mollusks. The French refer to these deposits as *"Trottoirs."* They have been mentioned by many writers as occurring around the various headlands and bays of the Mediterranean, along some of the French coasts, around the Gulf of Naples, and along portions of the coast of Algeria, as well as in many other localities along the Mediterranean and Adriatic coasts. The writer also observed them rimming some narrow rock terraces along the rocky coasts of Guam, Tinian, and Saipan in the Mariana Islands.

Sublittoral Zone

This zone is the shallow water area not uncovered at low tide. It is characterized by fairly uniform temperature and salinity conditions, only the upper part being strongly agitated by wave action. The upper limit of the sublittoral or neritic zone is low tide level, which obviously is a variable factor.

Apparently, the sublittoral zone is the most favorable area for the growth of coralline algae. In the tropical seas, numerous writers have described the tremendous development of coralline algae around the margins of the reef area exposed at normal low tide and extending downward from that level to a depth of perhaps 25 to 30 meters. The term "*Lithothamnion* Ridge" or "Nullipore Ridge" has been applied to this feature around the margins of many reefs, but similar developments have been observed around wave-cut terraces and low coastal terraces of other types of coast lines.

The "*Lithothamnion* Ridge" of coral reefs represents probably the most spectacular development of coralline algae to be found anywhere in the world. The term is a misnomer—species belonging to the genus *Lithothamnium* are seldom present. At Bikini, for example, the beautifully developed "*Lithothamnion* Ridge" is built entirely by a few species of *Porolithon*. The term "Algal Ridge" would be more correct. The outer rim, against which the waves always beat with great force at low tide, is made up of a great development of rounded algal heads, usually compact colonies of highly branching forms of the genus *Porolithon*. This belt is commonly rather narrow. Exceptionally low tides may expose a strip 40 or 50 feet wide, sometimes more, but frequently less. Behind the outer rim, in the area of less agitated waters more commonly exposed at normal low tide, there may be a wide pavement built of encrusting types (plate 139). This crust may be quite smooth or fairly irregular, cut by the upper ends of surge channels and blow holes. These two zones make up the entire "*Lithothamnion* Ridge." The calcareous algae are commonly the only plants there, but corals and mollusks may be present in varying numbers.

The greatest development of the coralline algae extends from low tide level down to depths of 30 to 70 feet, depending on local conditions. Algae continue downward to much greater depths but develop less luxuriantly, decreasing in size and abundance. Finckh (1904, p. 134) records algae as frequently occurring at depths of over 100 fathoms around Funafuti Atoll. A. Agassiz (1888, p. 287) suggests dredging living coralline algae from considerably greater depths.

Coralline algae also grow abundantly in the northern and southern seas, stretching from tropical to polar waters. Thanks largely to the work of Foslie, the *Lithothamnium* that grow in the area around the Norwegian coast are well known, and there are many observations concerning them. These *Lithothamnium* grow from the low tide limit downward. Large areas and banks of algae are exposed at low tide along Trondheim Fjord and elsewhere along the coast of Norway. Of interest is the fact that algae of this type have been found in abundance growing in the Arctic Ocean, off the northwest coast of Norway, and also in the seas around Iceland and Greenland. These algae are not merely isolated plants, but, in many cases, they have built banks of appreciable size.

Numerous observations have been made of banks of calcareous algae developing along the coasts of France, England, and Ireland.

The upper limit for growth of coralline algae is tide level or slightly above. The lower limit is not definitely known. Maximum depths from which algae have been collected are shown in Table VIII.

TABLE VIII

MAXIMUM DEPTHS FROM WHICH CRUSTOSE CORALLINE ALGAE HAVE BEEN COLLECTED (DATA FROM LEMOINE, 1940)

Location	Depth in Meters
White Sea	125
Around Spitzbergen	36
Around Iceland	40-70
Greenland	47
Ellesmereland	60
N. E. Canada	56
Anticosti Island	54-104
Western Norway	54
In the Baltic Sea	20
In the North Sea	36
Around Faroe Islands	45
Coast of Scotland	17
English Channel	43
Coast of Brittany	40
Coast of S.W. France	30
West Coast of Morocco	33-110
Cape Verde Islands	91
Coast of Florida	55
West Indies	40±
Around Mediterranean	30-120
Around Red Sea	20-30
Around Black Sea	36
Around Indian Ocean	45-109
Tropical Pacific	28-156
Antarctic Ocean	27-50

Crustose coralline algae have been collected from depths of as much as 156 meters, and some small, dark-colored, encrusting forms have been found much deeper. In connection with work around Bikini Atoll, the author observed material brought up from depths of over 1,100 feet, which was alive but not very prosperous. Apparently, algae grow at appreciably greater depths in the tropics than in the polar regions. It would appear that around coasts where the waters are more or less clouded with sediments or are freshened by the presence of river waters, the lower limit is less than it is farther out in clear waters. To date, all species obtained from depths greater than 100 meters belong to the genus *Lithothamnium*.

The depth and agitation of the water appear to have considerable influence upon the growth form of the algae. Encrusting types are found at all depths, but the highly ramified or branching forms grow only close to the surface. These branching forms are most plentiful between tide level and a depth of 30 meters. None have been found at depths greater than 80 meters. Among the encrusting forms, the thickest crusts are formed in shallow waters, and the crusts become thinner with depth, possibly as a result of slower growth.

Influence of Depth on Structure and Reproduction

Studies by French and English botanists present a wealth of data concerning specimens collected at depths ranging from the surface to about 45 meters in the English Channel. All of the specimens studied showed a reduction with depth in the size of the cells, particularly the cells of the hypothallus. A reduction in the thickness of the crusts was also observed. The hypothallus from inter-tidal specimens of *Lithothamnium polymorphum* had cells measuring from 12μ to 22μ long, while the hypothallus from bottom samples of the same species did not show any cells greater than 15μ in length, and in many cases, the cells did not exceed 10μ in length. In the case of *Lithothamnium lenormandi,* the cells of the hypothallus, which measured from 12μ to 25μ at the surface, measured only 10μ to 15μ in specimens from depth. A reduction with depth in the size of the cells of the perithallus was less clearly demonstrated, but a slight reduction did seem to occur. The presence of conceptacles or spore cases was noted in most species down to their limit of growth, except in *Lithothamnium polymorphum* which was sterile at depths below 45 meters. In all cases, however, there was a reduction with depth in the number of conceptacles.

Similar observations concerning the effect of depth on structure and reproduction of the coralline algae have been recorded for areas in the Mediterranean and the Cape Verde Islands and for a few places in the tropics.

Ecological Distribution of Species

Each of the various stages of the littoral and sublittoral zones is characterized by the presence of certain species. Along the English Channel, for example, of the 14 species of crustose corallines observed, five were localized in the littoral zone and three were found in the sublittoral zone, while six were found in both stages. The littoral stage is inhabited by certain species which are not found at depth. The same thing has been noted for groups of non-calcareous algae. Observations by Taylor (1950) and others in the tropics show a similar ecological distribution of species. The author is strongly impressed with the idea that species distribution is affected profoundly not only by depth but by the agitation and the salinity of the water and the intensity of light.

Marine Environments

Nature of the sub-stratum. Those coralline algae which develop as crusts (Melobesieae) usually are firmly attached to the sub-stratum. Some are but slightly anchored and become detached later in life; a few are free all or most of their lives.

Crustose algae show no preference for any kind of rock, and they may grow on pebbles, rocks, molluscan shells, barnacles, or pieces of glass, porcelain, or iron. Some very small species grow on other species of algae, either calcareous or non-calcareous types. In areas lacking rocks, the algae may attach themselves to sand that is held together by *Serpula* and other organisms or even to just fairly firm sand. In the tropical seas, algae grow on coral and coral debris, particularly in and around coral reefs. In so doing, the algae frequently do valuable geological work in binding these fragments together into a solid reef mass.

Those Melobesieae that form thickets of fragile sheets or leaf-like growths or that have short, more or less ramified branches, usually grow free. They rest upon the sandy bottom or muds. Very rarely do they grow on rocks or on shells.

Types such as *Lithothamnium calcareum* and *Lithothamnium fruticulosum* grow abundantly in sandy areas of the Mediterranean, along the coast of Ireland, and along the west coast of France. Many occurrences have been reported at depths of from 6 to 12 meters; some have been found at considerably greater depths. In the tropics, similar growth types have been found on muddy and sandy bottoms.

On the rocky bottoms of the Mediterranean, large areas are covered by algae and have been called the "Coralligene" areas. *Lithothamnium haucki* and numerous other species have been collected from such environments at various areas along the southern coasts of France and along the coast of Spain. Some of these forms also grow equally well in areas largely covered by mollusks.

Water circulation. The Melobesieae require a constant circulation of water. They are never found in stagnant or impure waters and a luxuriant occurrence appears to coincide with a strong agitation of the water. Pollock (1927, p. 27) has observed around the Hawaiian Islands a localization of different forms of Melobesieae. Along the outer edge of the reef exposed to the violence of the waves were encrusting forms and strong compact branching masses. Those branching forms that develop free nodules are found in depressions behind the outer part or the reef or along the reef front where the force of the waves is relatively slight; where the nodules may be rolled but not washed away by high seas.

Those inter-tidal Melobesieae that develop thin or mammillated crusts are found in exposed stations along capes, promontories, and rocky islands. In tropical regions, they are found along the borders of reefs exposed to the entire violence of the waves.

Melobesieae tolerate a violence of the wave action which no other algae can support. Locally, as for example on the "*Lithothamnion* Ridge" of certain reefs, they are the only algae present. Observations along the coast of France and the English Channel have shown some areas where a similar occurrence has developed.

The observations of many algologists in various parts of the world have shown that there is a definite localization of species in relation to the strength of the currents and the exposure to the waves.

Similarly, the distribution and growth form of the algae are affected below the zone of wave movement by the action of currents. Such a localization has been clearly demonstrated by studies of the algae of the English Channel. In a few cases, the same species may develop different growth forms under different conditions of water agitation and depth. Thus,

Lithothamnium incrustans, which normally develops abundantly in the shallower wave-swept areas along the coast, develops a growth form with rounded mammelons, of very different appearance, when it occurs in the strong currents of the deeper portions of the channel.

Branching types. A study of specimens from the coasts of Norway convinced Foslie that those algae that formed banks of considerable extent lived primarily in areas of strongly agitated water. Most of these algae form rounded heads of tightly packed branches. The only branching species known from the coasts of France, along the English Channel and the adjoining Atlantic is *Lithothamnium calcareum,* which is found along the bottoms of many bays and channels at locations where there are strong currents.

Mme. Weber van Bosse, in connection with the famous *Siboga* Expedition report on the calcareous algae of the Malaysian region, mentioned numerous areas where there were great accumulations of nodular growths of Melobesieae growing loose on the bottom. Sometimes these algae covered many acres, but always they were in strong current areas. Mme. Weber van Bosse observed that the nodules were colored on all sides, indicating live, growing surfaces, and demonstrating that the nodules were moved around by the waves. Similar observations were made by the present author in connection with the study of the reefs around Bikini in the Marshall Islands.

Along the Norwegian coast, the violence of the currents is sometimes so great that the algal nodules show indications of abrasion and scratching and the outer portions of the branches are broken and worn. Foslie attributed the growth form that appears as a well-developed, globe-like mass of long, slender, compact branches as a response to such conditions.

Thus, it appears that branching species exist only where there is certain movement of the water and that they adapt themselves to the intensity of the currents by change in growth form. Branching forms are not found very close to coasts nor between islands where circulation of the water is restricted.

Salinity. The Melobesieae live in waters of normal salinity but can exist in waters of slightly lower salinity. Detailed studies of algae along the coast of France have shown that in a number of places, particularly around the mouths of the rivers where there is a reduction in the salinity of the water, especially at low tide, certain of the calcareous algae occur in waters with salinity appreciably below that of normal sea water. However, in most of these cases, the low salinity exists only during the low tide periods. In general, the observations made along the coasts of France indicate that while both subfamilies of the Corallinaceae can stand a slight freshening of the water, the articulated coralline algae can live in fresher water for a longer time than can the crustose coralline algae.

In the open oceans, the algae growing at the surface must withstand greater variations of salinity than those that live at depth. This fact has been noted both in the Arctic regions, as for example around the coast of Greenland where melting ice sometimes freshens the surface water, and in the shallow waters around reefs and platforms in the tropics where heavy rainfall may decrease the salinity of the surface water.

Several species of the Melobesieae appear to be able to survive, for at least a short time, a concentration of salt appreciably above that of normal sea water.

Influence of light. The life of algae is related to the intensity of light. Penetration of light depends on several factors such as the clarity of the water, the intensity and quality of the solar radiation, the clarity of the atmosphere, and the presence or absence of other organisms which screen or partially screen the algae from the light. In a general way, the intensity of light varies from the Equator toward the Poles, being greatest around the Equator. The clarity of the water usually is greatest in the open seas and least around the coasts, particularly near the mouths of large rivers.

The clarity of the water in the tropics has been remarked by many students. Objects growing on the bottom sometimes can clearly be observed to depths of as much as 15 or 20 meters.

Intensity of the light. The distribution of algae as affected by the intensity of light has shown some interesting aspects. However, the observations made to date are somewhat contradictory, so it seems advisable to consider the problem from several different points of view.

A comparison of the intensity of the light under which Melobesieae can exist as compared to the non-calcareous algae, gives quite different results in different regions. In certain seas, the Melobesieae seem to need more intense light than other algae—at least, they do not descend to as great depths as other types. Around the Faroe Islands and in certain areas along the coast of Greenland, for example, the calcareous algae may descend to depths of 36 to 40 meters, while the other algae present may descend to as much as 55 meters. Similarly, studies around the Balearic Islands off the Mediterranean coast of Spain have shown that the calcareous algae do not descend to depths below 100 meters, while certain non-calcareous algae may grow at depths of as much as 180 meters. In other regions, however, the Melobesieae go to considerably greater depths than other algae. In the English Channel, other algae do not grow below about 35 meters, while the Melobesieae have been collected down to 45 or 50 meters.

Evidence as to the maximum amount of light that the Melobesieae can stand is also somewhat contradictory, probably because it is impossible to separate the effects of light and heat. As was mentioned in the discussion on the depth of water, algae may be found exposed at low tide. When they are exposed to the direct sun, they tend to become bleached and lose their pink or purplish color. This effect has been attributed by various students to the intensity of light, to heating, and to dehydration.

In a study of some of the tropical reefs, the author has been particularly interested in the "*Lithothamnion* Ridge" and has noted that at least the upper portions of the ridge are always exposed at low tide, while at exceptionally low tides, the algae may be considerably uncovered for a duration of several hours. During that time, they are exposed to the full tropical sun, yet they do not seem to lose their color. These areas are practically always kept moist and presumably are cooled by the spray from the breakers. These observations suggest that the algae will stand the light but may not stand the heating or dessication.

According to Setchell (1924), who made extensive studies around Samoa, the algae are affected less by the direct light of the sun than they are by the reflected light from the white surface of the reef. He observed that a certain number of species lived on dark basaltic rocks but were absent on the adjoining coral reefs, or, if those species were found on the reef, they were in places where they were protected from reflected light. These observations seem partly to reconcile some of the contradictory opinions of other authors.

In general, it seems that the tropical Melobesieae will support a much stronger intensity of light than will those Melobesieae of the temperate regions. Since most of the tropical crustose coralline algae show a chemical composition containing more magnesia than those of the cold waters, one wonders if this slightly different composition may affect the intensity of light which these algae can stand and which they absorb. Lemoine in 1911 called attention to the fact that the proportion of magnesium carbonate is larger among the tropical species of the *Lithothamnia* than among those species that live in temperate waters.

A number of observations indicate that there is a distribution of species according to intensity of light. For example, along the coasts of France, England, and Ireland, the species that will grow in the most shaded portions of grottos are quite different from those living nearby in the exposed areas.

In attempting to evaluate the importance of the intensity of light, it is difficult to separate the several factors involved, such as light intensity, depth of water, and so forth. The net result, however, is definitely a distribution of species both with depth and with intensity of light.

A diminution in the intensity of light also affects the structure and frutification of the algae. Along the French coast, *Lithothamnium lenormandi* growing in dark grottos is sterile, while in the same grotto, some other species of algae are able to fructify while still others are also sterile. The algae that do fructify, however, usually show a reduction in the size of the conceptacles. The cells of specimens growing in a dark grotto are often smaller in size than is normal for the same species growing outside in stronger light.

Temperature. The crustose coralline algae occur in all the seas of the world from latitude 80°31′ N., where *Lithothamnium lenormandi* has been found at a depth of 18 meters, down to latitude 73° S., where *Lithothamnium coulmanicum* has been found at a depth of 33 meters. However, the temperature of the water has a profound influence on the distribution of the genera. *Porolithon, Goniolithon,* and *Lithoporella* are essentially tropical, and they are absent in the Mediterranean area today. *Archaeolithothamnium* is limited to the tropics and sub-tropics and is represented by only one species in the Mediterranean. *Mesophyllum* and *Pseudophyllum* are known only from the warm seas.

Lithothamnium and *Lithophyllum* are universally dispersed, being represented by certain species in all areas from which the Melobesieae have been found. However, the most luxuriant growths, the largest number of species, all the very large forms, and most of the large branching forms of *Lithophyllum* are found in the tropics. Going toward the Poles, the number and variety gradually decrease, and the specimens become smaller, until, in the cold polar waters, only a few species, developing as thin crusts, are found. As nearly as can be judged from the fossil record, *Lithophyllum* has had a similar distribution throughout its known geologic history.

The temperature distribution of *Lithothamnium* is just the opposite. Today, a majority of the species occur in cool to cold waters. The largest specimens, most of the large branching forms, and practically all the well developed "*Lithothamnium* banks" are found in the polar or cold temperate seas. Foslie (1895) described many of these northern forms and mentioned their luxuriant growths along the coasts of northern Norway, Iceland, and Greenland. Going toward the Equator, the number, variety, and size of the forms decline, until, in the tropics, only a few species occur, mainly as small thin crusts. However, this was not true during Late Cretaceous and Eocene times when numerous species of *Lithothamnium* lived in the tropics and developed a variety of growth forms, including large branching types. Apparently, with the development of *Lithophyllum* and its offspring, *Goniolithon* and *Porolithon,* in the warm waters, *Lithothamnium* gradually has been pushed into the cooler waters.

Numerous observations have been made of the temperature of the water in which calcareous algae have been collected in European and tropical waters. The Charcot Expedition observed Melobesieae growing very abundantly and exposed at low tide where the temperature of the surface of the water showed oscillation only between 1.9°C and 1.1°C. Lemoine observed that in the Galapagos Islands, temperature changes of up to 10°C occurred in the period of a day, and laboratory experiments along the Adriatic coasts have shown that algae are quite adaptable and that many species can stand appreciable and rapid changes of temperature. In a bay in the Gulf of Naples, algae that were living in marine waters with a temperature of 24°C supported an elevation of temperature up to 45°C in eight hours without appearing to suffer any serious consequences.

Growth Rate

The enormous quantity of coralline algae which may be found in certain marine areas has led many people to wonder as to their rate of growth. There is, however, surprisingly little real information on this subject. Observations along the French coast indicate growth of *Lithophyllum incrustans* and *Lithothamnium lenormandi* at a rate of from 2 to 7 mm per year with a growth of from ½ to 1 mm per month in summer. In the tropical regions, the association of algae and corals on the reefs indicates that the two must have a comparable rate of growth. Extensive studies have been made of the growth rate of corals. The observations of Gardiner in the Maldives indicates that some corals may grow as much as 26½ mm per year, while Mayor observed a growth of up to 24 mm per year at Samoa.

Many observations show that encrusting coralline algae have developed on mollusks and have constructed fairly extensive crusts of moderate thickness during the lifetime of the mollusk which is usually only a matter of a few years. Similar growth on anchor chains and other parts of wrecked vessels of known date gives clues to rate of growth and indicates a comparable growth rate with that of corals.

Ecological Distribution of Articulated Coralline Algae

The articulated corallines (Corallineae), like the crustose corallines, are widely distributed in all modern seas, with their greatest development in the warmer seas. They develop as small bushy tufts, usually less than 10 cm high, formed of fine stems which are composed of cylindrical articles or flattened leaf-like segments. Articulated coralline algae commonly are strongly calcified, but because of their extremely small size, they seldom contribute much to the formation of limestones in spite of their abundance. However, some late Tertiary limestones from the Pacific islands have been observed which contain as much as 8 to 10 percent of Corallineae remains.

Geographical Distribution

Corallina and *Jania* are the only genera found in cool and temperate waters. *Amphiroa* occurs in the Mediterranean. In tropical regions, *Corallina* and *Jania* are frequently associated with *Amphiroa* on the rocks and reefs. Other genera also appear in the tropics, but they are much less common.

Ecological Distribution

The articulated corallines are often very abundant in the regions having tides of considerable amplitude. In these areas, they usually occur in the lower part of the littoral stage and in the upper part of the sublittoral zone, particularly on long, rocky coasts. Where the tides are feeble, articulated coralline algae may be observed growing on rocks and reefs in quite shallow water.

Different species of the articulated algae have adapted themselves to different environmental conditions. Some species develop in partly exposed positions; others require moderate depth; still others require even greater depth.

The upper limit of *Corallina officinalis* in the English Channel is low tide level. In some areas, this species has been observed to be exposed at low tide, however, it is practically always in places where the algae are kept well moistened by spray. Apparently, the algae can stand a slight emergence and drying, although forms that have developed under such conditions usually have a growth form somewhat different from the normal habit.

The author has observed luxuriant growths of articulated algae, belonging to the genera *Corallina* and *Amphiroa,* growing on rocks above low tide level on the rocky east coasts of Saipan and Tinian. There, however, they were kept wet by heavy spray and occasional surges from the breakers.

Corallina officinalis is particularly abundant in exposed stations where there is considerable agitation of the water. Where the water is calm, this species develops an elongated form and covers the rocks with a mantle of long slender branches.

Other species of the corallines prefer more sheltered environments. This statement is particularly true for the genera *Amphiroa* and *Jania,* although some species of *Amphiroa* have adapted themselves to varying conditions. It has been noted, however, that those species which occur in varied environmental conditions are quite variable in their growth forms and in their general appearance. Thus, *Amphiroa fragilissima* forms compact tufts composed of dense growths of short articles when it grows in exposed places, while the individuals that develop in sheltered areas are much longer and thinner.

Salinity. Corallina officinalis has been observed growing in estuaries where the water has an appreciably lower salinity than normal sea water. This species has also been observed in places where, during certain times of the year, there is an appreciable freshening of the water.

Light. In general, the articulated corallines seem to need fairly brilliant light. They practically never occur in grottos or very shaded areas, and while they will grow where there is some clouding of the water, they do not grow as profusely as where the light is brilliant.

Temperature. French botanists have made observations on the temperature changes that algae of this type can tolerate. These studies indicate that some of the Mediterranean species, at least, will stand considerable ranges in temperature.

Growth Rate

There are a few observations on the growth rate of the articulated corallines. The available information indicates that these algae may grow more rapidly than the crustose forms. Along the west coast of France, a tuft of this material grew 10 to 16 cm in a period between three and four years, giving a growth rate of 3 to 4 cm per year. In the tropics, the growth rate is apparently even more rapid.

Ecological Distribution of Green Algae

Codiaceae

All known Recent representatives of the family Codiaceae are marine. Most of them live in warm waters, that is, in the tropical and warm temperate areas, but there has been relatively little information published with regard to their ecological distribution in these areas. Probably the best known group that is of interest to us is *Halimeda.* Taylor, in his recent work on the plants of Bikini (1950), gives data for the specimens which he collected. His studies indicate that at Bikini Atoll the greatest development of the genus *Halimeda* is in the lagoons, although some fairly dense growths were noted by dredging at considerable depths, that is from 50 to 90 meters along the reef front. In the lagoons, certain species are characteristic of shallow depths (the areas exposed or nearly exposed at low tide and extending down to depths of 10 meters), while other species are characteristic of deeper waters. Where one species is found through a considerable range of depth, in the deeper waters it is usually represented by smaller, bushier forms than those seen at shallower depths. In many of the lagoon areas, the bottom is covered with a luxuriant growth, forming a *Halimeda* meadow. These meadows seem to be common even at depths as great as 50 or 60 meters. In many areas, dredging to collect bottom samples of the sediments revealed that at one time the lagoon bottom had been covered by and buried to considerable depths with *Halimeda* fragments. Along the ocean side of the reefs, *Halimeda* is found as occasional clumps in clefts in the cliff or around and between large coral knobs, but apparently, this species does not grow abundantly until considerable depth is attained, at least beyond the range of surf and heavy wave circulation. The present author found much the same distribution of *Halimeda* in his studies of the reef floras of the Marshall and Mariana Islands, and also in the Palaus. The greatest development of *Halimeda* that he noted was in the southern Palaus on the wide rock terraces and sunny lagoons to the west and northwest of Peleliu. There, at low tide, one could walk out for a considerable distance on a very gently sloping reef flat, which probably represented a wave cut rock terrace slightly overgrown by corals and algae. At normal low tide, at depths ranging from about a foot and a half to 15 or 20 feet, much of the surface seemed to be covered densely by a rich growth of *Halimeda,* including several species growing together. The adjoining islands show considerable accumulation of *Halimeda* limestones in the Recent and Pleistocene deposits.

Dasycladaceae

In 1920, Pia summarized the information then available on the ecology of the Dasycladaceae. He showed that these green algae normally occur at depths ranging from about low tide level down to 10 or 12 meters and that the most luxuriant growths are found just below tide level to depths of around 5 or 6 meters. Taylor (1950) mentioned that the few common types found at Bikini have rather seasonal developments, and that at certain times, they may occur in such abundance as to slightly color the reef flats at low tide. His observations also indicate that the dasyclads are restricted to waters less than 10 meters deep, with the greatest development close to low tide level. The present author observed rather similar development of these algae at Okinawa and Guam. At Guam, dasycladaceans occur rather commonly just below tide level, usually on the underside of rocks or coral heads, becoming more common on the bottom at deeper levels, with the greatest development apparently taking place just a few meters below low tide level.

Recent Dasycladaceae are strictly marine and are limited to tropical and warm temperate areas.

Ecological Distribution of the Charophytes

Present day charophytes live entirely submerged in shallow, quiet or very slowly moving bodies of fresh or brackish water. All the available evidence indicates that the group has been limited to similar environments since at least Early Devonian times. It is true that Recent marine sediments have occasionally yielded calcified parts of the oogonia, and similar fossils have been recorded associated with marine fossils from marine sediments in the Eocene in Florida and from various Devonian formaitons. However, most of the students of this subject believe that such occurrences have resulted from the transportation of the small, light, hollow bodies away from their normal habitat and that the charophytes have never lived in a typical marine environment. Cases have been cited of Recent charophytes growing associated with typical marine algae. A closer examination of the occurrences, however, usually shows that both types of algae have strayed from their normal habitat and are living together in a brackish-water environment. Thus, Peck (1957, p. 4) quotes Preston Cloud on the occurrence of charophytes growing with normally marine dasycladacean algae in the brackish waters of the Everglades near the southern tip of Florida, and Olsen (1944, p. 101) shows that the species of charophytes living in the Baltic Sea, which are often considered to be examples of charophytes growing in marine waters, do not occur in waters having a salinity greater than 18 parts per million.

Olsen, in his splendid work on Danish Charophyta (1944), discussed the various habitats in which the charophytes occur in Denmark. He classified the habitats as follows: fresh waters, including lakes, small sheets of water, and streams; and brackish waters, including the Baltic Sea, fjords, bays, some coastal lagoons, and coastal swamp areas. Olsen's work showed that the factors involved in the distribution of the various species of the charophytes appear to be the depth of the water, the pH of the water, the salinity of the water, and the amount of calcium carbonate in solution.

The depth of water in which charophytes will grow ranges from about half a meter to a maximum of about 9 meters. A few cases have been reported of charophytes occurring in water as deep as 25 or 30 meters, but most of these algae certainly occur in depths less than 9 meters. Probably the depth is very largely influenced by the light needs of the particular species.

Very little data are available concerning the distribution of the Charophyta with regard to temperature. Most of the studies of charopytes that have been made were carried out in the north temperate zone. Charophytes develop in enormous numbers in the glacial lakes and abandoned stream channels so common in the areas of former glaciation in the northern Mississippi Valley, New England, Great Britain, Denmark, southern Sweden, and northern Germany, as well as in Russia and Siberia. In warm countries such as India and Siam, however, charophytes occur in sufficient abundance to become a factor in problems of water pollution and purification.

The acidity or alkalinity of the water (the pH) is definitely another factor of importance. According to Olsen, pH conditions are of decisive importance regarding the distribution of charophytes. These algae have not been observed in constantly acid waters, and most of the species have been reported only from alkaline waters. However, charophytes do live in waters in which, during the course of the growing season, there is considerable fluctuation in the pH factor. A few species are known that develop almost exclusively in pure fresh water. A considerable number of species live in essentially fresh water but can develop and thrive in slightly saline waters. At the far extreme, are a number of species which occur only in brackish water, with a limiting amount of about 18 parts per million NaCl for normal growth and development. A few species of charophytes have been observed which, under either natural or experimental conditions, can for a short time stand a salinity up to about 26 parts per million.

The amount of lime in the water appears to be a factor of some importance also. Unfortunately, not enough detailed work has been done to give more than an approximate picture. From the available data, however, it would appear that there are some species which prefer waters where the content of lime is small, but that most of the species need, or at least prefer, waters in which appreciable quantities of lime are present.

Ecological Distribution of the Blue-Green Algae

Members of this group inhabit the greatest variety of environments of any of the phyla of algae. Schizophyceae live in marine, brackish, and fresh waters; in springs, both hot and cold, and both pure and highly mineralized; in salt lakes; and in moist soils.

Most of the marine forms grow in the inter-tidal zone, although a few species are planktonic. Most of the marine forms are free-living, but quite a number grow epiphytically on other marine algae.

The greatest number of the described species of blue-green algae live in fresh waters where they inhabit practically every type of fresh-water environment. Some blue-green algae can live even in rapidly moving or highly agitated waters. They occur in both permanent and temporary or intermittent bodies of water. Many of the forms found in the permanent bodies of water are planktonic and are quite seasonal in their developments. They form the green or yellowish-green scums so commonly observed on small ponds, lakes, and bodies of stagnant water. Some of the fresh-water species are sub-aerial, growing on the surfaces of moist rock, rock ledges, moist logs, and in moist soils. Quite a number of species are found in moist soils, from at or near the surface down to depths of a meter or more.

The members of the blue-green algae that have attracted the most interest are those that grow in springs, particularly high-temperature springs. Algae of this type have been found in spring waters where the temperature is only a few degrees below the boiling point, and the pastel shades observed in many hot spring deposits are due largely to their presence. Actually, as is discussed elsewhere in this treatise, some of these algae are of considerable importance in the formation of the calcareous tufas and siliceous sinters deposited by hot springs. Blue-green algae also occur in normal spring waters and in spring waters having fairly heavy concentrations of mineral matter.

Members of this group also are found in large numbers in lakes in the Arctic and Antarctic regions, where the temperatures during much of the year are quite low. Their presence has also been reported for short growing seasons in snow banks and on ice-covered slopes in Arctic or high mountain regions.

REFERENCES

Agassiz, A., 1888, Three cruises of the U. S. Coast and Geodetic Survey Steamer *Blake* in the Gulf of Mexico, in the Caribbean Sea, and along the Atlantic Coast of the United States from 1877 to 1880: London.

Chapman, V. J., 1940, Marine algal ecology: Botanical Review, v. 12, no. 10, p. 628-672.

Doty, M. S., 1957, Ecology of marine algae, annotated bibliography: Geol. Soc. America Mem. 67, v. 1, p. 1041-1050.

Finckh, A. E., 1904, Biology of the reef-forming organisms at Funafuti atoll, Section VI *in* The atoll of Funafuti: Royal Society of London, p. 125-150.

Foslie, M., 1895, The Norwegian forms of *Lithothamnion*: Det. Kgl. Norske Vidensk. Selsk. Skrifter 1894, Trondheim, p. 29-208.

——————, and Printz, H., 1929, Contributions to a monograph of the *Lithothamnia*: Kgl. Norske Vidensk. Selsk. Museet, Trondheim, 60 p., 85 pls.

Hedgpeth, J. W., and others, 1957, Treatise on marine ecology and paleoecology: Geol. Soc. America Mem. 67, v. 1 (Ecology), 1280 p.

Lemoine, Mme. Paul, 1911, Structure anatomique des Melobesiees: Annales Inst. Oceanographique, Paris (Masson et cie), v. 2, no. 2, 213 p., 5 pls.

——————, 1940, Les algues calcaires de la zone neritique: Soc. de Biogeographie (Contribution a l'etude de la repartition actuelle et passee des organismes dans la zone neritique): Paris, VII, p. 75-138.

Murray, G., 1891, The distribution of marine algae in space and in time: Proc. Trans. Liverpool Biol. Soc., v. 5, p. 164.

Olsen, Sigurd, 1944, Danish Charophyta: Det Kgl. Danske Vidensk. Selsk., Biol. Skrifter, v. 3, no. 1, 240 p., 2 pls.

Peck, R. E., 1957, North American Mesozoic Charophyta: U. S. Geol. Survey Prof. Paper 294-A, 42 p., 8 pls.

Pia, Julius, 1920, Die Siphoneae Verticillatae vom Karbon bis zur Kreide: Abhandlungen Zool.-Bot. Gesell. Wien, v. 11, pt. 2, p. 1-263, 8 pls.

Pollock, J. B., 1928, Fringing and fossil coral reefs of Oahu: B. P. Bishop Museum Bull. 55, 56 p.

Setchell, W. A., 1924, American Samoa: Part I. Vegetation of Tutuila Island, p. 1-188, 46 figs., pls. 1-20. Part III. Vegetation of Rose Atoll, p. 225-261, figs. 47-57, pls. 32-37: Carnegie Inst. Washington, Dept., Marine Biol., v. 20, Publ. no. 341.

——————, 1926, Nullipore versus coral in reef formation: Am. Philos. Soc. Proc., v. 65, no. 2, p. 136-140.

Taylor, W. R., 1950, The plants of Bikini: Univ. of Michigan Press, 227 p., 79 pls.

Weber van Bosse, Anna, and Foslie, M., 1904, The Corallinaceae of the *Siboga* Expedition: *Siboga* Expedition Repts., Brill-Leiden, Holland, v. 61, 110 p.

Systematic Descriptions
Phylum RHODOPHYCOPHYTA Papenfuss, 1946
(= Rhodophyta)

The Red Algae

The Rhodophycophyta represent the structurally highest group of algae. As in most of the major groups, however, this phylum includes forms with a wide range of structural developments. These structural features show considerable parallelism in each of the major groups.

Recent red algae are distinguished from the other major groups on the basis of the three following features:

1. Color. They contain a red pigment (phycoerythrin) and sometimes also a blue pigment (phycocyanin) in addition to the green chlorophyll. In most of the Rhodophycophyta, the phycoerythrin is present in such quantities that it masks the other pigments and gives the plants a distinctive red or purplish-red color.

2. Mode of reproduction. The red algae differ primarily from other algae by their sexual reproduction, in which non-flagellated male gametes are transported to the female carpogonium.

3. Lack of flagellated spores. The Rhodophycophyta also differ from all other algae (except the Myxophyceae) in lacking flagellated asexual spores.

These features cannot be recognized in fossil forms, however, and, consequently, cannot be used in classifying fossils.

The overwhelming majority of Recent red algae are strictly marine, although a few fresh-water forms are known. Under normal conditions, all of the marine species are sessile, and, in most cases, death soon ensues if a thallus becomes detached and free-floating. Among the crustose coralline algae, loose, broken fragments may develop into branching, although commonly infertile, plants. Marine species of red algae are found in all oceans, including the polar seas. However, most of them live in warm waters. Thirty-four percent of the Recent marine species are found in extratropical waters of the northern hemisphere, 22 percent are found in tropical waters, and 44 percent are found in the imperfectly known extratropical waters of the southern hemisphere (Setchell, 1915). The geographical distribution of the marine species is generally correlated with the surface temperature of the ocean. The increase in the temperature of the surface water as one passes from polar to tropical regions brings a change in the genera and species present. Most species of the Rhodophyceae are confined to zones of amplitude of approximately 5°C of the summer temperature, but certain species extend over zones representing a 10°C amplitude, and a few are known (Setchell, 1915) in zones with an amplitude of 20°C.

As indicated in the section on ecology, there is also great variation in the vertical distribution of Rhodophycophyta at a given locality. Some species grow only in the intertidal zone, and even there, a distinct zonation may occur. In most cases, there is but little vertical zonation among the sublittoral red algae (Børgesen, 1905). The maximum depth at which sublittoral algae will grow depends primarily upon the amount of light penetrating the water. This, in turn, depends upon the latitude and the turbidity of the water. Algae in the North Atlantic rarely grow below the 30-meter level (Børgesen, 1905; Hoyt, 1920; Printz, 1926), and the algae found at the lowest levels are almost exclusively Rhodophycophyta. In Florida (Taylor, 1928) and in the Mediterranean (Funk, 1927), where the water is clear and the sun almost directly overhead, algae have been found in abundance at the 75- to 90-meter level. At these deep-water stations, there are Chlorophycophyta and Phaeophycophyta intermingled with Rhodophycophyta. The greatest depth at which these algae have been found is about 200 meters (Printz, 1926).

A majority of the littoral marine red algae grow upon rocks or upon some other inanimate substratum, but there are also a number of species that grow upon other algae (Rhodophycophyta, Phaeophycophyta, or Chlorophycophyta). In most cases, a single species is restricted to a single host species.

REFERENCES

Børgesen, F., 1905, The algae-vegetation of the Faeröese coasts *in* Botany of the Faeröes: pt. 3, Copenhagen, p. 683-834, 12 pls., 14 figs.

Funk, G., 1927, Algae of Gulf of Naples: Pubbl. Stazione Zool. Napoli, v. 7 (supplemento), p. 1-507, 20 pls., 50 figs.

Hoyt, W. D., 1920, Algae of Beaufort, N. C.; U. S. Bureau of Fisheries Bull., v. 36, p. 371-556, 36 pls., 47 figs.

Printz, H., 1926, Algae of Norway: Skr. Norske Vidensk. Akad. i Oslo, Mat.-Nat. Kl., no. 5, p. 1-273, 10 pls., 29 figs.

Setchell, W. A., 1915, The law of temperature connected with the distribution of the marine algae: Ann. Missouri Bot. Gard., v. 2, p. 287-305.

Taylor, W. R., 1928, Algae of Florida: Carnegie Inst. Wash. Publ. no. 379, p. 1-219, 37 pls.

Class RHODOPHYCEAE Ruprecht, 1901
Order CRYPTONEMIALES Schmitz *in* Engler, 1892
Family CORALLINACEAE (Lamouroux) Harvey, 1849

The family Corallinaceae, which began in the Late Mesozoic, includes practically all the lime-secreting red algae of the present time. During the Paleozoic and Mesozoic, however, there were two other families belonging to the class Rhodophyceae, the Solenoporaceae and the Gymnocodiaceae, which were quite important. Both of these families appear to have become extinct by the end of the Mesozoic, their place in the Cenozoic being taken by the Corallinaceae. The ancestry of the Corallinaceae is not well known, but it is thought that members of this family descended from certain of the Solenoporaceae. All the Recent coralline genera appear rather suddenly in the fossil record, mostly during the Cretaceous or very early Tertiary. During the Cenozoic, the family was very important because of its abundant representation and because of the rock-building activities of some of its members.

Today, representatives of the family Corallinaceae are found in all seas, including tropical, temperate, and polar waters. The greatest variety of genera and species, and, by far, the largest number of individuals today live in the warm waters, but there are appreciable numbers of this family which are typical of even the polar seas, and in certain localities off the coasts of Norway, Iceland, and Greenland, coralline algae live in such abundance as to build sizeable banks.

Most of the genera and species of the Corallinaeae are strongly calcified. The calcium carbonate is deposited within and between the cell walls, with the result that the fossils show the cells and the structure of the tissue itself. This type of calcification permits definite identification of genera and even species.

Because of the type of calcification, the coralline algae are the most satisfactory of all the calcareous algae to study. Fossil coralline algae have been observed by geologists and paleontologists for over a century in rocks ranging in age from the Cretaceous to the Recent, but it was not until about 1911 that a detailed study of them was made. In that year, Madame Lemoine completed her great work on the anatomical structure of these algae and, for the first time, showed that the genera and species could often be identified from fragments of the tissue alone, even if the fragments were infertile. During the period from 1916 to 1940, she published a number of papers describing fossil coralline algae from many localities and put fossils of this type on a well-known scientific basis.

In addition to the anatomical and descriptive work, Madame Lemoine also showed that fossil coralline algae have both stratigraphic and ecological significance and that they have considerable possibilities as time fossils. Before such fossils can be used for accurately dating the rocks, however, it is necessary that a larger number of the species be described and their time ranges determined. Considerable progress has been made in this direction during the last ten years by European, Asiatic, and American geologists and paleontologists. The work done by the Military Geology Branch of the U. S. Geological Survey on the mandated islands of the Pacific during the last 15 years has added greatly to our knowledge in this field. In most of the Pacific islands that were studied, there are thick sequences of Cenozoic limestones which are richly fossiliferous, and the fossils include abundant foraminifera associated with the coralline algae. Consequently, the rocks can be dated on the basis of the foraminifera, and this dating permits determination of the age of the various species of algae and gives an idea of their time ranges.

The botanist working with Recent algae uses the growth habits of the plant, the structure of the tissue, the size and shape of the cells, and the structure, size, and position of the conceptacles to separate the various genera and species. Unfortunately, the paleobotanist, endeavoring to determine the fossil algae in the rocks, seldom has an opportunity to use all of these data. Usually, the specimens he has are only small parts of the plants, fragments of the crust, or pieces of the branches which, in many cases, are separated from the basal crusts. Most of the fragments obtained are sterile. Consequently, he is often forced to determine the genera and name the species without having nearly as much data as would be desired.

Chemical Composition

Because of their highly calcified character, the coralline algae have attracted considerable attention from the chemists, and many chemical analyses of these algae have been made. Several charts showing chemical compositions are given earlier in the text in the chapter on the chemical composition of the algae.

Terminology

A complete glossary giving the technical terms used in describing the various types of algae and algal limestones is appended at the end of the text. However, a few of the common and more important terms may profitably be presented here.

In discussing algae, the entire plant or plant tissue is usually spoken of as the *thallus*. Among the Corallinaceae, the tissue of the thallus commonly is differentiated into two parts, the *hypothallus* and the *perithallus*. The *hypothallus* forms the basal portion of the plant and frequently the central part of the branches. Also, in case of injury, the scar tissue that develops usually is hypothallic tissue. The *perithallus* is above the basal hypothallus in crustose forms and is outside the medullary hypothallus in branching forms.

The plants reproduce by *spores,* which develop in spore cases called *sporangia.* In most of the genera of the corraline algae, the sporangia are collected into large cases called *conceptacles.*

Determination of Genera and Species

The various genera can be separated on the basis of characteristic development in the type and structure of the hypothallus and perithallus, and on the basis of the structure, size, and arrangement of the conceptacles. Species are usually differentiated on the basis of differences in the size of the cells, the size and shape of the conceptacles, and peculiarities in the texture and structure of the tissue.

Basic Structural Features

The basic structural features are the hypothallus, the perithallus, and the conceptacles.

Among the crustose coralline algae, the basal hypothallus may develop into any one of three different types. These types are illustrated on Plate 1. In the first type, or *simple hypothallus,* we find that the bottom layer consists of a layer of large cells which gradually bend upward. In other words, the simple hypothallus consists merely of a number of curved layers of large cells (plate 1, figure 1).

The second type, or *co-axial hypothallus,* consists of arched or curved layers of cells which in vertical section appear to form semicircles or arcs (plate 1, figure 3).

The third type, or *plumose hypothallus,* consists of a long thick tissue in which the cell layers appear to start at the center and then curve more or less evenly both upward and downward, producing a feathery or plumy structure in vertical section (plate 1, figure 2).

When typically developed, these three kinds of hypothallus are quite characteristic and easily distinguished. In addition to them, however, we do find in most genera one or two species in which the hypothallus is very poorly developed. Usually, in these cases, the hypothallus has been reduced to a single, or at most a few, irregular or horizontal layers of large cells.

Normally, the cells of the hypothallus are considerably larger than are those of the perithallus. There are a number of exceptions to this statement, however, particularly in the genus *Lithophyllum.* The importance of the hypothallus varies greatly both between different species within a genus and between different genera. At one extreme, we have a monostromatic plant, such as *Lithoporella,* in which the plant consists entirely of a single layer of exceptionally large hypothallic cells, except around the conceptacles where several layers of small cells may be present. At the other extreme, we find certain species of *Lithothamnium* or *Lithophyllum* in which the hypothallus is absent. However, in most of the cases, the hypothallus is fairly well developed, and in quite a number of species, it normally may be thicker than the perithallus.

In many branching forms of coralline algae, there is a distinct differentiation of the tissue into a thick central medullary hypothallus and a thin outer marginal perithallus. In the genus *Mesophyllum,* however, this differentiation of tissue in branching forms is not clear. The same is true for some branching forms of *Archaeolithothamnium,* and the tissue is not strongly differentiated among most branching species of *Lithothamnium,* although there are some exceptions to this statement. In *Lithophyllum,* the medullary hypothallus is distinct and of the co-axial type. In some species of *Archaeolithothamnium,* a co-axial medullary hypothallus is present. This statement is particularly true in a number of the Late Cretaceous and very early Tertiary species. A co-axial hypothallus also is usually well developed among the branching species of *Goniolithon* and *Porolithon* and among practically all of the articulated coralline algae.

As Madame Lemoine pointed out in her classic work of 1911, there are two quite distinct basic types of perithallic tissue among the coralline algae. In type 1, the basic structure results from the presence of threads of cells which, in crusts, develop from the top of the hypothallus and continue upward as distinct threads. The cells that make up these threads are separated by cross partitions. The cross partitions in adjoining threads of cells may or may not be at corresponding levels. Commonly, they are not. The walls surrounding the cell threads are thicker and much more distinct than the cross partitions that separate the cells. As a result, these cell threads show up as the most conspicuous element in the perithallic tissue.

In type 2, the tissue appears to be composed of horizontal layers of cells. While the cell threads can be distinguished, it is observed that the cells are of equal size, that the cross partitions in adjoining threads occur at corresponding positions, and that the cross partitions appear to be stronger than the vertical partitions. Consequently, in cross section, the tissue appears to be composed of distinct horizontal layers of cells.

In addition to these two distinct basic types of tissue, the perithallus is often said to be regular or irregular. In a regular tissue, the sizes of the cells are quite similar, and the cell layers or cell threads are of approximately equal height and diameter. In an irregular tissue the cell layers or cell threads may vary greatly in size, and the individual cells within the layers or threads may show considerable variation also.

In addition to the above features, the perithallic tissue in certain genera contains *megacells.* These are cells much larger than the average cells of the perithallic tissue. The megacells may occur singly, in vertical clusters, or in horizontal lenses.

Examples of the various types of perithallic tissue are shown in Plate 2.

The different types of perithallic structures are features that are quite characteristic of genera and of species within genera. For example, tissue of type 1 is characteristic of *Archaeolithothamnium* and of *Lithothamnium.* The second type of tissue is characteristic of *Lithothamnium,* while perithallic tissue, usually of type 2, containing megacells is distinctive of *Goniolithon, Fosliella,* and *Porolithon* (plate 3).

Among the coralline algae, the sporangia usually are collected into conceptacles. The one exception is in the genus *Archaeolithothamnium* where the sporangia occur as lenses or layers with the sporangia isolated within the tissue (plate 2, figure 1).

There are two distinct types of conceptacles observed among the coralline algae: those which have a single large aperture in the center of the roof for the escape of spores and those in which there are a number of apertures in the roof.

The conceptacles are of three kinds: male, female, and asexual, or as they are often called, tetrasporic (plate 5). The male and female conceptacles usually have a single aperture, but among the asexual conceptacles, we find both of the types already mentioned, those with a single aperture and those with multiple apertures. The multi-apertured type is characteristic of the genera *Lithothamnium* and *Mesophyllum,* while the single-apertured type is found among the genera *Lithophyllum, Porolithon, Goniolithon,* and *Lithoporella,* as well as among practically all of the articulate coralline algae.

Key to the Tribes and Genera of the Crustose Coralline Algae

I. Sporangia collected into conceptacles III
Sporangia not collected into conceptacles II

II. Tissue many layered with hypothallus and perithallus .... *Archaeolithothamnium*

III. Conceptacles perforated by a few or many pores IV
Conceptacles perforated by a single pore V

IV. Roof of sporangial conceptacles perforated by few to many pores .... Tribe Lithothamnieae
- 1. Thallus self sustaining, not parasitic 2
  - 2. Hypothallus a single layer of cells, at least in part; thallus epiphytic; hypothallic cells in section square or somewhat horizontally elongated .... *Melobesia*
  - 2. Hypothallus of many layers of cells 3
    - 3. Hypothallus of curved rows of cells 3A
    - 3. Hypothallus co-axial (arched rows or layers of cells) .... *Mesophyllum*
      - 3A. Sporangial conceptacles superficial or submersed .... *Lithothamnium*
      - 3A. Sporangial conceptacles deeply immersed .... *Clathromorphum**

V. Roof of sporangial conceptacles perforated by a single pore .... Tribe Lithophylleae
- 1. Megacells present 2A
- 2. Megacells absent 2B

*These genera have not been found as fossils and are not described in this book

2A. Hypothallus consists of a single layer of cells 3
- 3. Hypothallic cells vertically and obliquely elongated .... *Hydrolithon**
- 3. Hypothallic cells square or nearly so .... *Fosliella**

2A. Hypothallus consists of several to many layers of cells 4
- 4. Megacells in lenses or horizontal clusters .... *Porolithon*
- 4. Megacells in vertical rows or singly .... *Goniolithon*
- 4. Megacells in both horizontal and vertical clusters .... *Paraporolithon*

2B. Hypothallus composed of cubic cells 3
- 3. Thallus of several layers of cells .... *Heteroderma**
- 3. Thallus of many layers of cells, not epiphytic .... a
  - a. Hypothallus commonly coaxial but sometimes of curved layers of cells; perithallus of layers of cells .... *Lithophyllum*
  - a. Hypothallus of curved layers of cells; perithallus of vertical threads of cells .... *Tenarea*

2B. Hypothallus of one or two layers of obliquely elongated cells 4
- 4. Thallus characteristically epiphytic or epizoic; commonly expanding locally to two or more layers with cells nearly equidimensional .... *Dermatolithon*
- 4. Thallus prostrate, epiphytic or epizoic, often superimposed; single layered except immediately around conceptacles; cells vertically elongated .... *Lithoporella*

2B. Hypothallus of large polygonal cells 5
- 5. Medullary hypothallus coaxial .... *Aethesolithon*

---

* These genera have not been found as fossils and are not described in this book.

TABLE IX

DIAGNOSTIC FEATURES OF THE COMMON GENERA OF CRUSTOSE CORALLINE ALGAE

	Structural Features									
Genus	Hypothallus, Type 1	Hypothallus, Type 2	Hypothallus, Type 3	Perithallus, Type 1	Perithallus, Type 2	Megacells absent	Megacells present	Sporangia not in conceptacles	Conceptacles with numerous apertures	Conceptacles with a single aperture
Archaeolithothamnium	X			X		X		X		
Lithothamnium	X			X		X			X	
Mesophyllum		X			X	X			X	
Lithophyllum		X			X	X				X
Goniolithon		X			X		X			X
Porolithon		X			X		X			X
Paraporolithon		X			X		X			X
Lithoporella			X			X				X
Dermatolithon			X		X	X				X
Melobesia			X		X	X			X	
Tenarea	X			X		X				X

Subfamily MELOBESIEAE

Genus *Archaeolithothamnium* Rothpletz, 1891

Plate 2, figure 1; plate 9, figures 1-3; plate 131.

Description—

Tissue formed of many layers of cells, differentiated into hypothallus and perithallus. The basal hypothallus is formed of curved layers of cells. A distinct medullary hypothallus occurs in some branching forms. In crustose forms, the perithallus is commonly quite regular. The most distinctive feature of the genus is that the sporangia are not collected into conceptacles, but form lenses or layers embedded in the tissue.

Remarks—

The tissue strongly resembles that of *Lithothamnium*. In numerous cases, infertile crusts of one genus cannot be distinguished from infertile crusts of the other. The author has observed that in thin sections well-preserved tissue of *Archaeolithothamnium* commonly will have a brownish tint, while the tissue of *Lithothamnium* will be dark gray or black, but this generalization is not always true.

Generic range —

Jurassic to Recent. Reached the zenith of its development during the Late Cretaceous and Eocene.

Geographic distribution—

All Recent species are confined to tropical and subtropical marine waters, and apparently fossil species had similar restrictions.

Genus *Lithothamnium* Philippi, 1837
Plate 1, figures 1, 2; plate 4, figure 1.

Description—

Tissue composed of numerous layers of cells, differentiated into hypothallus and perithallus. Commonly, the basal hypothallus consists of threads of cells which start horizontally, then curve upward as in *Archaeolithothamnium*. In quite a number of forms, however, the hypothallus may be relatively thick, consisting of threads of cells which curve both upward and downward from the middle horizontal layer, forming a plumose structure in vertical section. The perithallus consists of vertical threads of cells. The tissue may be quite regular with cells of nearly equal length and horizontal partitions occurring at about the same level, or, more often, it may be irregular with adjoining cells of unequal length. The sporangia are collected into conceptacles, which have numerous apertures in the roofs.

Remarks—

This genus contains a great variety of growth forms ranging from thin, encrusting forms to large, thick, branching types. Today, *Lithothamnium* occur in all shallow seas. Most Recent tropical species are thin crusts; the large, highly branched forms occur in the Arctic regions. This does not seem to have been true during the Eocene when both thin crusts and highly branching types occurred in the warm water floras.

Generic range—

Upper Jurassic ? or Lower Cretaceous to Recent.

Geographic distribution—

All seas, but greatest development today in cool to cold waters.

Genus *Mesophyllum* Lemoine, 1928
Plate 4, figure 2.

Description—

Structurally, this genus lies between *Lithothamnium* and *Lithophyllum*. It has tissue similar to that of *Lithophyllum*, with a perithallus composed of layers of cells and, commonly, a co-axial hypothallus, and it has multi-apertured conceptacles similar to those of *Lithothamnium*.

Remarks—

Branching species commonly show pronounced growth zones suggestive of *Lithothamnium*. Fertile specimens often show numerous, relatively large conceptacles which actually occupy an appreciable percentage of the volume of a branch.

Generic range—

Eocene to Recent. May have reached its zenith during the upper Eocene.

Geographic distribution—

Essentially tropical and subtropical seas today.

Genus *Lithophyllum* Philippi, 1837

Plate 1, figure 3; plate 2, figure 2; plate 5, figures 1-3; plate 8, figure 1; plate 10, figures 1-3.

Description—

This genus also has a tissue normally differentiated into a hypothallus and a perithallus. Characteristically, the basal hypothallus is co-axial, that is, formed of regularly curved or arched layers of cells. In some cases, however, it consists of a few irregular or curved rows of cells. Rarely, it is absent. The perithallus is formed of regular layers of cells. Branching species have a well-developed co-axial medullary hypothallus, surrounded by a thinner marginal perithallus. The sporangia are collected into conceptacles which are pierced by single large openings in each roof for the escape of spores.

Remarks—

Today, *Lithophyllum* has its greatest development in the warmer seas where the majority of the species and all the large forms occur. Probably the same was true during the past.

Infertile specimens often cannot be distinguished from infertile specimens of *Mesophyllum,* although, commonly, the latter show more pronounced growth zones, especially in branching forms.

Generic range—

Late Cretaceous to Recent.

Geographic distribution—

All seas, but greatest development in the tropics.

Genus *Goniolithon* Foslie, 1900

Plate 3, figure 1; plate 8, figure 2.

Description—

Resembles *Lithophyllum* in conceptacles and hypothallus. The perithallus is similar to that of *Lithophyllum* except for the development of megacells, which normally occur in vertical clusters of 2-5. The presence of these megacells makes the tissue rather irregular.

Remarks—

Species with both crustose and strongly branching habits are known.

A few widely distributed Recent species are found in the tropical Pacific.

Generic range—

Miocene to Recent.

Geographic distribution—

The tropical oceans.

Genus *Porolithon* Foslie, 1909

Plate 3, figures 2, 3; plate 7, figures 1, 2.

Description—

Porolithon closely resembles *Lithophyllum* in type of conceptacles and hypothallus. The perithallus differs from that of *Lithophyllum* in having numerous lenses of megacells, more or less regularly scattered throughout the tissue. In addition, lateral pores connecting adjoining cells are common.

Remarks—

The genus contains both crustose and branching forms. It is abundantly represented today in the tropical Pacific by a relatively small number of widely distributed species.

Generic range—

Late Miocene to Recent.

Geographic distribution—

Tropical seas.

Genus *Paraporolithon* Johnson, 1957

Plate 11, figure 3.

Description—

This genus resembles *Lithophyllum* in type of conceptacles and hypothallus but differs in having small, horizontal lenses of megacells similar to those of *Porolithon*. *Paraporolithon* also has short vertical columns of megacells similar to those of *Goniolithon*.

Remarks—

Possibly an ancestral form for both *Porolithon* and *Goniolithon*.

Generic range—

Lower Miocene.

Geographic distribution—

Saipan and Guam, Mariana Islands.

Genus *Aethesolithon* Johnson, 1961

Plate 10A, figures 1-3.

Description—

Plants crustose and branching. Tissue irregular, with irregular layers or lenses of cells. Cells rounded to polygonal, frequently large. The branching forms show a well-developed medullary hypothallus of large polygonal cells, and a marginal perithallus of irregular layers of rounded to polygonal cells. Conceptacles small and highly arched, probably with a single aperture.

Remarks—

This genus, with its irregular tissue and large polygonal cells, is quite distinctive. It differs appreciably from the typical coralline algae which have relatively small cells, normally rectangular in longitudinal section.

Generic range—

Middle and upper Miocene.

Geographic distribution—

Guam and the Philippine Islands.

Genus *Lithoporella* Foslie, 1909

Plate 11, figures 1-2; plate 12, figures 1, 3.

Description—

Thalli form very thin crusts. Each thallus consists of a single layer of large, elongated cells, except around conceptacles where the thalli thicken and several layers of small cells may be present. The conceptacles are similar to those of *Lithophyllum,* each having a single large aperture in the roof for the escape of spores.

Remarks—

Lithoporella are very common and widespread. The thalli grow attached to other calcareous algae, corals, Foraminifera, Bryozoa, and other organisms.

Species are separated on the basis of cell and conceptacle size, but determining the species is more difficult than in other genera of crustose coralline algae because of the great range in cell size even in one individual and the wide overlap in cell dimensions among described species. The size and shape of the conceptacles are better criteria. However, there is quite a size range in the conceptacles, and it unfortunately happens that the majority of the fossil fragments observed in sections are infertile.

Classification is further hampered by the great differences in cell lengths, even in the same specimen, and by the range in size of conceptacles, where present. These ranges are given in the resume of the characteristics of the principal modern species listed below.

Species	Cells	Conceptacle diameters
L. melobesioides Foslie	25-85μ by 15-30μ	600-1000μ
L. atlantica Foslie	32-60μ by 18-40μ	500- 800μ
L. conjuncta Foslie	36-55μ by 14-30μ	400- 800μ

These species show such great and overlapping variation that they could easily be considered as representing one variable species.

The same wide variation is true of the observed fossils, and because most of them fall within the range of the Recent *Lithoporella melobesioides* Foslie, they are commonly attributed to that species, which thus attains a time range from Eocene to Recent.

There has been some question among phycologists and paleontologists whether these monostromatic genera represent early, primitive developments or whether they represent a later, more or less degenerative evolution from thicker forms. The recent finding of Late Jurassic specimens with almost identical structure suggests they represent an early development among the coralline algae.

Generic range—

Upper Jurassic to Recent.

Geographic distribution—

All seas, but greatest development in warm waters.

Genus *Dermatolithon* Foslie, 1899

Description—

Plants develop as thin crusts that are circular or irregular in outline. They grow on other algae, shells, coral, or other hard objects. Thalli may grow one upon another. Each thallus only a few layers of cells thick, but contains both hypothallus and perithallus. Hypothallus consists of one or two layers of long cells which are vertically and obliquely elongated. Perithallus built of a few layers of nearly cubic cells. Conceptacles slightly to strongly convex, each with a single aperture in the roof.

Generic range—

Eocene to Recent.

Geographic distribution—

Most warm and temperate shallow marine waters.

Genus *Melobesia* Lamouroux, 1812

Plate 12, figure 2.

Description—

Thallus forms a very thin crust, one to several layers of cells thick. Cells cubic or horizontally elongated. Conceptacles have several to many pores in the roofs.

Remarks—

Infertile specimens of this genus cannot always be distinguished from infertile specimens of *Dermatolithon*.

Generic range—

Eocene to Recent.

Geographic distribution—

Temperate and tropical marine areas, especially littoral zones.

Genus *Tenarea* Bory, 1832

Description—

Thallus crustose; hypothallus simple. In the perithallus, the cell layers are not apparent or are irregular, and the cell threads commonly are conspicuous. Conceptacles are strongly arched, each with a single aperture.

Remarks—

Structurally, this genus appears to be intermediate between *Lithophyllum* and *Lithothamnium,* having a tissue similar to that of the latter and conceptacles similar to those of the former.

Tenarea is a widespread tropical and subtropical genus today. It is the main constituent of the widespread crustose pavements known as "trottoirs" along the Mediterranean.

Generic range—

Pliocene (possibly late Miocene) to Recent.

Geographic distribution—

Fossil forms have been found in the Mediterranean and Caribbean regions, and in the Mariana Islands. Today the genus is found in most warm marine waters.

Subfamily CORALLINEAE

Genus *Amphiroa* Lamouroux, 1812.

Plate 4, figure 3; plate 13, figure 3; plate 14, figure 5.

Description—

Plants are clusters of segmented fronds which branch dichotomously or trichotomously at regular intervals. Conceptacles are lateral. Segments are cylindrical to flattened or show a thicker center and thinner margins. Within the well-developed hypothallus of individual segments, one or more tiers of long cells commonly alternate with single tiers of shorter ones. Marginal perithallus moderately to well developed.

Remarks—

Representatives of this genus are abundant in modern warm and temperate seas. Pleistocene and Recent forms commonly have alternations of layers of long and short cells in the medullary hypothallus. This alternation also occurs in forms from earlier epochs but is much less common.

Generic range—

Late Cretaceous to Recent.

Geographic distribution—

Widespread in warm and temperate marine waters.

Genus *Arthrocardia* Decaisne (emend. Areschoug), 1852

Description—

Fronds fragile; branching pinnate-cymoid; nodes unizonal. Medullary filaments of segments straight, with cells in transverse zones of equal length. Conceptacles terminal, in cymoid clusters, upright or with pores apical.

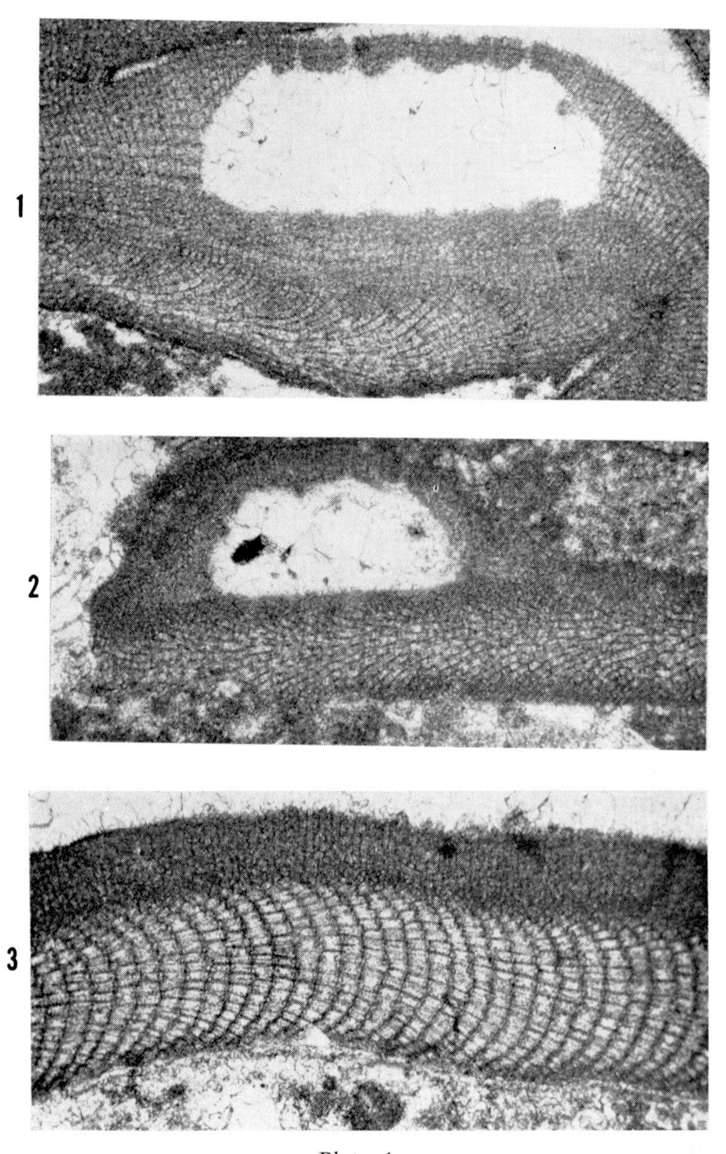

Plate 1
Coralline Algae — Types of Hypothallus
(Published by permission, U. S. Geological Survey)

Figure 1. Simple hypothallus of curved threads of cells (type 1) (also shows a conceptacle chamber with many apertures) (x100). *Lithothamnium* cf. *L. bofilli* Lemoine. Eocene of Ishigaki.

Figure 2. Plumose hypothallus (type 3) and a multi-aperture conceptacle chamber (x100). *Lithothamnium* species, Eocene of Saipan.

Figure 3. Coaxial hypothallus (type 2). *Lithophyllum prelichenoides* Lemoine (x100), Miocene of Saipan.

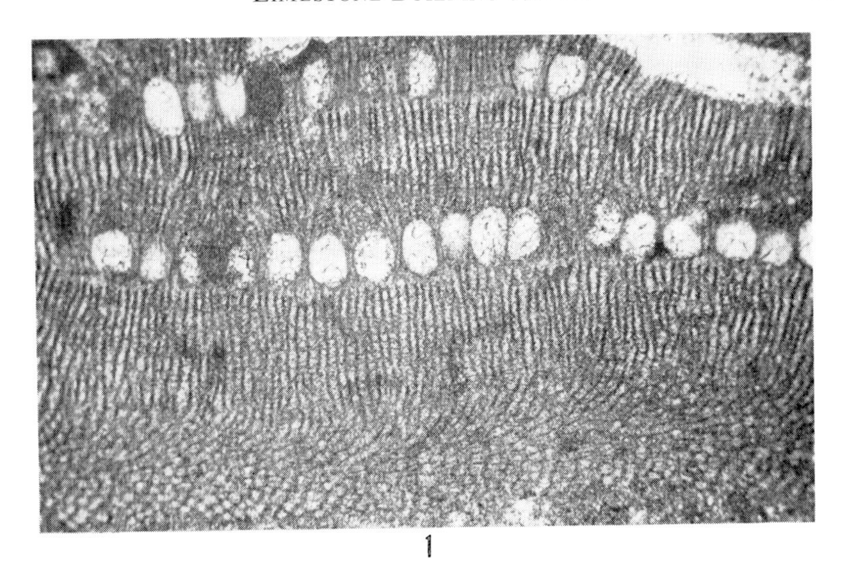

1

2

Plate 2
Coralline Algae — Types of Perithallus
(Published by permission, U. S. Geological Survey)

Figure 1. Type 1. Pronounced vertical threads of cells (horizontal partitions thin and inconspicuous). Some hypothallus at base (x100). Sporangia imbedded in the tissue. *Archaeolithothamnium* species, Eocene, Saipan.

Figure 2. Type 2. Horizontal partitions strong, giving a layered appearance (x100). *Lithophyllum* species, Eocene of Palau.

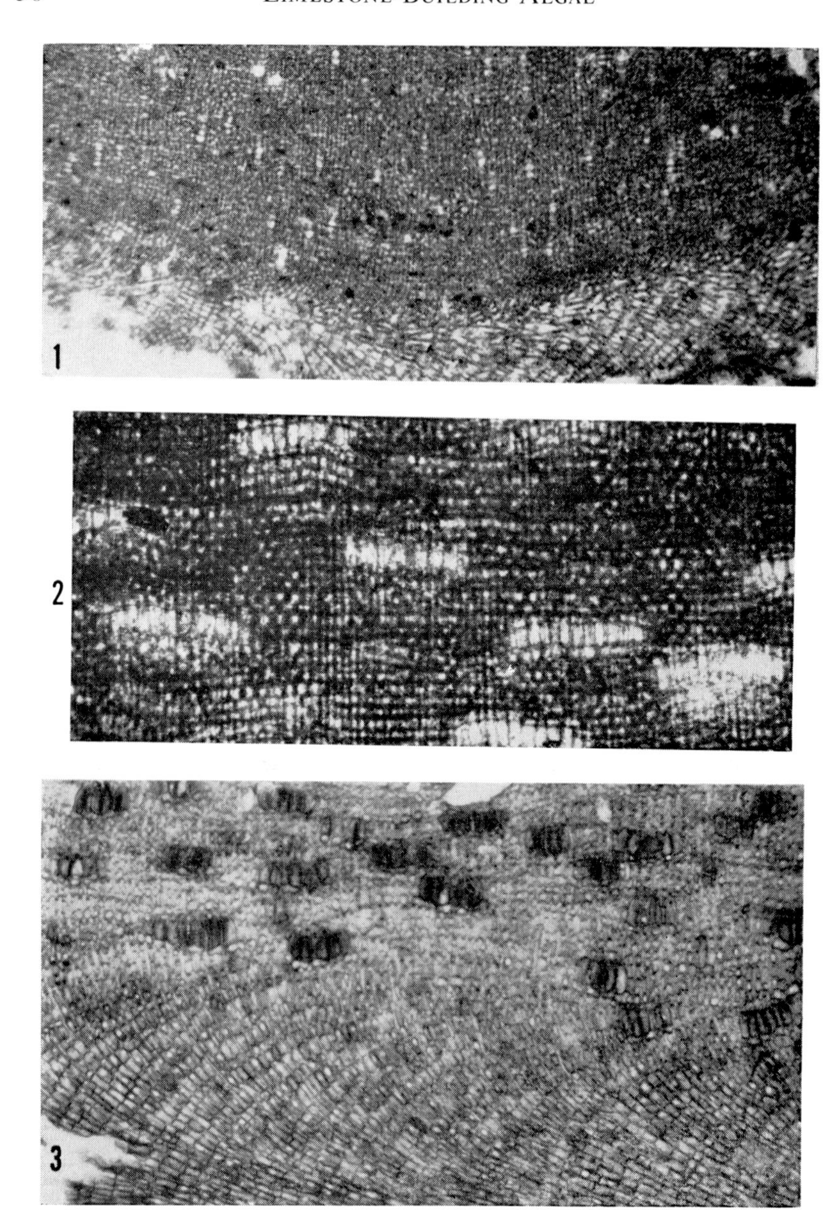

Plate 3

Genera *Goniolithon* and *Porolithon*

Figure 1. *Goniolithon* cf. *G. fosliei* (Heydr.) Foslie. Recent, Guam. Vertical section (x50).

Figure 2. *Porolithon craspedium* Foslie. Recent, Bikini Atoll. Detail of a vertical section showing clusters of megacells (x125).

Figure 3. *Porolithon gardineri* Foslie. Recent, Bikini Atoll (x125).

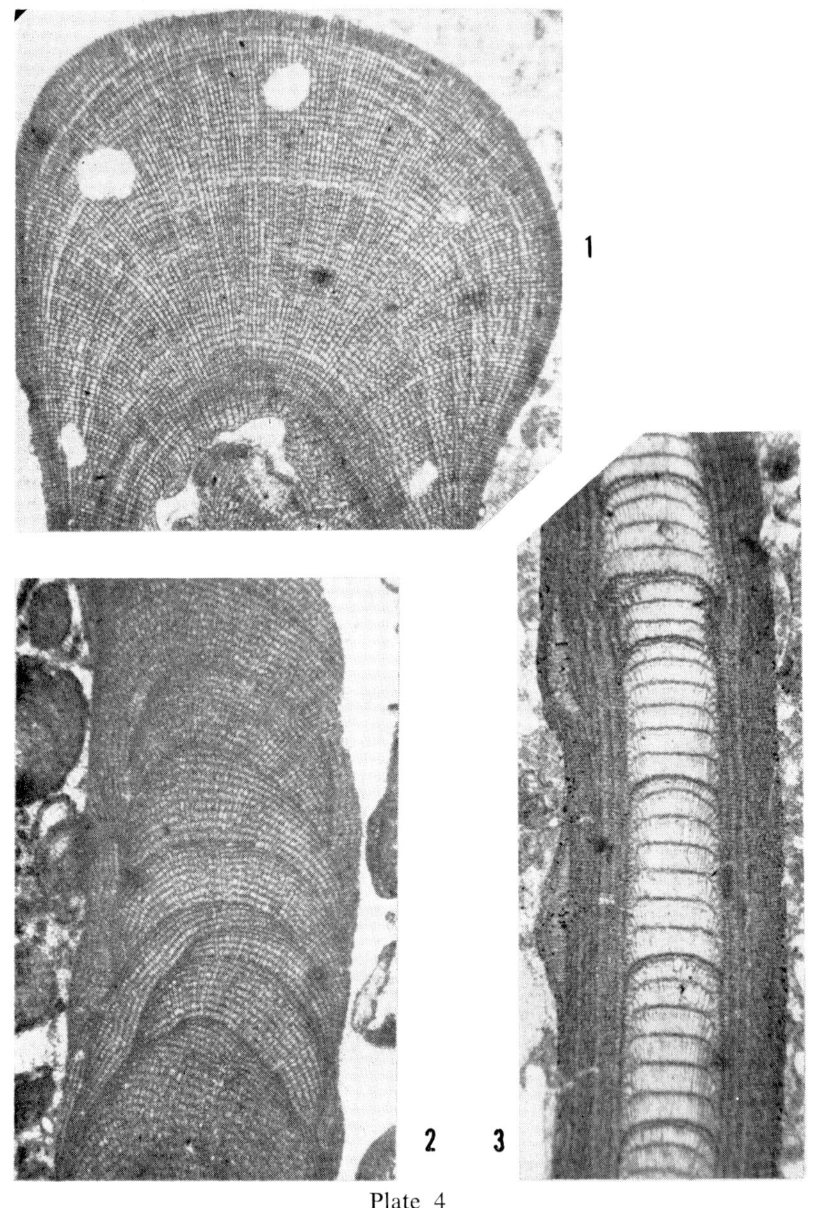

Plate 4
Corallinaceae — Structure of Branches
(Published by permission, U. S. Geological Survey)

Figure 1. Small branches or mammillae formed of regular perithallic tissue of type 1. (x50). *Lithothamnium* species. Pliocene of Miyako, Ryukyu Islands.
Figure 2. Spine-like branch of perithallic tissue with strong lenticular growth zones (x50). *Mesophyllum guamense* Johnson, Lower Miocene of Guam.
Figure 3. Long slender branch with well-developed medullary hypothallus and marginal perithallus (x40). *Amphiroa* species, Miocene of Guam.

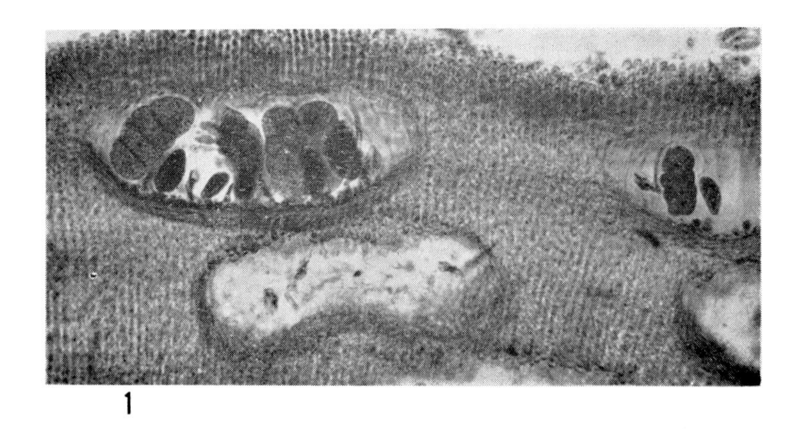

1

2

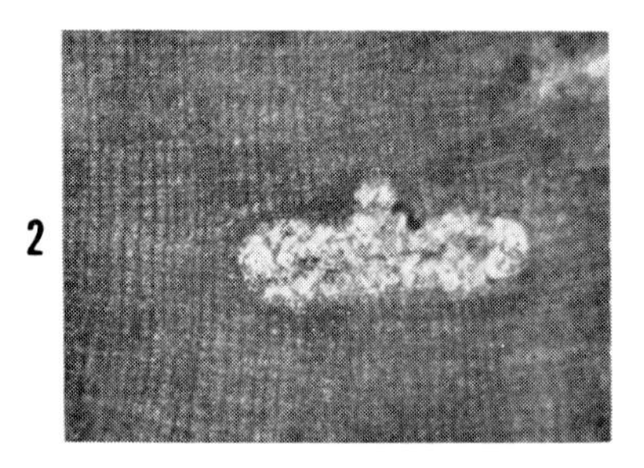

3

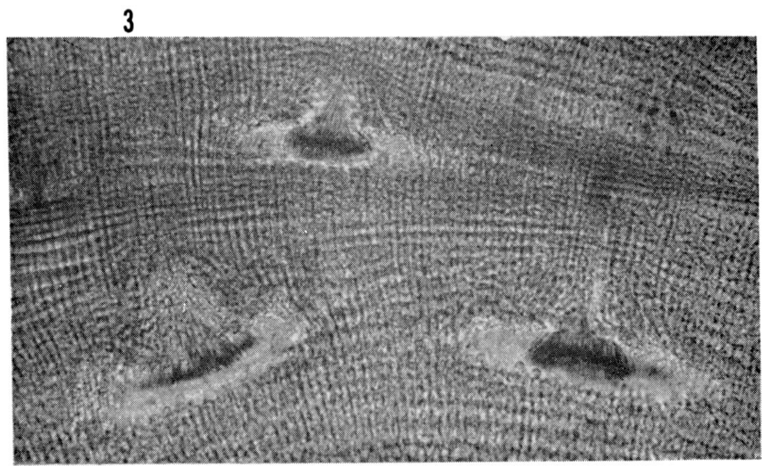

Plate 5

Coralline Algae — Conceptacles

Figure 1. Conceptacle, single-aperture type with tetraspores (x150).
Figure 2. Conceptacles with single aperture (x100).
Figure 3. Male conceptacles (x150).

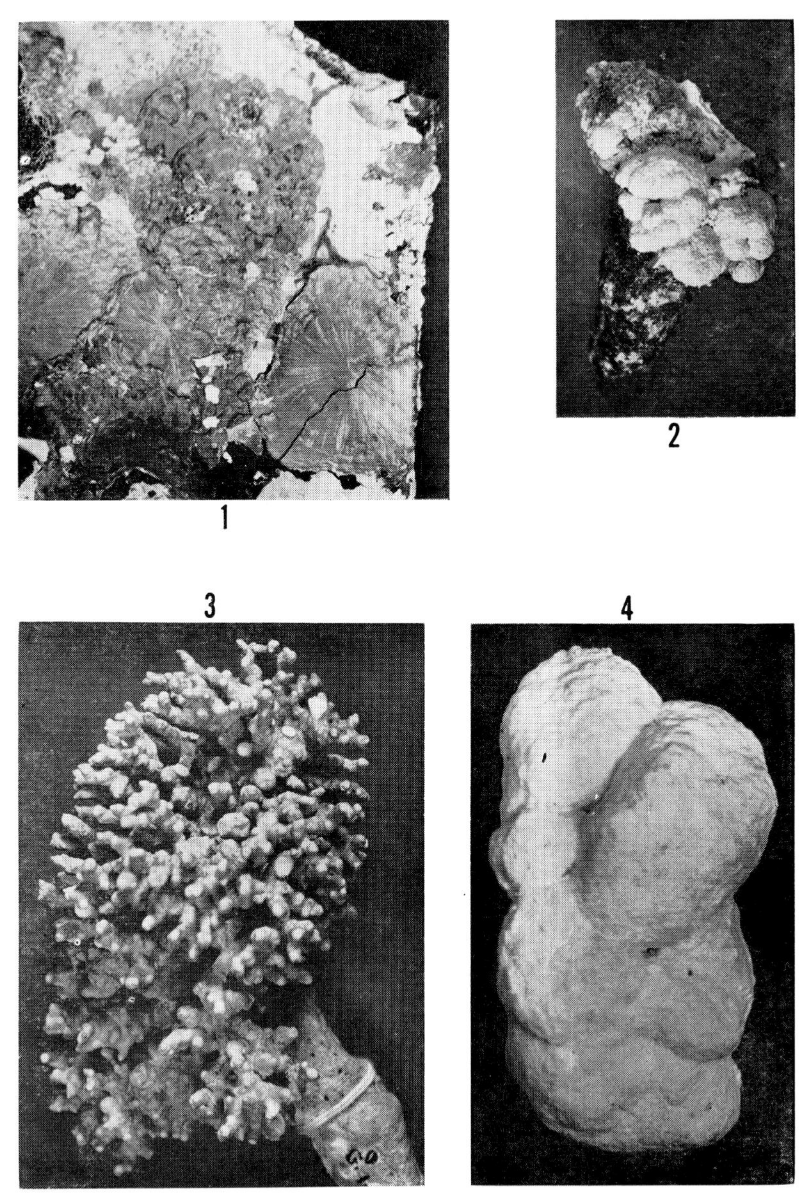

Plate 6
Growth forms of Coralline Algae
Figure 1. Very thin crusts on a piece of glass.
Figure 2. Small nodular masses on a piece of coral.
Figure 3. Short branches on a piece of coral.
Figure 4. Irregular nodular mass. All x1. Recent, Guam.

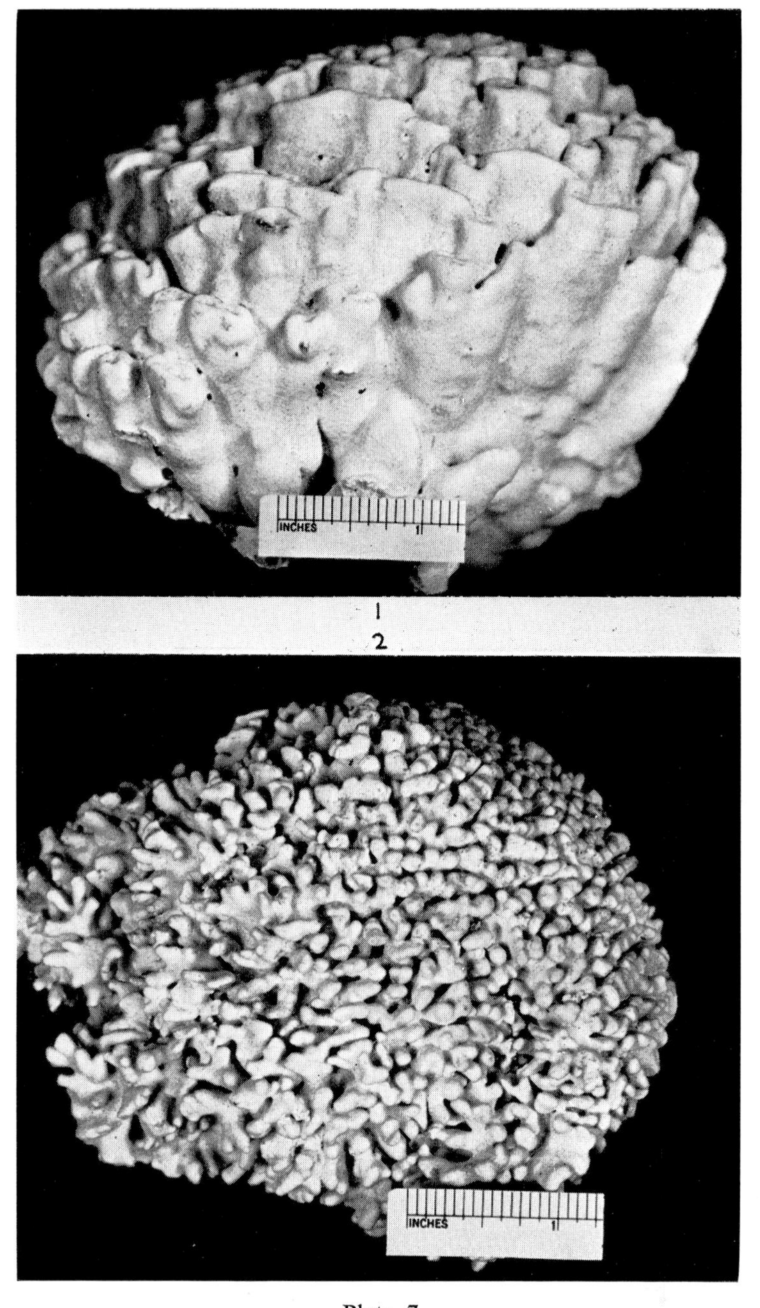

Plate 7
Coralline Algae — Growth Forms
(Published by permission, U. S. Geological Survey.)

Figure 1. A massive, compact plant with partly fused, large digitate branches *(Porolithon craspedium).* Recent, Bikini Atoll.

Figure 2. A compact plant with closely packed branches *Porolithon gardi neri Foslie.* Recent, Bikini Atoll.

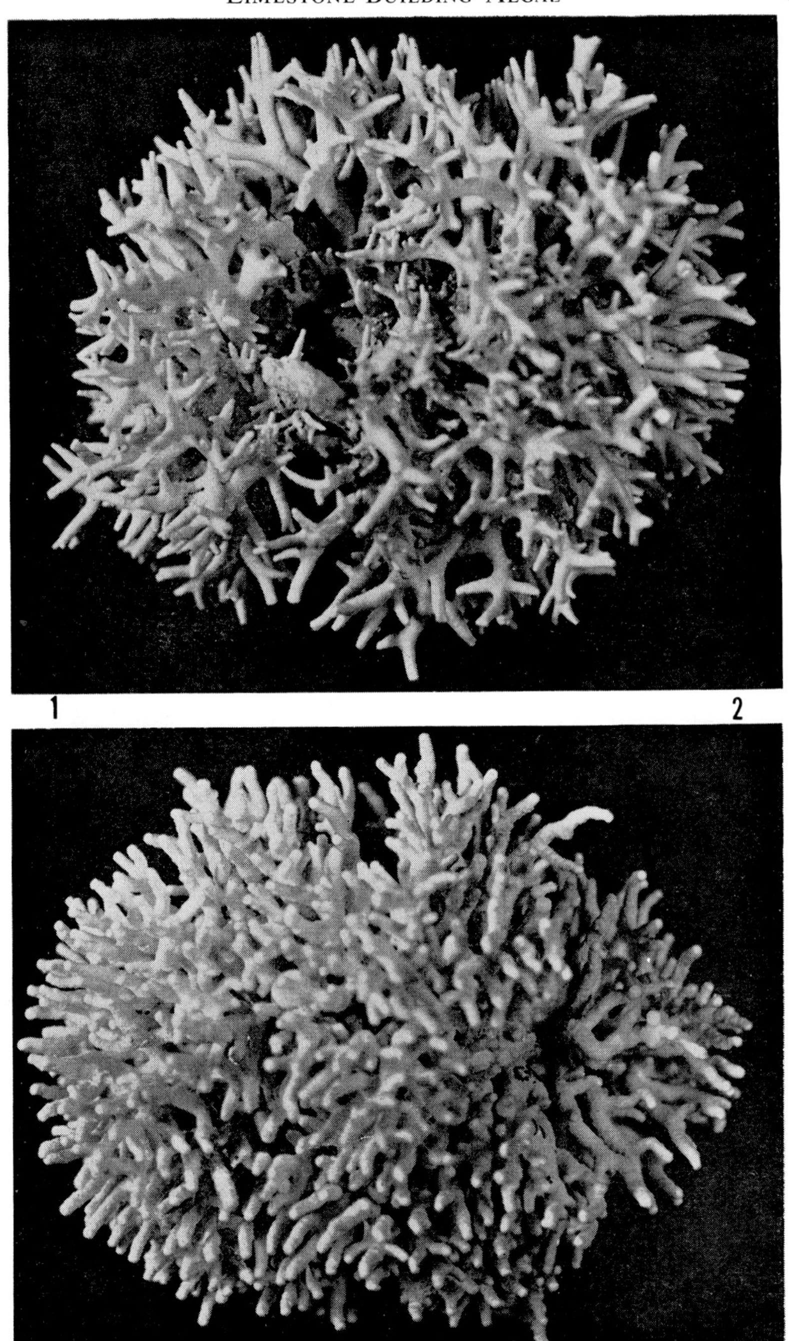

Plate 8
Coralline Algae — Growth Forms
Loose Branching Masses.
Figure 1. *Lithophyllum moluccense* (x1). Recent, Guam.
Figure 2. *Goniolithon frutescens* (x1). Recent, Guam.

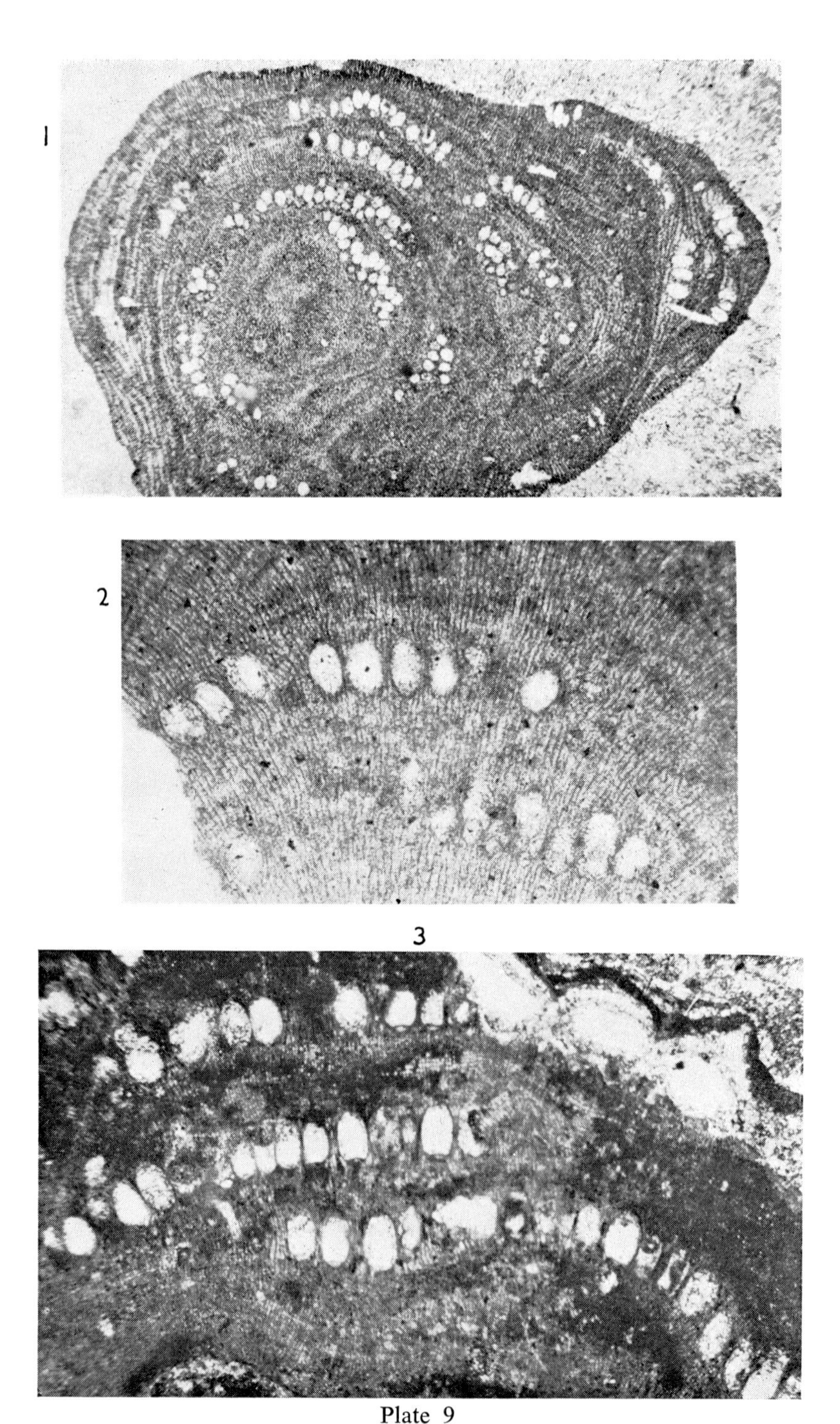

Plate 9

Genus *Archaeolithothamnium*

Figure 1. *Archaeolithothamnium lugeoni* Pfender (x40). Eocene of Peru. Shows growth zones in a section of a nodular crust and irregular layers of sporangia.

Figure 2. *Archaeolithothamnium floridium* Johnson and Ferris. A detail (x75) of perithallic tissue and sporangial chambers. Eocene of Florida.

Figure 3. *Archaeolithothamnium lauensum* Johnson and Ferris (x100). Miocene, Lau, Fiji.

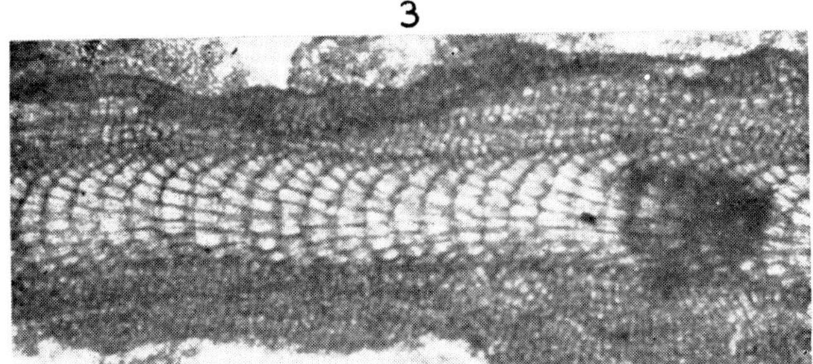

Plate 10

Genus *Lithophyllum*

Figure 1. Vertical section (x75) showing typical coaxial hypothallus and regular perithallus. Miocene, Lau, Fiji.

Figure 2. Slightly oblique section (x75) of perithallus and conceptacle chamber. Miocene, Lau, Fiji.

Figure 3. Section of thin crust (x75) with coaxial hypothallus surrounded by perithallic tissue. Eocene, Borneo.

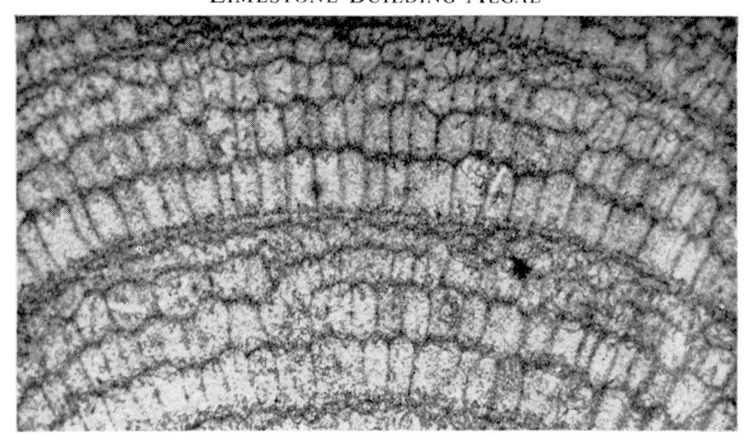

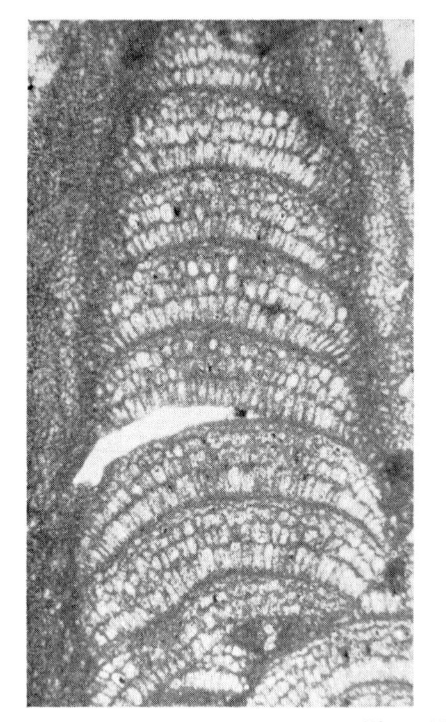

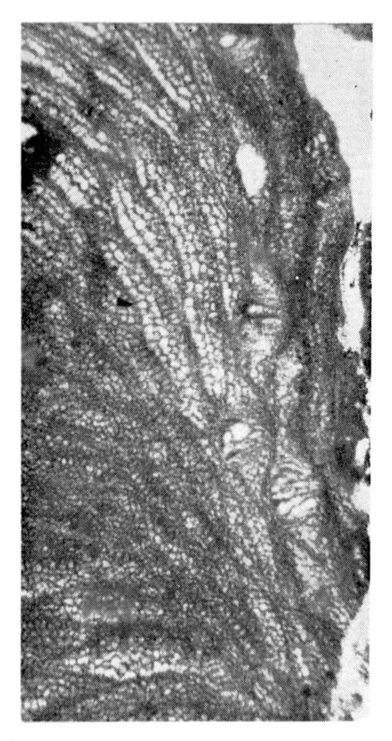

Plate 10A
Genus *Aethesolithon*

Figure 1. A slightly oblique vertical section (x50) of *Aethesolithon problematicum* Johnson. Miocene, Bonya limestone of Guam. Shows the well-developed medullary hypothallus.

Figure 2. A vertical section of *A. grandis* Johnson (x50) showing the irregular perithallic tissue with conceptacles.

Figure 3. A detail of the medullary hypothallic tissue (x125) in vertical section.

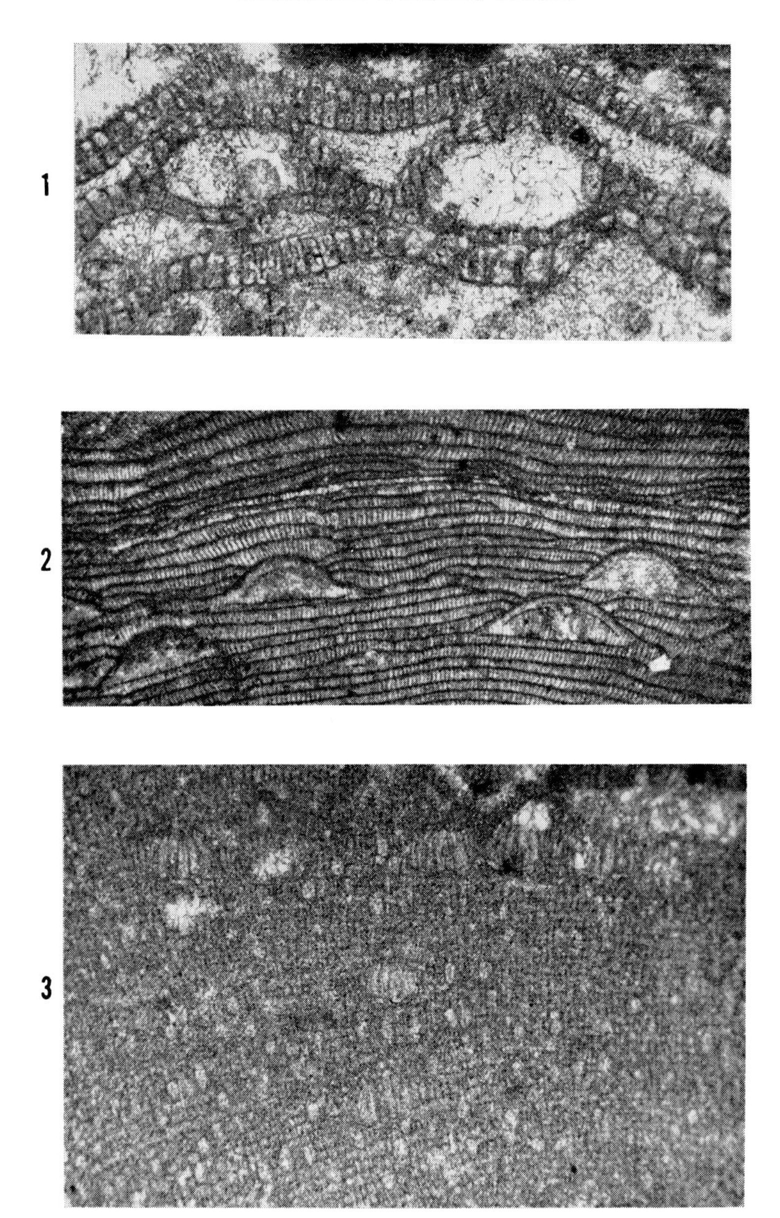

Plate 11
Genera *Lithoporella* and *Paraporolithon*
(Published by permission, U. S. Geological Survey)

Figure 1. *Lithoporella melobesioides* Foslie. Several loose thalli and a conceptacle chamber. (x100). Miocene of Guam.

Figure 2. *Lithoporella* sp. (x50). A number of superimposed thalli, four with conceptacle chambers.

Figure 3. *Paraporolithon saipanense* Johnson (x100). Miocene, Saipan.

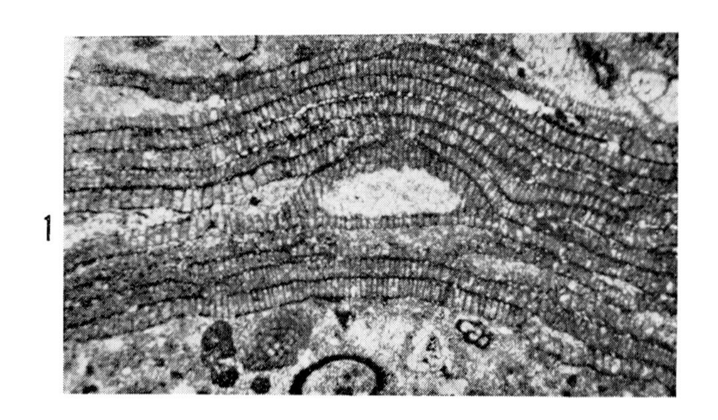

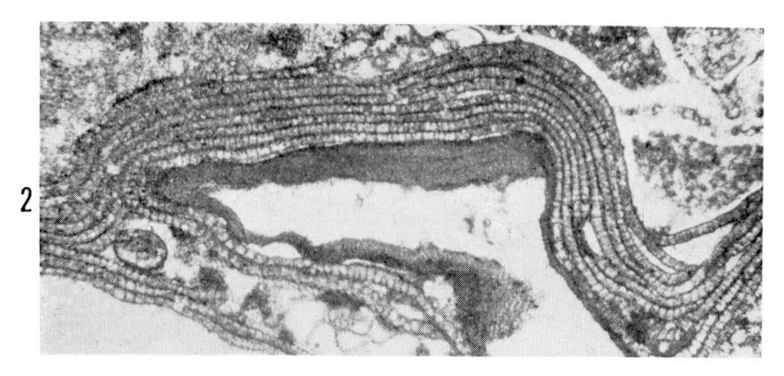

Plate 12

Genera *Lithoporella* and *Melobesia*

Figures 1 and 3. *Lithoporella melobesioides* Foslie. 1. Miocene of Saipan (x40). 3. Miocene of Guam (x100).

Figure 2. *Melobesia* sp. Pleistocene, Saipan (x40).

Remarks—

Quite similar to *Amphiroa*. Differs in having terminal conceptacles, and in having the arched layers (zones) of cells that form the medullary hypothallus all of equal length. Commonly the top of each arched layer is flattened.

Arthrocardia is rather rare in the fossil record.

Generic range—

Late Cretaceous to present.

Geographic distribution—

Tropical seas.

Genus *Calliarthron* Manza, 1937
Plate 14, figure 3.

Description—

This genus differs from other known genera of articulated coralline algae in that the cells of the hypothallus are flexuous and interlacing. The conceptacles are located along the lateral margins of the segments.

Remarks—

The first fossil representative of the genus to be recorded was described by Johnson (1957, p. 237) from the lower Miocene of Saipan.

Generic range—

Lower Miocene and Recent.

Geographic distribution—

Recent species appear to be limited to shore areas in the temperate zone. Some Tertiary species were sub-tropical.

Genus *Corallina* Linnaeus, 1758
Plate 14, figures 1, 4; plate 15, figures 1-4.

Description—

Plants are clusters of segmented stems which branch at close intervals, ordinarily in a plane. Branching typically pinnate (dichotomous to trichotomous or irregular). Perithallus weakly developed and inconspicuous, the greater part of the individual segments consisting of hypothallic tissue. Segments mainly clavate, flattened cylindrical, or flattened *Halimeda*-like, but varying widely in shape. Conceptacles lateral or terminal.

Remarks—

Abundant and widespread at present in warm and temperate seas and has been found as far north as latitude 70°.

Generic range—

Eocene to Recent.

Geographic distribution—

Widely distributed in warm and temperate marine waters.

Genus *Jania* Lamouroux, 1812
Plate 14, figure 2.

Description—

Plants consist of masses of slender, dichotomously branching fronds. Each frond is a series of slender segments formed of tiers of hypothallic cells surrounded by a narrow perithallus that characteristically is restricted to a single layer of small rectangular cells. The hypothallic cells tend to be wider in proportion to length than in most genera of articulate corallines. In many instances, the hypothallic cells are elongately wedge-shaped in section, and the successive tiers of cells tend to meet along irregular lines.

Remarks—

Living *Jania* occurs widely in the tropic and temperate seas, where it is represented by many species.

Generic range—

Late Cretaceous to Recent.

Geographic distribution—

Widespread in tropical and temperate seas.

REFERENCES

Foslie, M., and Printz, H., 1929, Contributions to a monograph of the *Lithothamnia*: Kongelige Norske Vidensk. Selsk. Museet, Trondheim, 60 p., 85 pls.

Ishijima, W., 1954, Cenozoic coralline algae from the western Pacific: Tokyo (privately printed), 87 p., 49 pls.

Johnson, J. Harlan, 1954, Fossil calcareous algae from Bikini atoll: U. S. Geol. Survey Prof. Paper 260-M, p. 537-546, pls. 188-197.

———, 1957, Fossil algae from Saipan: U. S. Geol. Survey Prof. Paper 280-E, p. 209-246, pls. 37-60.

Lemoine, Mme. Paul, 1911, Structure anatomique des Melobesiees: Annales Inst. Oceanographique, Paris (Masson et cie), v. 2, no. 2, 213 p., 5 pls.

———, 1939, Les algues calcaires fossiles de l'Algerie: Ser. Carte Geol. de l'Algerie Mem. 9 (Paleontologie), 128 p., 3 pls.

Manza, A. V., 1940, A revision of the genera of articulated corallines: Philippine Jour. Sci., v. 71, no. 3, p. 239-315, 20 pls.

Maslov, V. P., 1956, Fossil calcareous algae in the U. S. S. R.: Trudy Inst. Geol. Nauk, S.S.S.R., no. 160, 301 p., 86 pls.

Pfender, Juliette, 1926, Les Melobesiees dans les calcaires Cretaces de la Basse-Provence: Soc. Geol. France Mem., new ser., Paris, v. 3, no. 2, 30 p., 10 pls.

Pia, Julius, 1926, Pflanzen als Gesteinsbildner: Berlin, 355 p., 166 figs.

———, 1927, Thallophyta *in* Hirmer, M., Handbuch der Palaobotanik: Berlin and Munich, p. 1-136.

Setchell, W. A., 1926, Nullipore versus coral in reef formation: Am. Philos. Soc. Proc., v. 65, no. 2, p. 136-140.

Suneson, S., 1937, Studien über die enturcklungsgeschichte der corallinaceen: Lunds Univ. Arsskr. N. F. adv. 2, v. 33, no. 2, 101 p., 4 pls.

Taylor, W. R., 1950, The plants of Bikini: Univ. of Michigan Press, 227 p., 79 pls.

Weber van Bosse, A., and Foslie, M., 1904, The corallinaceae of the *Siboga* Expedition: *Siboga* Expedition Repts., Brill-Leiden, Holland, v. 61, 110 p.

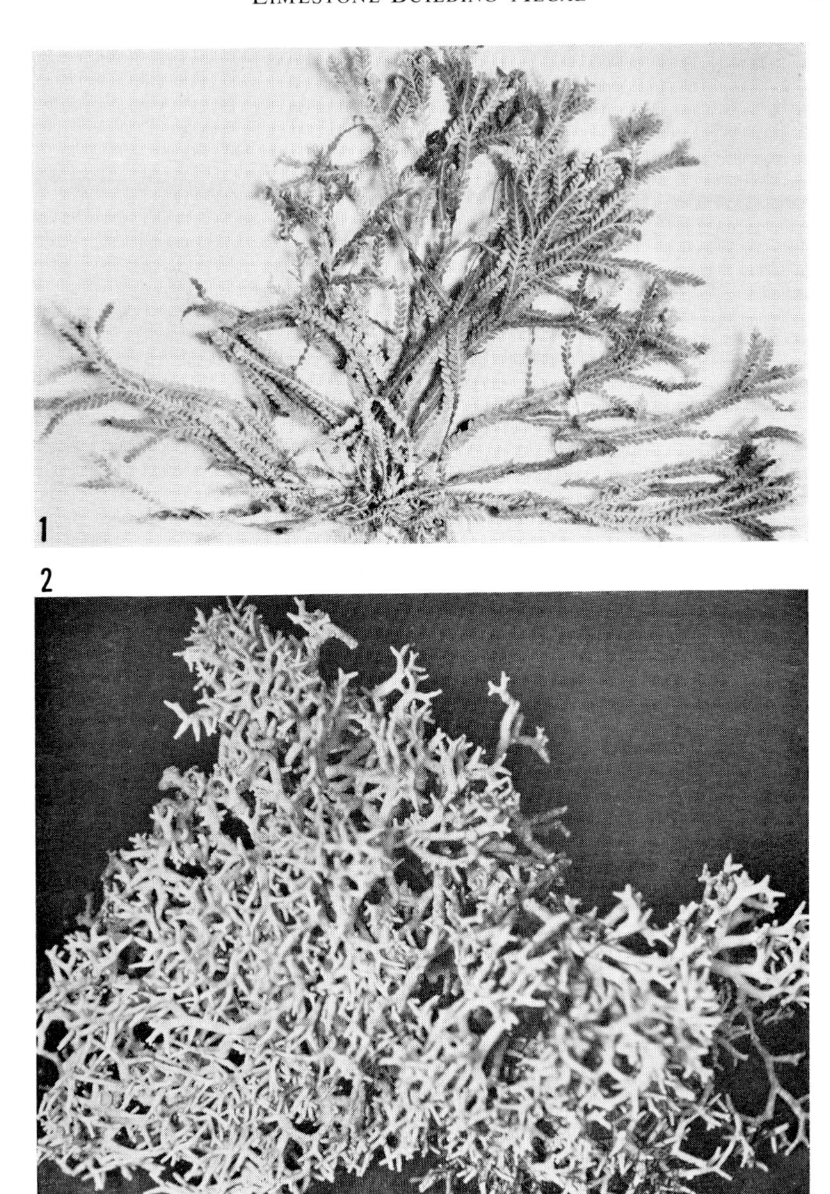

Plate 13
Articulated Coralline Algae
Figure 1. A Recent plant from Agana Reef, Guam (x1).
Figure 2. *Amphiroa* sp. Angaur, Palau.

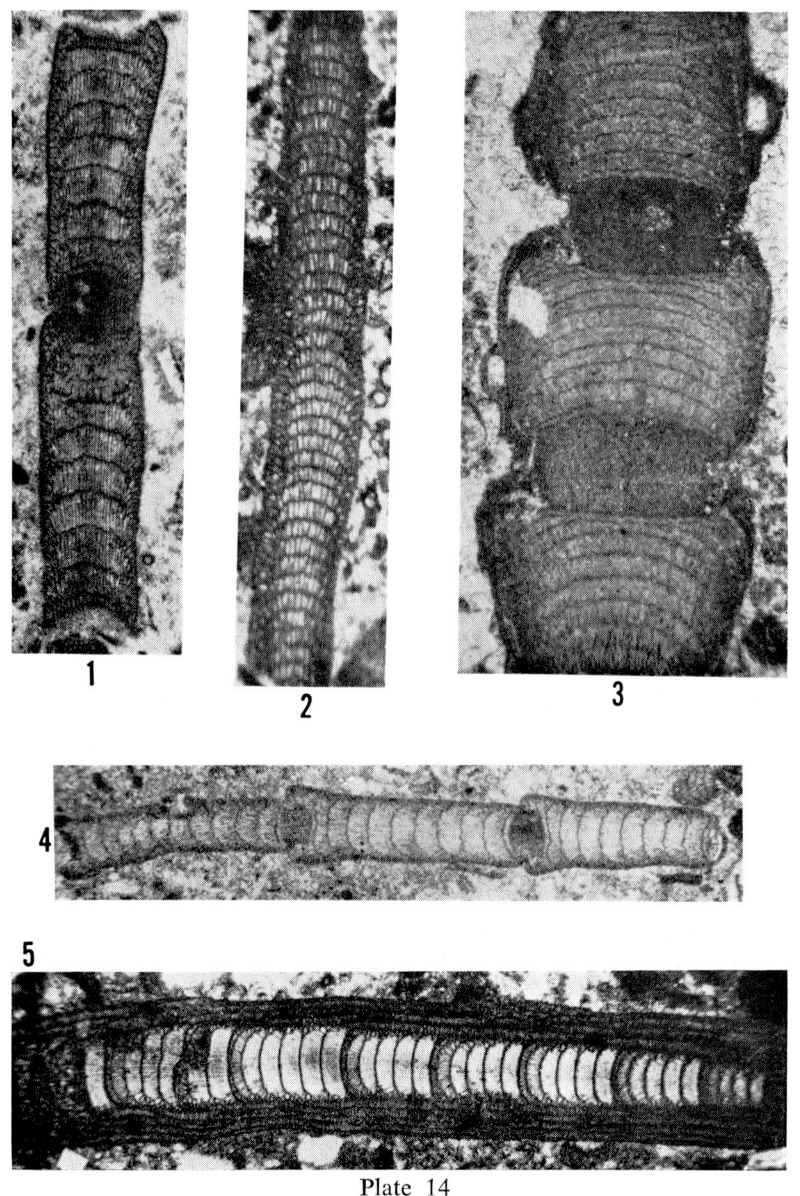

Plate 14

Articulated Coralline Algae

Figure 1. *Corallina prisca* Johnson (x100). Upper Eocene, Saipan.
Figure 2. *Jania* sp. Long section (x75). Miocene, Saipan.
Figure 3. *Calliarthron antiquum* Johnson (x75). Three segments with nodes between, and a conceptacle cavity on upper right margin. Miocene, Saipan.
Figure 4. *Corallina neuschelorum* Johnson (x25). Miocene, Saipan. Four segments with the nodes between.
Figure 5. *Amphiroa fragilissima* Lmx. (x40). Pleistocene, Saipan. A longitudinal section with medullary hypothallus and marginal perithallus. Hypothallus shows attenuations of 3-5 layers of long cells to one layer of short cells.

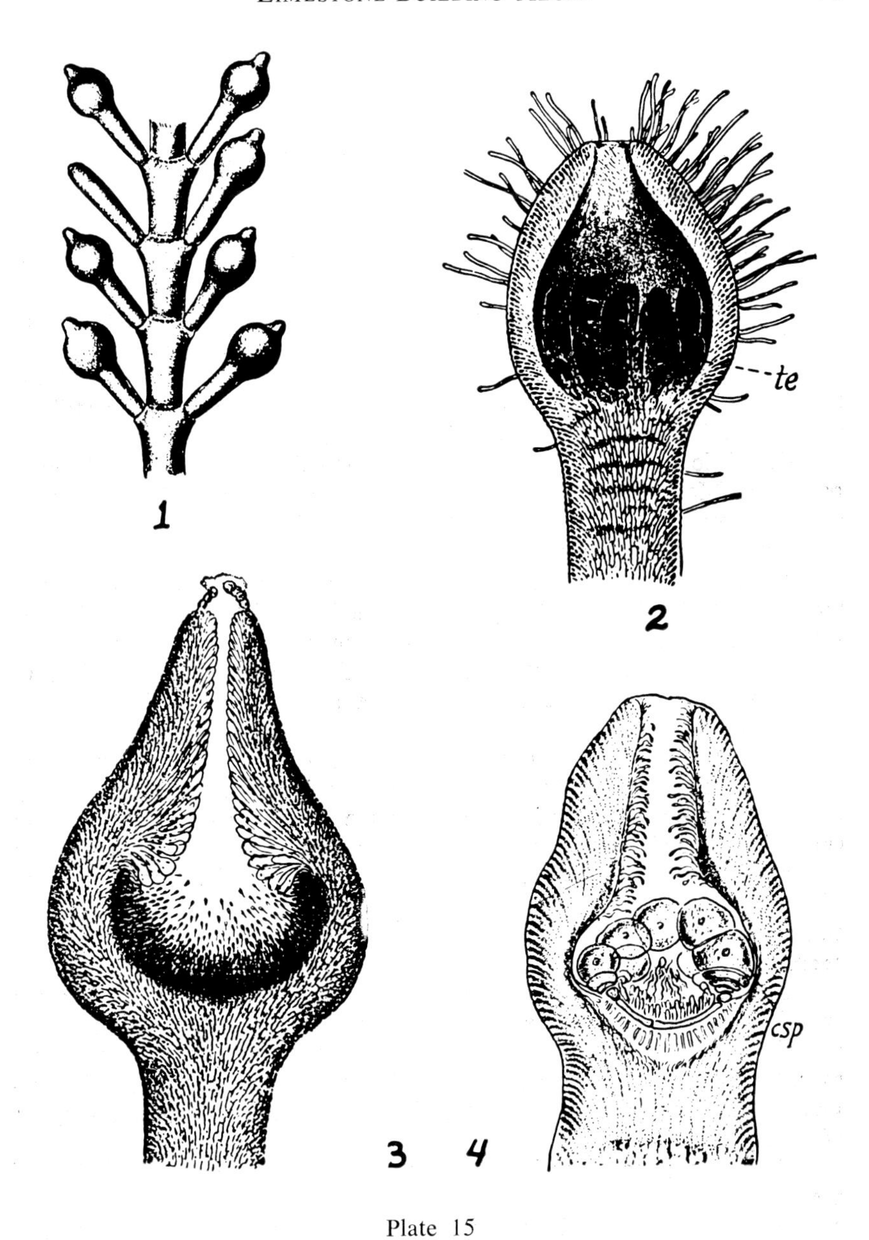

Plate 15

Conceptacles, Articulated Coralline Algae

Figures 1-4. *Corallina mediterranea* Arsch. (after Oltmanns). 1. Branch with male conceptacles (x15). 2. Mature tetraspore conceptacle, long section (x75). 3. Longitudinal section, male conceptacle (x110). 4. Longitudinal section, female conceptacle (x100).

Family SOLENOPORACEAE Pia, 1927

This family is closely related to the Corallinaceae. During the Paleozoic and Mesozoic eras, the Solenoporaceae appear to have occupied the same ecological niche that is now occupied by the Corallinaceae. The genera of the latter family are supposed to have developed during the Mesozoic era from members of the Solenoporaceae. The general structure of the tissue is quite similar in the two families, although there are some striking differences. These differences are listed below.

1. Little is known regarding the reproductive organs of the Solenoporaceae. Most of the specimens studied have shown no indications of such structures, and it has been presumed that, in most cases at least, the sporangia were external and uncalcified and were not preserved with the fossils. It is true that a few specimens, particularly of Silurian age, have shown some structures that suggest possible isolated sporangia or, in one case, even small conceptacles. However, the supposed sporangial structures that have been observed are not uniform in appearance, size, or shape, with the result that there is considerable difference of opinion as to their exact nature.

2. The cells of the Solenoporaceae, in general, are much larger, particularly much longer, than those of the Corallinaceae. However, among the articulated corallines of today, there are some specimens which have cells of similar size and length, and in the genus *Lithoporella,* there are species with cells that attain an equal size. In cross section, most members of the Solenoporaceae show cells having a polygonal, or rarely a circular, outline in contrast to cells of the Corallinaceae which commonly are rectangular or circular in outline.

3. Among the Solenoporaceae, the cross partitions separating the cells in the cell threads are much more poorly developed and far less conspicuous in sections. In some cases, they appear to be absent. However, there is considerable question if this feature is not a result of incomplete calcification and poor preservation rather than an original absence of the structure.

4. Among the Corallinaceae, there normally is a differentiation of the tissue into hypothallus and perithallus. Such differentiation of tissue has rarely been observed among the Solenoporaceae, and when seen, is usually very poorly developed. Among the Paleozoic Solenoporaceae, a few forms have been observed in which there appears to be a hypothallus consisting of one or two layers of cells much larger than those of the rest of the tissue. During the Mesozoic, some branching forms occurred which show a slight differentiation of tissue, but, in general, this feature is absent among the Solenoporaceae.

Most of the Solenoporaceae develop as rounded nodular masses, although a few encrusting forms have been observed, particularly in the Late Paleozoic and Mesozoic. Nodular masses bearing irregularly rounded protuberances or short warty branches are known, particularly from the Jurassic, and truly branching forms have been found in limited numbers in Jurassic and Cretaceous beds.

During the Paleozoic era, representatives of the Solenoporaceae appear to have been rather scarce and limited in number, variety, and distribution, although in a few small areas, considerable numbers of a few species did occur. It is very rare, however, that they occur in sufficient numbers to really be considered as builders of limestone. In a few cases, as for example, in the Silurian of Gotland, Solenoporaceae appear in the role of reef builders to the extent, at least, that they occur in fair numbers in the various reefs and bioherms. However, their role in reef building during the Paleozoic was very limited as compared with that of other organisms. Thus, in the Silurian and Devonian reefs that have been studied, the stromatoporoids and the corals were vastly more important than were the Solenoporaceae. In general, one gets the impression that during the Paleozoic, Solenoporaceae were not nearly as important as the coralline algae are today.

During the Mesozoic, Solenoporaceae became more important, particularly during the Jurassic period. They occur abundantly in many regions during the Jurassic and locally were important rock builders. During this period, they appear to have attained their greatest size and, probably, the greatest number of individuals, although the variety of forms known from that period is still quite limited. They definitely were of importance as rock builders in a number of regions during that time. Of interest also is the fact that in the Jurassic beds, many specimens have been found which still show the original red or pink color. This color, as in the case of Recent forms, speedily fades when exposed to strong light.

The Solenoporaceae continue into the Cretaceous, and, particularly during the Lower Cretaceous, they seem to have been quite widespread and to have been represented by a considerable variety of types which also exhibit a number of growth forms. The true coralline algae, especially the crustose forms belonging to the genera *Archaeolithothamnium* and *Lithothamnium,* appear and become locally abundant during the Cretaceous. Apparently, they competed actively with the Solenoporaceae, and we find the Solenoporaceae rapidly decreasing in numbers and importance during the latter part of the Cretaceous. A few forms survived into the Paleocene, but most of them became extinct by the end of the Cretaceous, and all of them were gone by the Eocene.

The Solenoporaceae appear to have lived under the same general ecological conditions that the Corallinaceae do today. At least, this is the general picture that one gets from the study of the algae themselves and from studies of associated fossils.

Genus *Solenopora* Dybowski, 1877

Plate 16; plate 17; plate 19, figures 2, 4; plate 119, figures 1, 2; plate 129.

Description—

Solenoporaceae in which the most apparent structures in vertical sections are the vertical or slightly radiating cell threads. The cross partitions separating the cells in the threads are irregularly spaced and commonly are thinner and less conspicuous than the vertical walls of the threads. Little or no differentiation of tissue occurs. Growth form commonly rounded nodular masses.

Remarks—

The cross partitions between cells in threads were thin and probably poorly calcified, so sometimes they are not preserved. A few branching forms appear in the Late Paleozoic.

Generic range—

Cambrian to Cretaceous.

Geographic distribution—

Nearly world wide.

Genus *Parachaetetes* Deninger, 1906

Plate 18, figures 1-2; plate 19, figures 1, 3.

Description—

Solenoporaceae with very regular tissue. In vertical sections, the cells appear to be in regular horizontal layers and vertical (or slightly radiating) threads. The cross partitions separating the cells in the threads are thick and regularly spaced and at the same levels in adjoining threads. As a result, the tissue in vertical or radial sections commonly has a grid-like appearance.

Remarks—

Not as abundantly represented by species or individuals during most of the Paleozoic as *Solenopora,* but more important during the Mesozoic.

Generic range—

Ordovician to Cretaceous.

Geographic distribution—

Nearly world wide.

REFERENCES

Brown, Alex, 1894, On the structure and affinities of the genus *Solenopora,* together with a description of new species: Geol. Mag., New Ser., London, Dec. IV, v. 1, p. 145-151 and 195-203, pl. 5.

Deninger, K., 1906, Einige neue Tabulaten und Hydrozoen aus mesozoischen Ablagerungen: Neues Jahrb. f. Mineralogie, v. 1, p. 61-70, pls. 5-7.

Dybowski, W., 1877, Die Chaetetiden der ostbaltischen Silur-Formation: Dorpat, 134 p., illus.

Høeg, O. A., 1933, Ordovician algae from the Trondheim area (Norway): Norske Akad. Oslo. Skr. Mat.-Natur. Kl., no. 4, p. 1-15, pls. 1-3.

Johnson, J. Harlan, 1946, Late Paleozoic algae of North America: American

Midland Naturalist, v. 36, no. 2, p. 264-274, 2 pls.
———, 1960, Paleozoic Solenoporaceae and related red algae: Colorado School of Mines Quarterly, v. 55, no. 3, 77 p., 23 pls.
———, and Konishi, Kenji, 1959, Studies of Silurian (Gotlandian) algae: Colorado School of Mines Quarterly, v. 54, no. 1, p. 29-30, pls. 1-2; p. 136-146, figs. 1-11.
Maslov, V. P., 1956, Fossil calcareous algae in the U. S. S. R.: Trudy Inst. Geol. Nauk, S.S.S.R., no. 160, p. 66-72, pls. 15-22.
Peterhans, E., 1929, Algues de la famille des Solenoporacees dans le Malm du Jura balois et soleurois: Soc. Paleontologique Suisse, Mem., v. 49 (1929-1930), no. 1, p. 1-15, 7 pls.
Pia, Julius, 1930, Neue Arbeiten über fossile Solenoporaceae und Corallinaceae: Neues Jahrb. f. Min., Geol. u. Pal., Referate, Stuttgart, III, p. 122-147.
———, 1939, Sammelbericht über fossile algen — Solenoporaceae 1930 bis 1938: Neues Jahrb. f. Min., Geol. u. Pal., Referate 1939, no. 3, p. 731-760.
Sinclair, G. W., 1956, *Solenopora canadensis* (Foord) and other algae from the Ordovician of Canada: Trans. Royal Soc. Canada, v. 50, ser. 3, p. 65-81, pls. 1-4.
Yabe, H., 1912, Über einige gesteinsbildende Kalkalgen von Japan und China: Science Report Tohoku Imp. Univ., Sendai, ser. 2, Tokyo, v. 1, p. 1-8, pls. 1-2.

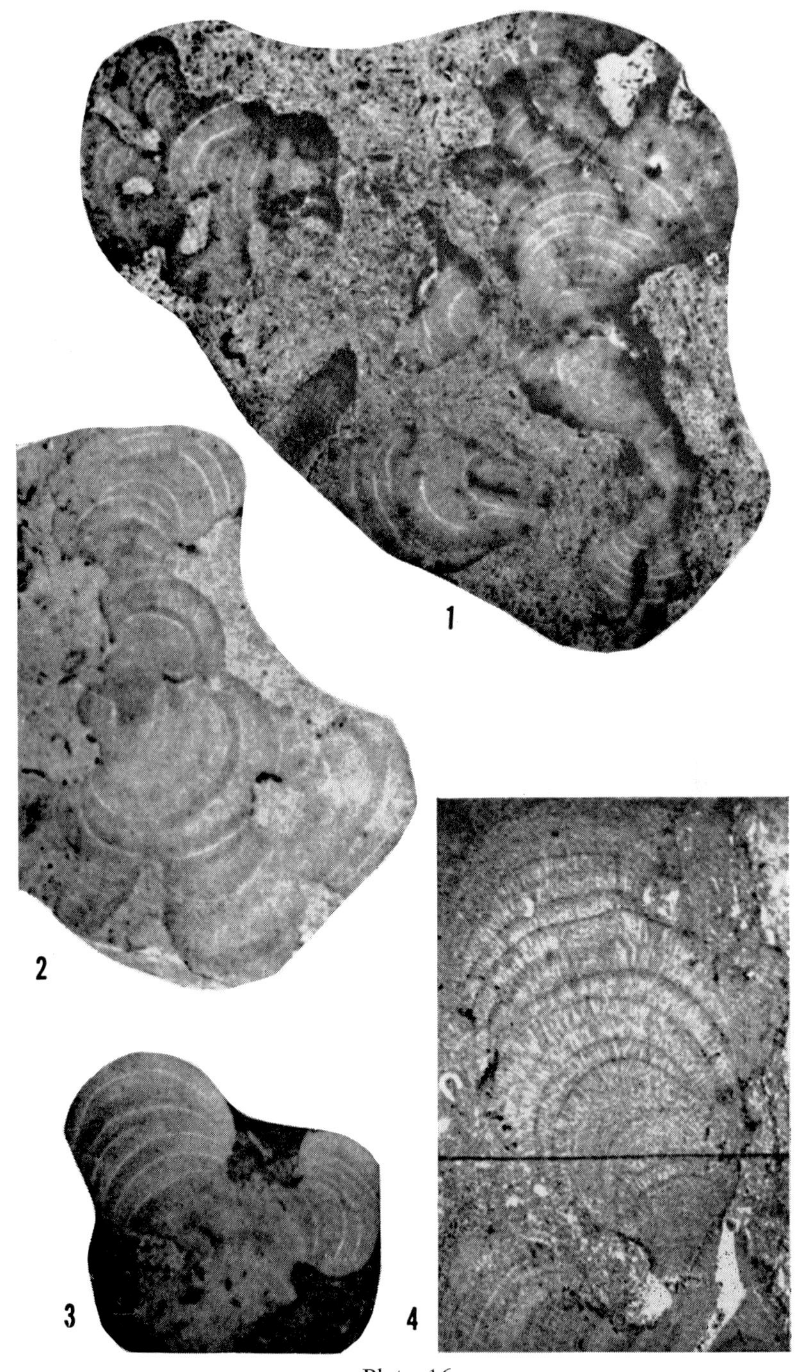

Plate 16
Genus *Solenopora*
Figures 1-4. *Solenopora spongioides* var. *iuchvii* Maslov. Lower Silurian, U.S.S.R. (From Maslov, 1956.) 1-3. Polished slabs of limestone containing thalli of *S. spondioides* (x25). 4. A section (3.5).

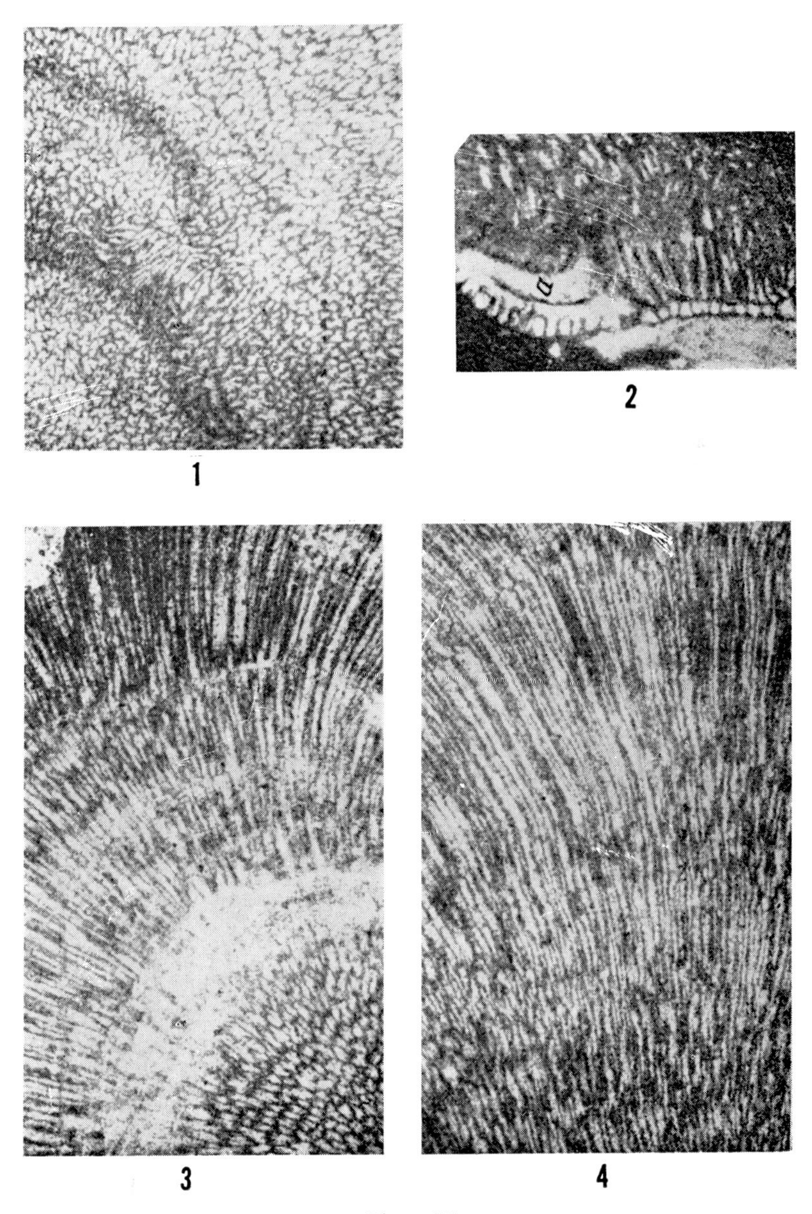

Plate 17

Genus *Solenopora*

Figures 1-4. *Solenopora* cf. *S. nigra* Brown, Silurian, Esthonia. (From Maslov, 1956.) 1. A cross section (x45) showing irregular outlines of cell walls. 2. A nearly perpendicular section (x40) showing the hypothallus consisting of a single layer of large cells. 3-4. Nearly vertical sections (x45) of the tissue.

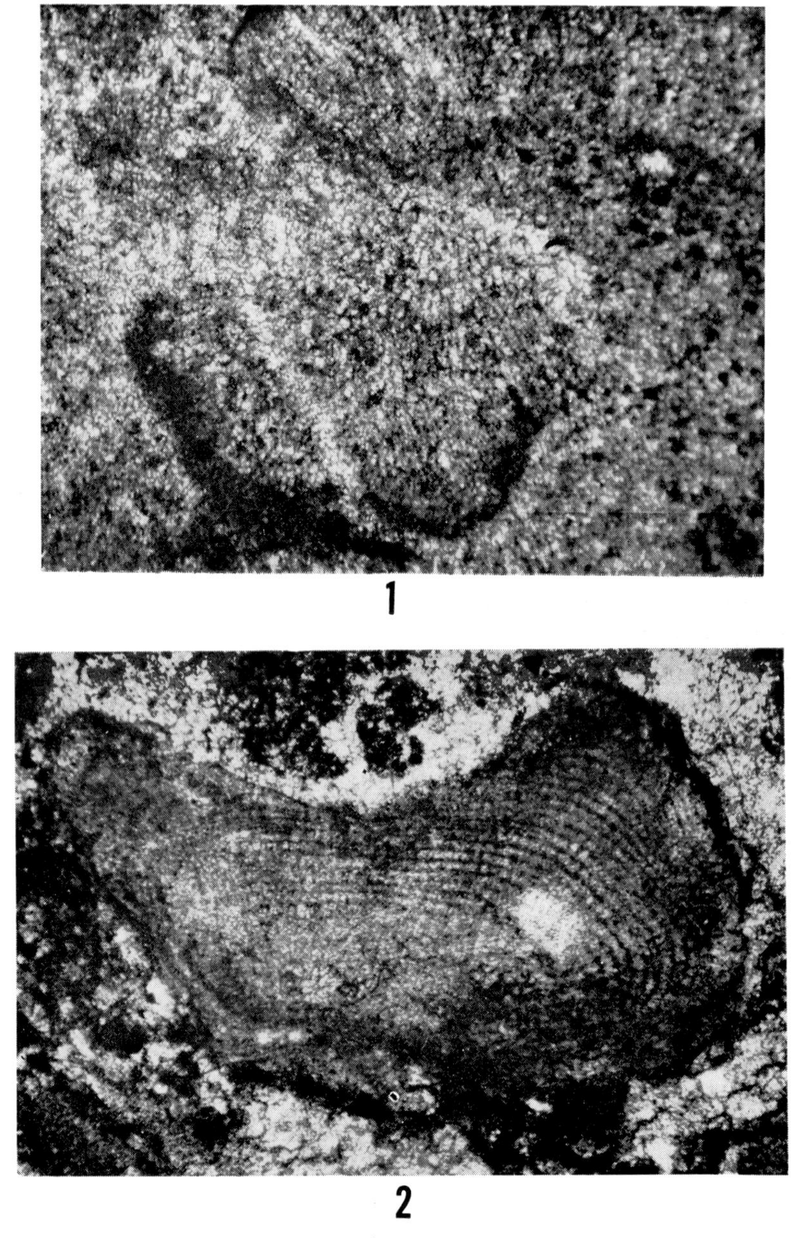

Plate 18
Genus *Parachaetetes*

Figures 1-2. *Parachaetetes* sp. Lower Silurian (?) Gazelle formation, Siskiyou County, northern California (Laudon collection). 1. Oblique section (x40). 2. Almost longitudinal section (x50) showing the development of what appear to be horizontal cell layers instead of the vertical cell rows.

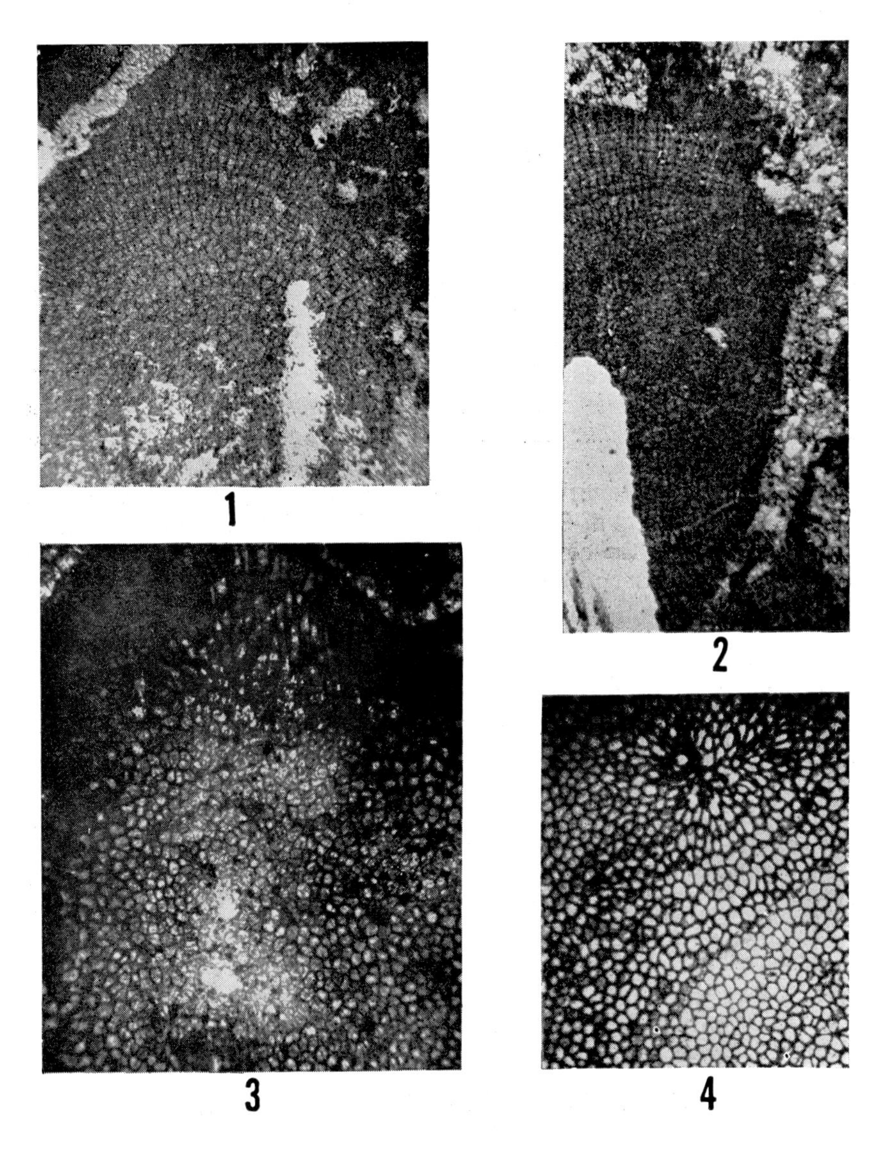

Plate 19

Genera *Parachaetetes* and *Solenopora*

Figure 1. *Parachaetetes glenwoodensis* Johnson. Mississippian, Glenwood Springs, Colorado (x25).

Figure 2. *Solenopora similis* Paul. Mississippian, near Glenwood Springs, Colorado (x40).

Figure 3. *Parachaetetes glenwoodensis* Johnson. Mississippian, Saskatchewan, Canada. Section (x40) across a hemispherical thallus.

Figure 4. *Solenopora garwoodia* Hinde. Mississippian, England. Shows "stellate" arrangement of cells (x25). (After Garwood.)

Family GYMNOCODIACEAE Elliott, 1955

This family is an extinct group of red algae having segmented or unsegmented thalli. The segments or units are of very varying size, form, and degree of calcification. The sporangia are internal.

The family was erected by Elliott to include two genera: *Gymnocodium* Pia, 1920 (as emended by Elliott) and *Permocalculus* Elliott, 1955. Representatives of *Gymnocodium* are known only from the Permian, but species of *Permocalculus* have been found in both Permian and Cretaceous beds.

Pia named the genus *Gymnocodium* in 1920 and, in 1937, compared its structure with that of Recent specimens belonging to the genus *Galaxaura*. He came to the conclusion that, since the structural features were very similar, *Gymnocodium* was probably a red algae, and he placed it in the family Chaetangiaceae.

Recent species of *Galaxaura* are feebly calcified, and the calcification is rather peculiar in the sense that it usually affects a thin zone of tissue just under the outer layer, or, in other words, it is slightly subdermal.

In 1955, Elliott, working with large collections of *Gymnocodium* material from the Near East, made a detailed comparison of the structural features of the genera *Gymnocodium* and *Galaxaura*. He agreed with Pia that the structural features of the two had many corresponding aspects, but he felt that there were sufficient differences between them for *Gymnocodium* to be placed in a different family. Hence, his creation of the family Gymnocodiaceae. His study involved a restudy of previously described species of *Gymnocodium,* as well as consideration of new material. In the course of the study, he found that the material could be subdivided into two groups on the basis of structural features, so he emended and somewhat restricted the old genus *Gymnocodium* and erected a new genus *Permocalculus.*

Genus *Gymnocodium* Pia, 1920 (emend.)
Plate 20, figures 1-6.

Description—

The genus was redescribed by Elliott as follows:

> "Gymnocodiaceae represented by hollow calcareous segments, which are cylindrical, oval, or cone-shaped, and circular or oval in cross section. Rarely bifurcating. The walls are perforated by pores which radiate oblique-distally and widen markedly outward. The interior of segments may be empty or may show calcified traces of the plant fibers that filled it in life in the form of longitudinally oblique streaks. The sporangia are ovoid, in terminal segments. The segments are usually smaller

than those of *Permocalculus,* and the perforations usually coarser."

Remarks—

Gymnocodium is interpreted as a plant, very similar in life to *Galaxaura,* consisting of numerous thalli, each composed of many segments which sometimes bifurcate, while the thalli grow upward in a cluster from the point of attachment.

The genus was created by Pia about 1920 based on the type species *Gyroporella bellerophontis* Rothpletz, 1894. Previously, Pia had interpreted the fossils as being dasycladaceans and referred them to the genus *Macroporella.* The name *Gymnocodium* indicates that Pia considered them to be green algae, differing from the dasyclads, and presumably belonging to the family Codiaceae. However, as mentioned above, after studying much more material and making more detailed comparative structural studies, he transferred *Gymnocodium* to the red algae in 1937.

Fossils belonging to this genus are very numerous in certain localities and have a widespread geographic distribution. Locally, as in northern Italy, they are so abundant as to be limestone builders (Accordi, 1956).

Generic range—

Permian, particularly Middle and Upper.

Geographic distribution—

Northern Italy, Yugoslavia, Greece, Hungary, Turkey, Persia, the Salt Range of India, the East Indies, Japan, and Texas, New Mexico, and Kansas.

Genus *Permocalculus* Elliott, 1955
Plate 21, figures 1-4.

Description—

Thallus irregular, segmented. Segments of variable form: spherical, ovoid, or barrel-shaped, or elongated, irregularly finger-like, or with pinching and swelling units. Calcification varying from very thin to massive or solid. Pores small and cortical. Sporangia cortical or medullary.

Remarks—

Very similar to *Gymnocodium* but commonly having larger units or segments, with finer (smaller) pores and more irregular calcification.

Generic range—

Upper Permian to Cretaceous.

Geographic distribution—

Northern Italy, Austria, Serbia, Turkey, northern Iraq, Salt Range of India, Japan, and West Texas and southwestern New Mexico.

REFERENCES

Accordi, Bruno, 1956, Calcareous algae from the Upper Permian of the Dolomites (Italy) with stratigraphy of the "Bellerophon-zone": Jour. Palaeontological Soc. India, v. 1, p. 75-84, pls. 6-12.

Elliott, G. F., 1955, The Permian calcareous alga *Gymnocodium:* Micro-paleontology, v. 1, no. 1, p. 83-91, pls. 1-3.

———, 1956, *Galaxaura* (calcareous algae) and similar fossil genera: Jour. Wash. Acad. Sci., v. 46, no. 11, p. 341-343, 2 figs.

Johnson, J. Harlan, 1951, Permian calcareous algae from the Apache Mountains, Texas: Jour. Paleontology, v. 25, p. 21-30.

Konishi, Kenji, 1952, Occurrence of *Gymnocodium,* a Permian alga, in Japan: Paleontological Soc. Japan, Trans. Proc., new ser., no. 7, p. 215-221.

Pia, Julius, 1912, Neue Studien über die triadische Siphoneae Verticillatae: Beitr. Pal. Geol. Österreich-Ungarns, v. 25, p. 25-81.

———, 1920, Die Siphoneae Verticillatae vom Karbon bis zur Kreide: Zool.-Bot. Ges. Wien, Abh., v. 11, no. 2, p. 1-263.

———, 1926, Pflanzen als Gesteinsbildner: Berlin, 355 p.

———, 1927, Thallophyta *in* Hirmer, M., Handbuch der Paläobotanik: Munich and Berlin.

———, 1937, Die wichtigsten Kalkalgen des Jungpaläozoikums und ihre geologische Bedeutung: Congr. Av. Etude Strat. Carb., 2nd (Heerlen, 1935), C. R., v. 2, p. 765-856.

Rao, S. R. N., 1948, Palaeobotany in India; No. 6—*Gymnocodium* cf. *bellerophontis* from the *Productus* beds in the Salt Range, Punjab: Indian Bot. Soc. Jour., v. 26, no. 4, suppl., p. 249.

———, and Varma, C. P., 1953, Fossil algae from the Salt Range; I — Permian algae from the Middle *Productus* beds: Palaeobotanist, v. 2, p. 19-21.

Rothpletz, Aug., 1894, Ein geologischer Querschnitt durch die Ost-Alpen nebst Anhang über die sogenannte Glarner Doppelfalte: Stuttgart.

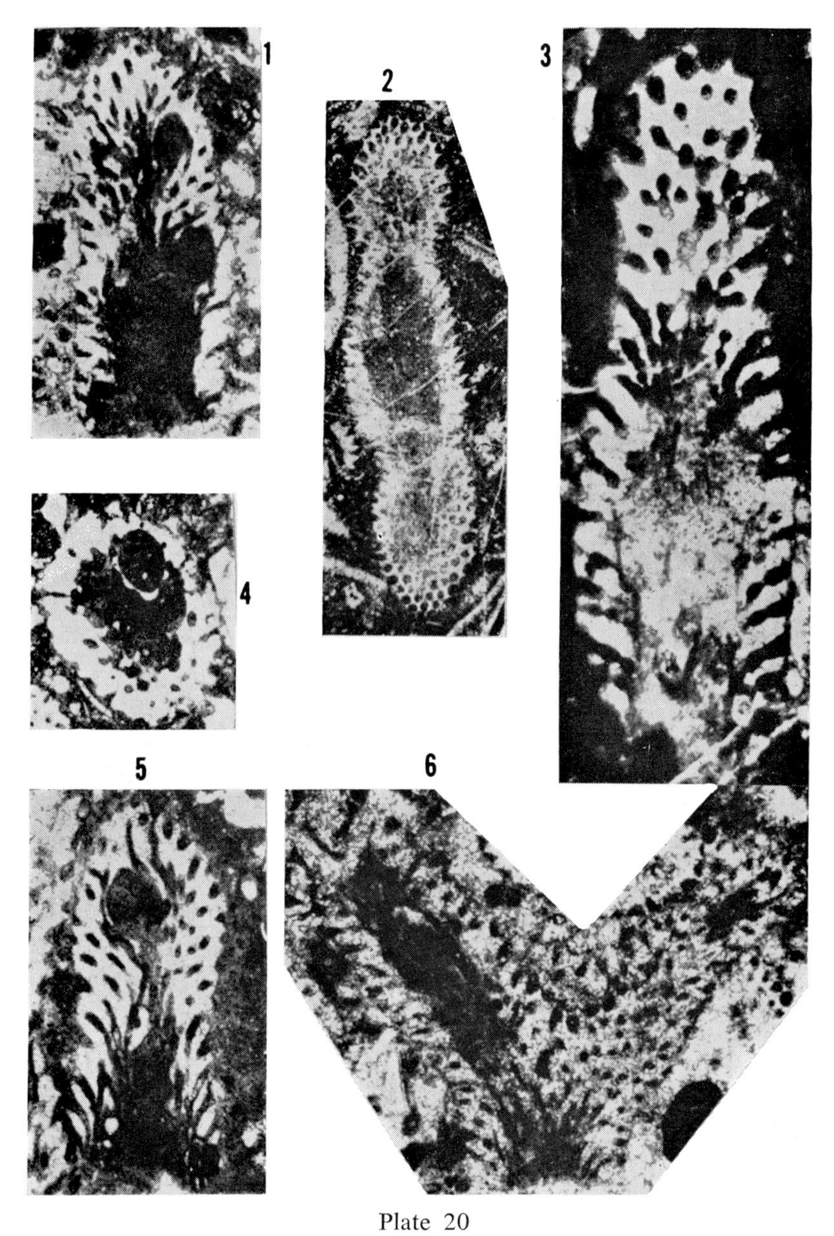

Plate 20
Genus *Gymnocodium*
Figures 1-6. *Gymnocodium bellerophontis* (Rothpletz). Upper Permian, the Dolomites, Austria. (2 x11, all others x28) (From Pia, 1937.)

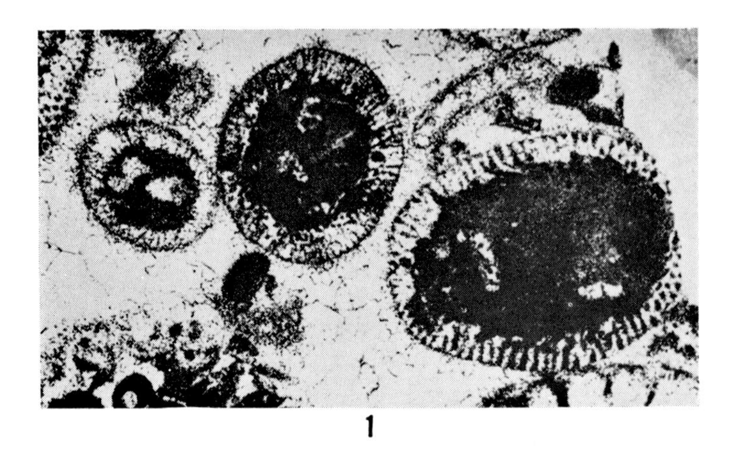

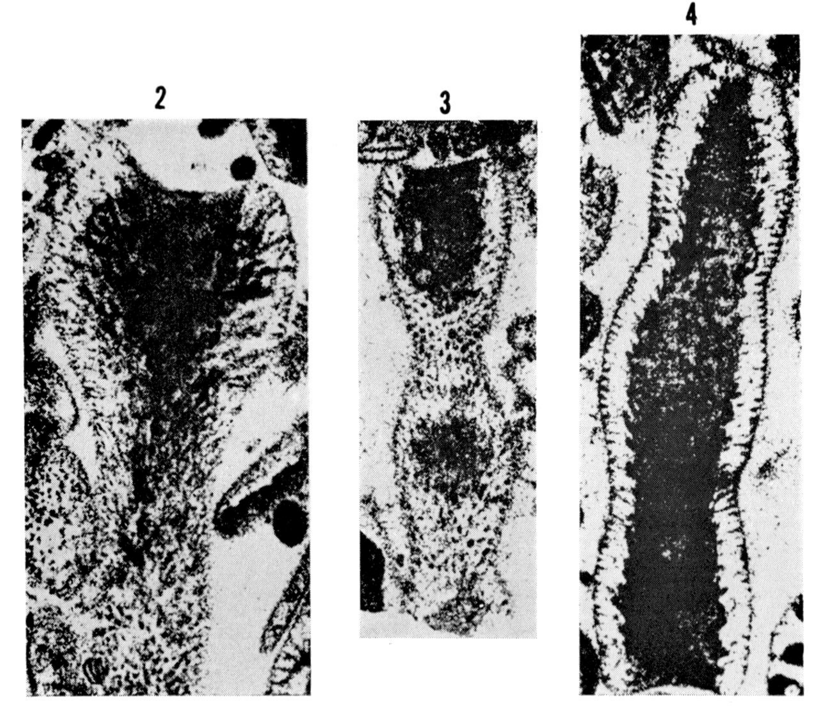

Plate 21
Genus *Permocalculus*
Figures 1-3. *Permocalculus plumosus* Elliott (from Elliott, 1955). Upper Permian, northern Iraq (all x28). 1. Three transverse sections. 2. Oblique tangential section. 3. A tangential section.
Figure 4. *Permocalculus digitatus* Elliott.

Red Algae of Uncertain Affinities

In addition to the genera with the Corallinaceae, the Solenoporaceae, and the Gymnocodiaceae, there are several other quite important genera which appear to belong to the red algae but which do not fit into any of the above families. Four such genera are described below.

Genus *Archaeolithophyllum* Johnson, 1956
Plate 22; plate 23.

Description—

Plants develop in two growth forms: (1) as small irregular platy or crustose masses which appear to have grown unattached on the sea bottom and which, in some cases, developed small protuberances or branches on the upper surface, and (2) as slender branching forms with cylindrical branches.

The tissue is clearly differentiated into a hypothallus and a perithallus. The hypothallus is co-axial and formed of large polygonal cells. Commonly, it is quite thick. The perithallus is formed of regular layers of small rectangular cells. Normally, it is much thinner than the hypothallus. The sporangia are collected into conceptacles which are nearly circular in ground plan and are highly arched. Each may have had only a single aperture.

Remarks—

This genus is surprisingly advanced structurally. The well-developed co-axial hypothallus and the conceptacles are features which did not become common among the red calcareous algae until late in the Mesozoic. The large polygonal hypothallic cells suggest affinities with the Solenoporaceae, while the small rectangular perithallic cells and the conceptacles resemble those of the Corallinaceae.

Generic range—

Pennsylvanian, especially the upper Des Moines and lower Missourian.

Geographic distribution—

Abundant and widespread in the Mississippi Valley, the Mid-Continent region, and in Texas.

Genus *Cuneiphycus* Johnson, 1960
Plate 24, figures 1-3.

Description—

Thallus slender, probably segmented, consisting of cylindrical or wedge-shaped branched members. Branches cylindrical or, more often, wedge-shaped. Tissue consists of layers of large, commonly elongated, rectangular or wedge-shaped cells. No suggestions of hypothallic tissue observed. Sporangia unknown.

Remarks—

The fossils observed in thin sections are commonly cylindrical or wedge-shaped fragments or segments that are sometimes branched. Superficially, they suggest badly worn pieces of articulated coralline algae, but they have longer and wider cells and a less complex tissue.

Generic range—

Pennsylvanian.

Geographic distribution—

Abundant in West Texas and adjoining parts of New Mexico. Recently found in Missouri and in the Wichita Falls region of Texas.

Genus *Komia* Korde, 1951
Plate 25, figures 1-3.

Description—

Thallus strongly branched. Hypothallus consists of small cluster of drawn-out filaments. Perithallus composed of dichotomously branching filaments which form a compact massive tissue.

Remarks—

This genus was described originally by Korde from the Middle Carboniferous of the Ural region of Russia. It has been known to the Humble Oil and Refining Company geologists in this country for many years as *"Desmoinesia."*

Generic range—

Middle Carboniferous of Russia, Lower Pennsylvanian of central Japan, Des Moines group (Pennsylvanian) of West Texas and New Mexico.

Geographic distribution—

Ural region of Russia, central Japan, West Texas, New Mexico.

Genus *Ungdarella* Maslov, 1950
Plate 26, figures 1-3.

Description—

Thallus branched but not articulated. Hypothallus consists of a single row of cells. Branches of perithallic tissue, consisting of simple or branching filaments. Filaments commonly branch at a large angle, making an irregular, almost contorted tissue. Sporangia unknown.

Remarks—

The coarse-textured tissue of this form suggests the Solenoporaceae, but the irregular tissue with the peculiar branching filaments is quite different, as is the strongly branching growth habit.

Generic range—

"Upper Carboniferous."

Geographic distribution—

Known only from the central Ural region of Russia.

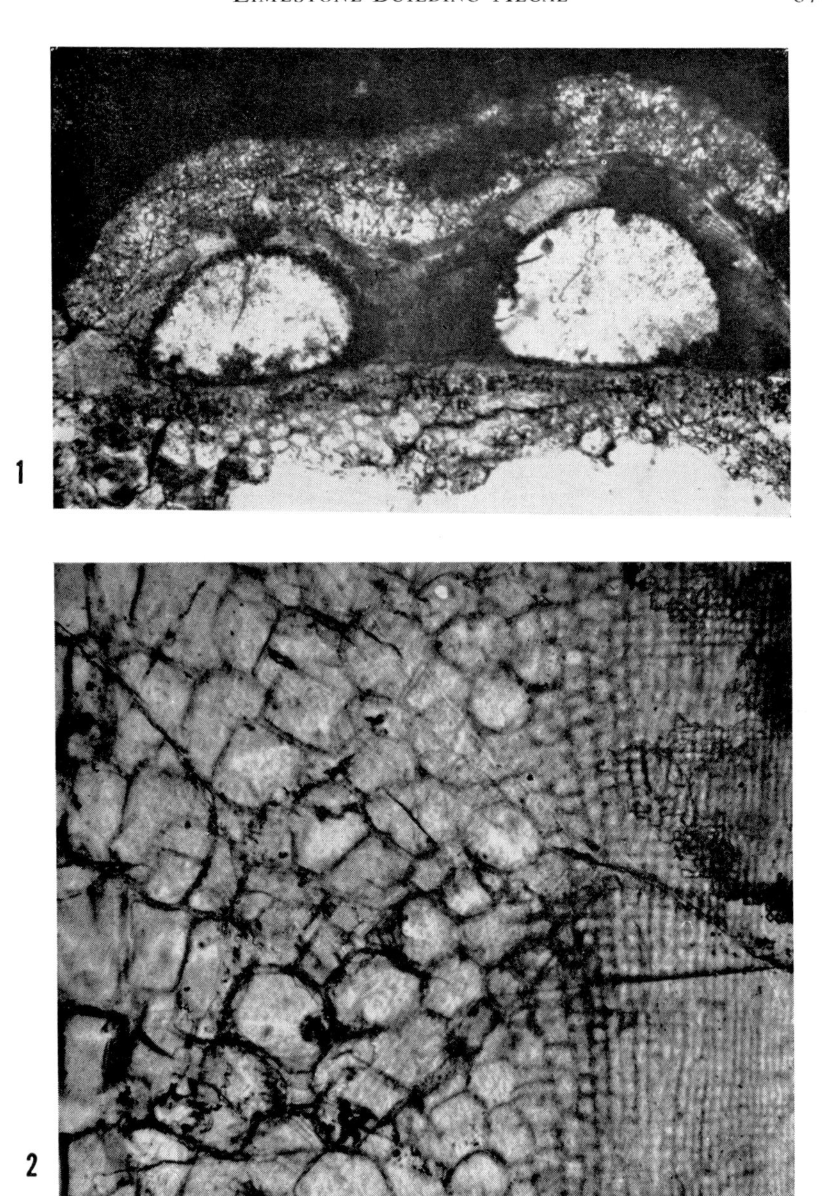

Plate 22
Genus *Archaeolithophyllum*
Figures 1-2. *Archaeolithophyllum missouriensum* Johnson. Pennsylvanian. (From Johnson, 1956.) 1. Slide shows two conceptacle chambers surrounded by perithallic tissue with a little of the hypothallus at the base (x50). From Carroll County, Missouri. 2. Detail of the tissue (x100). Hypothallus to left, perithallus to right. From Collinsville, Illinois.

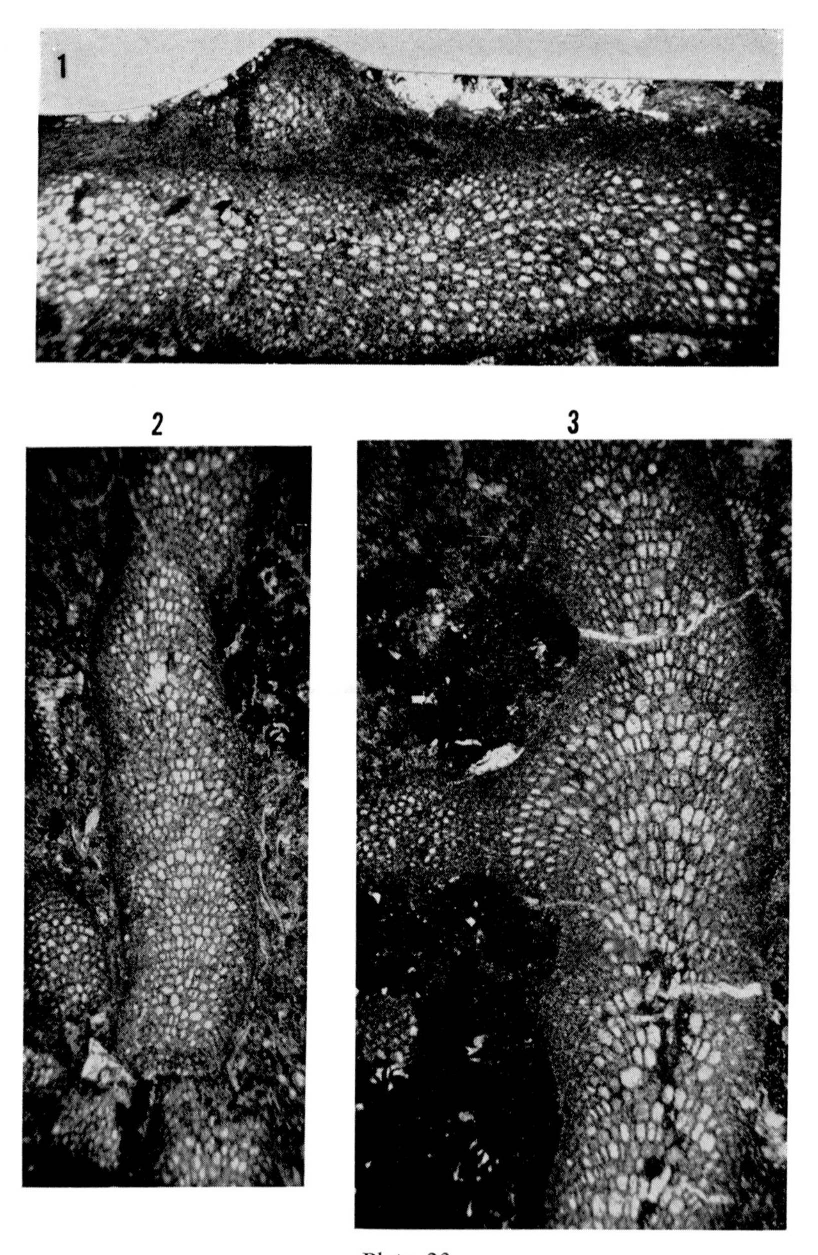

Plate 23
Genus *Archaeolithophyllum*

Figures 1-3. *Archaeolithophyllum* species. Pennsylvanian, Young County, Texas. 1. Section (x40) showing a bud developing on side of tissue, probably the beginning of a branch. 2. Longitudinal section (x33). 3. A specimen showing branching (x40).

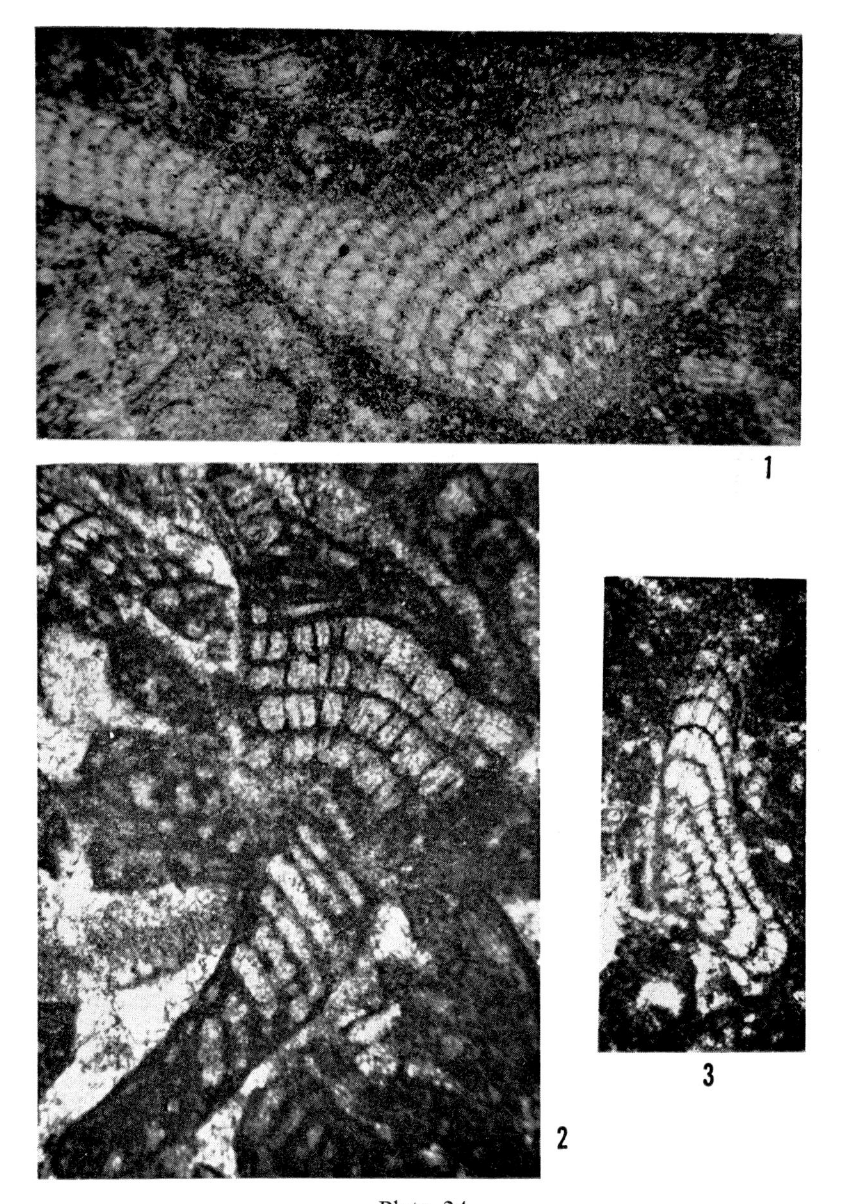

Plate 24
Genus *Cuneiphycus*

Figures 1-3. *Cuneiphycus texana* Johnson, n. sp. Pennsylvanian. 1. Vertical section (x40) through a branching segment. Marble Falls limestone. 2. Nearly complete segment (x40). Pow-Wow Canyon, Hueco Mountains, Texas. 3. An oblique section (x40) of a branching segment. Hueco Mountains, Texas.

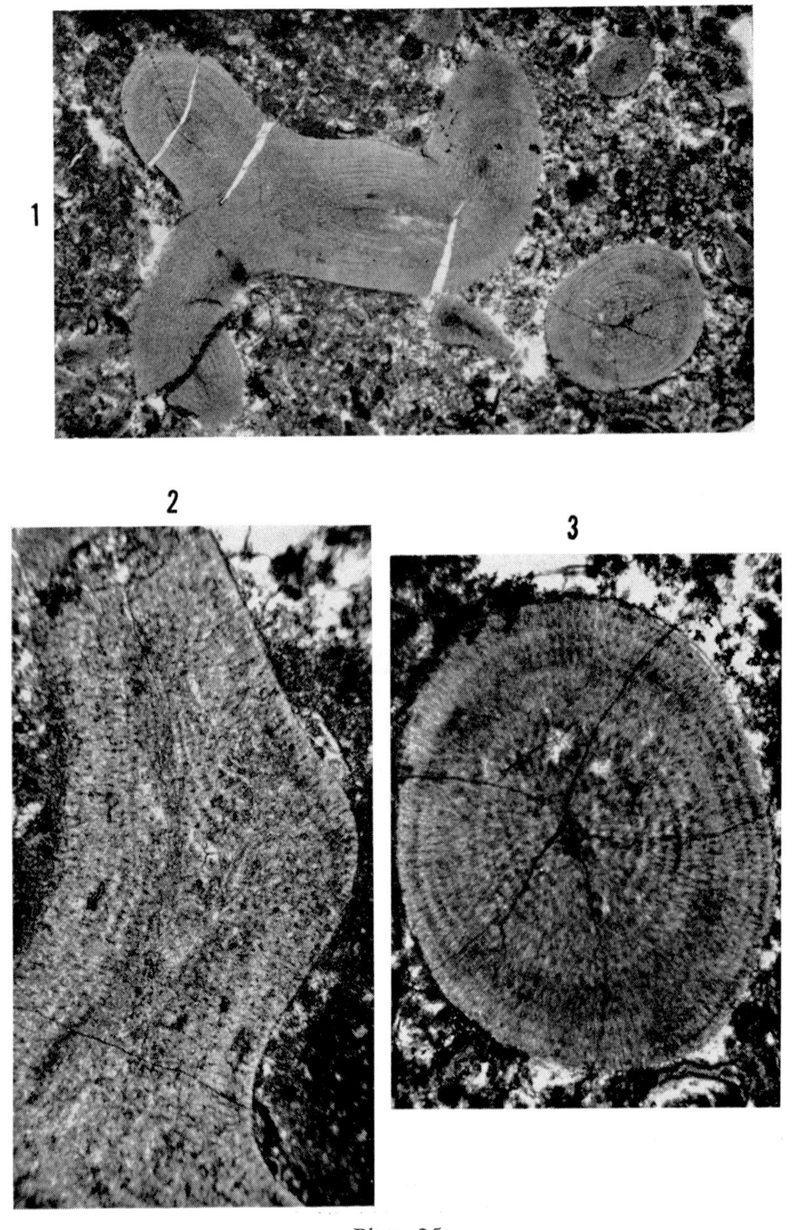

Plate 25
Genus *Komia*

Figures 1-3. *Komia* sp. ? Pennsylvanian, southern New Mexico. 1. Slide (x15) showing a nearly longitudinal section and two cross sections (to right). 2. A detail of a cross section (x15). 3. Detail of a long section near the tip (x50).

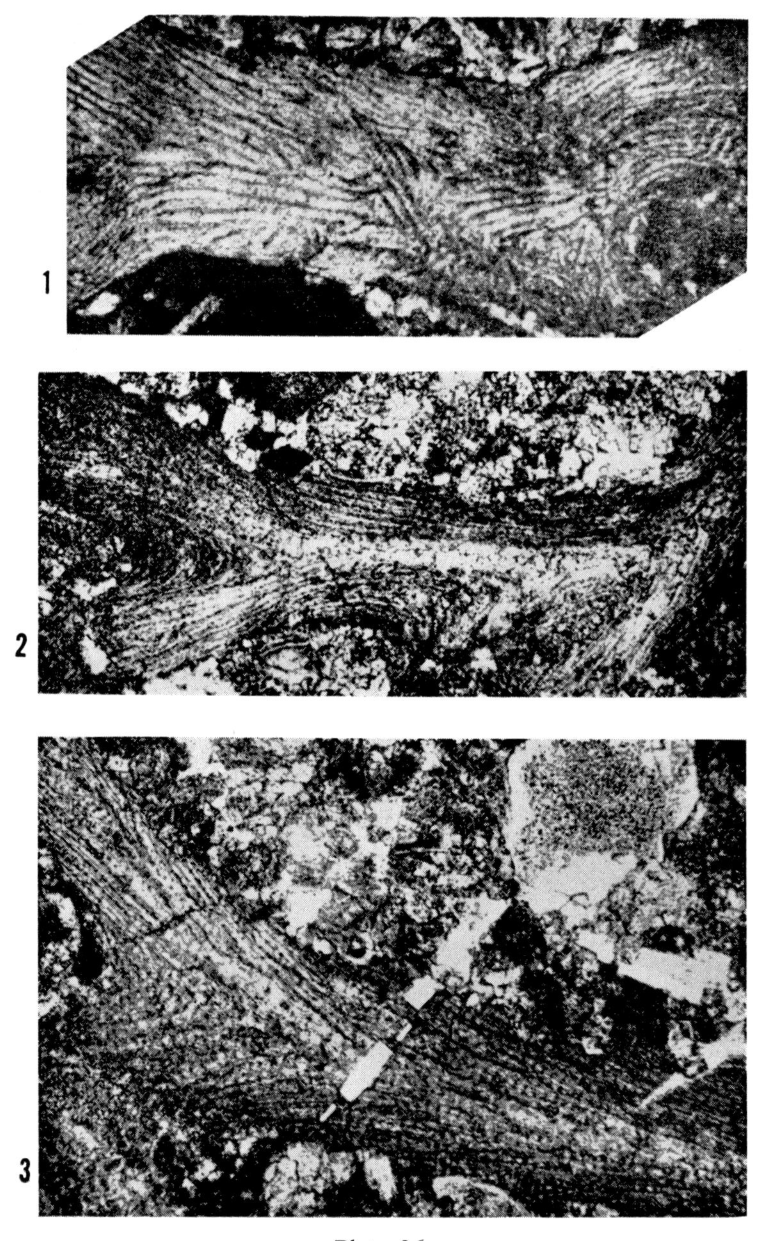

Plate 26
Genus *Ungdarella*
Figures 1-3. *Ungdarella uralica* Maslov. Carboniferous, Ural region, Russia. (From Maslov, 1956.) 1. Detail of tissue (x45). 2-3. Longitudinal sections (x26) showing branching of the thallus and arrangement of the cell threads.

REFERENCES

Johnson, J. Harlan, 1956, *Archaeolithophyllum*, a new genus of Paleozoic coralline algae: Jour. Paleontology, v. 30, no. 1, p. 53-55, 1 pl.

————, 1960, Paleozoic Solenoporaceae and related red algae: Colorado School of Mines Quarterly, v. 55, no. 3, 77 p., 23 pls.

Korde, K. B., 1951, Novye rody i vidy izvestkovykh vodoroslei iz kamennougolnykh otlozhenii severnogo Urala: Moskov. Obshch. Ispyt. Prirody, Trans., Otdel Geol., v. 1, p. 175-182, illus.

Maslov, V. P., 1956, Fossil calcareous algae in the U. S. S. R.: U. S. S. R. Acad. of Sci., Proceedings Institute Geological Sciences No. 160, 301 p., 86 pls. (for *Ungdarella*, p. 73-76, text-figs. 18-19, pl. 21, figs. 2-3; pl. 23, figs. 1-4).

Phylum CHLOROPHYCOPHYTA Papenfuss, 1946
(= Chlorophyta)

The Green Algae

This phylum is one of the largest and most important phyla of algae. The distinctive features of the Chlorophycophyta are: 1) the plants contain the pigments chlorophyll a, chlorophyll b, carotin, and xanthophyll (the same pigments found in the higher land plants), and the green algae contain these pigments in essentially the same proportions as the higher land plants; and 2) reproduction may be sexual or asexual. The motile reproductive cells are flagellated.

The phylum includes a great variety of structural types, ranging from microscopic unicellular forms to large plants which can be easily recognized megascopically. Both planktonic and sessile forms are found among the Chlorophycophyta.

Today, green algae live in marine, brackish, and fresh waters, with the greatest number of the known species residing in fresh water. Most of the fresh-water species that have been described are microscopic organisms, while a large proportion of the megascopic forms are marine.

This phylum is an old group, dating from far back in the Precambrian. One receives the impression that by Cambrian times, most of the major structural types of green algae had already developed and the number and variety of green algae present in Cambrian seas was probably as great as that found today.

Only two families belonging to the phylum Chlorophycophyta have developed the lime-secreting habit, and not all members of these two families are calcified. These families, the Codiaceae and the Dasycladaceae, are the only ones of interest to the geologist and paleobotanist as fossils or as limestone builders. Both of these families have a known geologic range extending from the Cambrian to the present.

During their long geologic history, some members of each of these families developed strong lime-secreting habits and, locally, became sufficiently abundant to be important as rock builders. The details of these activities will be given in connection with the discussion of the particular families.

Some of the forms which we class with the Porostromata may actually belong among the green algae. This statement is particularly true for the genus *Girvanella*.

Class CHLOROPHYCEAE Kützing, 1843
Order SIPHONALES Wille *in* Warming (1889) orth. mut.
Blackman et Tansley, 1902
Family CODIACEAE (Trevisan) Zanardini, 1843

This family is one of the most important groups of the rock-building green algae, and, in the course of its long history, it has been represented by quite a number of genera. The Codiaceae have a freely branched, tubular thallus, in which the tubular branches are interwoven to form a plant body of definite shape. In other words, the basic structure is a mass of branching tubular filaments, which may be closely or loosely packed.

In the Recent seas, the family is represented by 16 genera and about 125 species. Only a small percentage of these are calcified.

The calcification is interesting and rather characteristic. It starts at the outer surface or just inside the dermal layer and works inward. Very young growths commonly are uncalcified, while old ones may be thoroughly impregnated with lime. Commonly, the outer portion of a plant is more or less well-calcified, while the inner portion may not be calcified at all. This type of calcification is found in the fossil forms. Good structure is commonly observed in the outer portion of a fossil, but the structural features gradually fade as one works inward, and the center of the fossil, which frequently is filled with clear secondary calcite, may not show any of the original structure.

For convenience in study, the fossil codiaceans may be divided into two groups on the basis of growth form. The first of these groups includes crustose or nodular forms, composed of more or less tightly-packed branching tubular filaments. The second group contains erect branching or segmented forms. Among these branching forms, the branches, or segments, consist of tubular filaments commonly quite loosely packed. In most of these forms, the tubular elements in the center are large and roughly parallel to the axis of the plant. They branch into smaller and smaller tubes, which bend away from the center, ending in a dermal layer commonly composed of the tips of very fine tubular branches more or less perpendicular to the outer margin.

Those forms belonging to the first group are divided into genera on the basis of the character of the branching and into species according to the shape of the plant, the angle of branching, and the diameter of the individual branching tubes.

In the second group, the forms are separated into genera on the basis of whether the thallus is segmented or non-segmented, and whether the thallus is branched or unbranched. Species are separated largely on the dimensions of the plant, the segments, and the tubes.

All known Recent species of Codiaceae are marine, and most of them are restricted to the warm seas. There is strong evidence suggesting that the same was true in the past.

The family is known to have occurred as early as the Cambrian and, apparently, during that period was represented by a considerable variety of forms. Codiaceae appear to have occurred in abundance in the warm seas from at least the Ordovician to the Recent. By Ordovician times, the characteristic basic structures are known to have been present.

Lime-secreting codiacean algae throughout geologic time have occasionally developed in such numbers in certain localities as to be important rock builders. Limestones constructed largely of their remains are known from the Ordovician, Silurian, Devonian, Mississippian, Permian, Jurassic, and Cretaceous. During the Cenozoic, the most important representatives of the family belong to the genus *Halimeda,* which seems to have been an important rock builder whenever conditions were favorable.

Table X shows some of the characteristics of the more important genera of Codiaceae.

TABLE X
CHARACTERISTICS OF SOME GENERA OF CODIACEAE

Group 1
Crustose or nodular forms of closely packed branching filaments.

Genus	Character of branching	Tube diameter Uniform	Tube diameter Variable	Geologic Range
Bevocastria			X	Mississippian
Cayeuxia		X		Jurassic - Lower Cretaceous
Garwoodia		X		Mississippian to Triassic?
Hedstroemia		X		Ordovician to Mississippian
Ortonella		X		Silurian to Pennsylvanian?

Group 2
Tufts of erect branches, loosely packed, commonly segmented.

Genus	Not segmented	Segmented	Shape of segment or mass
Halimeda		X	Flattened, leaf-like
Microcodium	?		Ovate to nearly cylindrical
Ovulites		X	Ovate, tubular, hollow
Palaeocodium	X		Globular
Paleoporella		X	Cylindrical with rounded ends
Succodium		X	Cylindrical with rounded ends

Genus *Anchicodium* Johnson, 1946
Plate 27, figures 1-5.

Description—

Thallus develops as an irregular crust from which straight or nearly straight cylindrical stems may develop. These stems may branch or develop rounded protuberances. Some species show irregular constrictions; others are very regular. Each crust is composed of a spongy mass of rounded threads, the central portion of which tends to be poorly organized or pith-like. Toward the outer margin, the threads tend to become parallel. In some species, the threads end in tufts of fine branches that usually are nearly perpendicular to the outer surface. Calcification is variable, starting at the outer surface and working inward. Commonly, only a thin outer zone is calcified, but sometimes the entire tissue is affected. Calcified areas normally preserve the microstructure whereas the uncalcified areas are filled with clear calcite. Sporangia are unknown.

Remarks—

This genus is abundantly represented in the Late Pennsylvanian and Early Permian of the Mid-Continent region and of Japan.

Generic range—

Middle Pennsylvanian to Lower Permian.

Geographic distribution—

Kansas, Nebraska, Texas, and New Mexico in the United States; Honshu, in Japan.

Genus *Bevocastria* Garwood, 1931
Plate 28, figures 1-3.

Description—

Thallus consists of an irregular growth of tubes averaging 0.04 mm in diameter and exhibiting constrictions at fairly regular intervals. Tubes curved and undulating, bifurcated at irregular intervals, with a tendency for all the tubes to assume a parallel growth. Tubes arranged in irregular but roughly concentric faggot-like groups. Detailed structure and mode of growth very variable. Preservation commonly poor. No traces of sporangia known.

Remarks—

This genus is distinguished from *Rothpletzella* Wood (1948; = *"Sphaerocodium"* Rothpletz, 1890, pars) by its absence of the *Codium*-like expansions and the general mode of growth.

Generic range—

Lower Carboniferous.

Geographic distribution—

England and Scotland.

Genus *Cayeuxia* Frollo, 1938
Plate 29, figures 1-5.

Description—

Plants form rounded tufts or cushions, ranging in diameter from a few millimeters to more than a centimeter. The tufts consist of a mass of loosely-packed, radially-arranged, branching tubular filaments. The type of branching characterizes the genus: the original filament continues straight with a branch starting off at an angle of nearly 45°

for a short distance. The branch then turns and grows approximately parallel to the parent stem. Sporangia unknown.

Remarks—

Suggests *Garwoodia* but differs in the angle of branching and in having only a single branch coming off at a time instead of a tuft of branches.

Generic range—

Jurassic and Lower Cretaceous.

Geographic distribution—

North Africa, Hungary, Yugoslavia, Middle East, Texas.

Genus *Garwoodia* Wood, 1941
Plate 30, figures 1-2.

Description—

Molds of algal threads occurring in nodules and having a markedly radial arrangement. The threads, or tubes, are practically straight and of equal thickness for relatively long distances, so that a very characteristic appearance is seen in section. Branching occurs rather frequently. A new tube is given off almost at right angles to the parent, goes a short distance, then rebranches (again at right angles), producing a peculiar forked branching. Branches subequal in diameter. Reproductive organs unknown.

Remarks—

Wood (1941) restudied the available type specimen of the genus *Mitcheldeania* (= *M. nicholsoni* Wethered) and found it to represent an aggregate of at least three genera — *Girvanella, Ortonella* and *Bevocastria* (?). The other form of the genus, *M. gregaria,* with its peculiar double right-angled branching, which several generations of paleontologists had considered the peculiar and easily recognizable characteristic of *"Mitcheldeania,"* he named *Garwoodia.*

The genus *"Mitcheldeania"* was first placed among the Hydroctinidae of the Hydrozoa and was generally ignored until Garwood carefully restudied it about 45 years ago and redescribed it as a calcareous alga.

Generic range—

Lower Carboniferous to Triassic (?).

Geographic distribution—

England, France, Belgium, North Africa, India, Japan, British Columbia, Washington, Colorado, and Kansas.

Genus *Halimeda* Lamouroux, 1812
Plates 31, 32, 33, and 136.

Description—

Plants bushy, consisting of tufts of segmented branching stems or fronds. The segments may be broad and leaf-like, flattened, subcylindrical, or even subconical. The older segments become strongly calcified. Calcification proceeds inward from the outer surface and commonly is incomplete.

The segments are composed of tubular filaments. These are coarse in the center of the segment but branch into smaller and smaller tubes ending in clusters of fine, short tubes perpendicular to the surface.

Remarks—

Recent species are separated largely on the method of branching, the shape of segments, and the character of the node, features not available in fossils which almost always consist of individual segments. Consequently, it is seldom possible to identify fossil fragments specifically, especially in sections, although it is very easy to recognize them generically.

Halimeda occur abundantly in many shallow-water areas of the warm seas. Locally, they may be so abundant as to be limestone builders.

Generic range—

Late Cretaceous to Recent.

Geographic distribution—

The tropical oceans of the world today, and apparently much the same since Eocene times.

Genus *Hedstroemia* Rothpletz, 1913
Plate 34, figures 1-5.

Description—

Thallus forms rounded masses composed of tubes having a more or less radial arrangement. Branching common, into pairs or clusters of branches, the branches forming an acute angle with one another. Sporangia unknown.

Remarks—

Quite widespread. Locally, sufficiently abundant to make important contributions to limestone building.

Generic range—

Ordovician, Silurian, and Lower Carboniferous.

Geographic distribution—

Norway, Sweden, Baltic region, Ontario, and Japan.

Genus *Microcodium* Glück, 1912
Plate 35, figures 1-2.

Description—

The structures called *Microcodium* are subcylindrical to lobate, rarely ovoid or nearly spherical. They consist of large cuneiform cells which radiate out, like the petals of a flower, from a circular or elliptical central mass which shows little preserved structure.

Remarks—

There is considerable difference of opinion as to the nature of these objects. Some authors consider them to be inorganic. In the writer's opinion, they are definitely organic and probably are algal, but are of very uncertain systematic position.

Generic range—

Eocene to Miocene.

Geographic distribution—

Germany, France, Switzerland, Bikini, Saipan.

Genus *Ortonella* Garwood, 1914
Plate 36, figures 1-4.

Description—

Thallus forms small rounded nodules or nodular masses, each of which consists of a series of fine ramifying tubes which radiate from the center of the nodules. The tubes are straight or slightly undulating, completely and often widely separated, and circular in cross section. They vary slightly in size, but individual tubes show a nearly uniform diameter throughout. The tubes have a marked dichotomous branching with the angle of divergence of the branches usually about 40°. There appears to be a tendency for this branching to take place in several neighboring tubes at about the same distance from the center of the nodule. Sporangia are unknown.

Remarks—

The species are separated on the basis of diameter of tubes and angle of branching.

Generic range—

Silurian to Upper Carboniferous.

Geographic distribution—

England, France, Belgium, Japan, Colorado, New Mexico, Kansas, Missouri, Alberta.

Genus *Ovulites* Lamarck, 1816
Plate 37, figures 1-12.

Description—

Small, hollow, calcareous egg-shaped or tubular bodies, 1.75 to 4 mm long. At each end of the body, there is an irregularly circular opening, or, more rarely, two openings (but in the latter case at one end only). The walls are of variable thickness (25 μ to 300 μ), perforated by numerous fine pores (8 μ to 28 μ in diameter) which are arranged in no apparent order.

Remarks—

These were long problematical fossils, common in the Eocene beds of the Paris Basin. With the finding of specimens with several of the bodies held in position by encrusting bryozoans, however, they were finally recognized by Munier-Chalmas (1879) as being analogous in structure to Recent Codiaceae of the genus *Penicillus*.

Generic range—
Eocene.

Geographic distribution—
Paris Basin in France and Belgium, England.

Genus *Palaeocodium* Chiarugi, 1947
Plate 38, figures 1-2.

Description—
Thallus forms a hollow globular mass, about 3.5 cm in diameter, probably open at the top. The tissue is composed of moderately closely packed coarse threads arranged more or less perpendicular to the outer surface. The filaments average about 100 μ in diameter and branch frequently, keeping a nearly uniform thickness. Branching dichotomous. Calcium carbonate deposited around the threads forming a firm dense mass. Sporangia unknown.

Generic range—
Lower Carboniferous.

Geographic distribution—
Libya.

Genus *Palaeoporella* Stolley, 1893
Plate 39, figure 1; plate 117, figure 2.

Description—
Thallus articulated and composed of slender cylindrical internodes, bifurcated frequently but irregularly. Each internode strongly calcified (sometimes even in the medullary part), composed of a mass of long slender branching tubes, the extremities of which form a very fine cortical layer.

Remarks—
Although a superficial observation may suggest this genus to be a dasycladacean such as *Bornetella* [as Stolley originally thought (1893)] or *Cymopolia* (Stolley, 1896), the internal structure, which occasionally is preserved, indicates it to be a codiacean (Pia, 1926 and 1927).

Generic range—
Basal Ordovician* to Devonian (?)**.

Geographic distribution—
Western Europe and the Baltic region; Llano region of Texas.

* The Ellenburger group in Texas is considered to be either basal Ordovician or uppermost Cambrian.

** There is a doubtful occurrence from Eifel (Johnson and Konishi, 1958): otherwise, all occurrences are from Ordovician and lowermost Silurian.

Genus *Succodium* Konishi, 1954
Plate 40, figures 1-4.

Description—

A branching, calcareous codiacean composed of articulated segments. Each segment has rounded ends and is formed of three components: (1) a feebly calcified central medulla of longitudinal, ramified, medullary filaments; (2) a strongly calcified subcortical part of irregularly interwoven utricles; and (3) a thin outer cortical layer outlining outlets of tapering utricles. Gametangia-like expansions, densely disposed at the same level, form the boundary between the subcortical and cortical parts. (Konishi, 1954)

Remarks—

Distinguished from other genera of Permian codiaceans by its articulated thallus and nearly symmetrical stucture, as well as by the abrupt expansion of the utricles to form the gametangia-like structures.

Generic range—

Upper Permian.

Geographic distribtuion—

Japan.

REFERENCES

Allard, P., Gannat, E., Laplaiche, N., and Lefavais-Raymond, A. and M., 1959, Sur un niveau a *Microcodium* a la base du Tertiare de Bresse: Soc. Geol. France, C. R., 1959, fasc. 6, p. 150, 1 fig.

Barton, E. S., 1901, The genus *Halimeda: Siboga* Expedition Mon. 60, Leiden, Holland, p. 1-32, 4 pls.

Chiarugi, Alberto, 1947, *Palaeocodium saharianum*, n. gen., n. sp., nuova Codiaceae paleozoica del deserto libico: Palaeontographia Italica, v. 41 (n.s. v. 11), p. 121-130.

Emberger, L., 1944, Les plantes fossiles dans leur rapports avec les vegetaux vivants: Paris (Masson et cie), p. 44-99.

Frollo, M. M., 1938, Sur un nouveau genre de Codiacee du Jurassic superieur des Carpates orientales: Soc. Geol. France Bull., ser. 5, v. 8, no. 3-4, p. 269-271, 1 fig., 1 pl.

Garwood, E. J., 1931, The Tuedian beds of northern Cumberland and Roxburgshire east of the Luddel water: Geol. Soc. London Quart. Jour., v. 87, p. 97-152, pls. 7-14.

Glück, H., 1914, Eine neue gesteinbildende Siphonee (Codiacee) aus dem marinen Tertiar von Süddeutschland: Badische Geol. Landesanstalt Mitt., v. 7, p. 1, Heidelberg.

Høeg, O. A. 1927, *Dimorphosiphon rectangularis*, preliminary note on a new Codiaceae from the Ordovician of Norway: Avh. Norske Vid. Oslo, 1 Mat.-Naturv. Kl., no. 4, p. 1-15, pls. 1-3.

Johnson, J. Harlan, 1945, Calcareous algae of the upper Leadville limestone near Glenwood Springs, Colorado: Geol. Soc. America Bull., v. 56, p. 829-848, 5 pls., 1 fig.

————, 1946, Mississippian algal limestones from the vicinity of St. Louis, Missouri: Jour. Paleontology, v. 20, no. 2, p. 166-171, pl. 30.

————, 1946, Lime-secreting algae from the Pennsylvanian and Permian of Kansas: Geol. Soc. America Bull., v. 57, p. 1087-1120, 10 pls., 5 figs.

————, 1946, Late Paleozoic algae of North America: Am. Midland Naturalist, v. 36, no. 2, p. 264-274, 2 pls.

————, 1951, Permian calcareous algae from the Apache Mountains, Texas: Jour. Paleontology, v. 25, no. 1, p. 20-30, pls. 6-10.
————, 1952, Ordovician rock-building algae: Colorado School of Mines Quarterly, v. 47, no. 2, p. 29-56, pls. 1-12.
————, 1953, *Microcodium* Glück est-il un organisme fossile?: Acad. Sci. Paris, Compte Rendu, v. 237, no. 1, p. 84-86.
————, and Konishi, Kenji, 1956, Studies of Mississippian algae: Colorado School of Mines Quarterly, v. 51, no. 4, (Codiaceae p. 23-38, pls. 4-7).
————, 1958, Studies of Devonian algae: Colorado School of Mines Quart., v. 53, no. 2, (Codiaceae p. 40-46, pls. 10-13).
————, 1959, Studies of Silurian algae: Colorado School of Mines Quarterly, v. 54, no. 1, (Codiaceae, p. 39-43, pls. 8-13).
Konishi, Kenji, 1954, *Succodium* a new Codiacean genus and its algal associates in the Late Permian of southern Kyushu, Japan: Jour. Fac. Soc., Univ. Tokyo, sec. II, v. 9, pt. 11, p. 225-240, 2 pls., 1 fig.
Morellet, L., and Morellet, J., 1939, Tertiary siphoneous algae in the W. K. Parker Collection: British Museum (Nat. Hist.), 55 p., 6 pls.
Munier-Chalmas, E., 1879, Observations sur les algues calcaires confondues avec les foraminiferes et appartenants au groupe des siphonees dichotomes: Soc. Geol. France Bull., ser. 3, v. 7, p. 661-670, 4 figs.
Pia, Julius, 1926, Pflanzen als Gesteinsbildner: Berlin, 355 p.
————, 1927, Thallophyta *in* Hirmer,, M., Handbuch der Palaobotanik: Munchen und Berlin, p. 1-136.
————, 1937, Die wichtigsten Kalkalgen des Jungpalaozoikums und ihre geologische Bedeutung: 2nd Cong. pour l'avancement des etudes de stratigraphie Carbonifere, Compte rendu, Heerlen, Holland, p. 765-856, pls. 85-97.
Rothpletz, Aug., 1890, Über *Sphaerocodium Bornemanni*, eine neue fossile Kalkalge aus den Raibler Schichten der Ostalpen: Bot. Centralbl., v. 52, p. 9.
Stolley, E., 1893, Über silurische Siphoneen: Neues Jahrb. Mineralogie, v. 2, p. 135.
————, 1896, Über gesteinsbildende Algen und die Mitwirkung solcher bei der Bildung der Skandinavisch-baltischen Silurablagerungen: Naturwiss. Wochenschr., v. 11, p. 173.
Taylor, W. R., 1950, Plants of Bikini: Univ. Michigan Press, p. 76-93, pls. 39-51.
Wood, A., 1941, The Lower Carboniferous calcareous algae *Mitcheldeania* Wethered and *Garwoodia* gen. nov.: Geologists' Assoc. London Proc., v. 52, pt. 3, p. 216-226, 3 pls.
————, 1948, "*Sphaerocodium*" a misinterpreted fossil from the Wenlock limestone: Geologists' Assoc. London Proc., v. 59, pt. 1 p. 9-22, 4 pls.

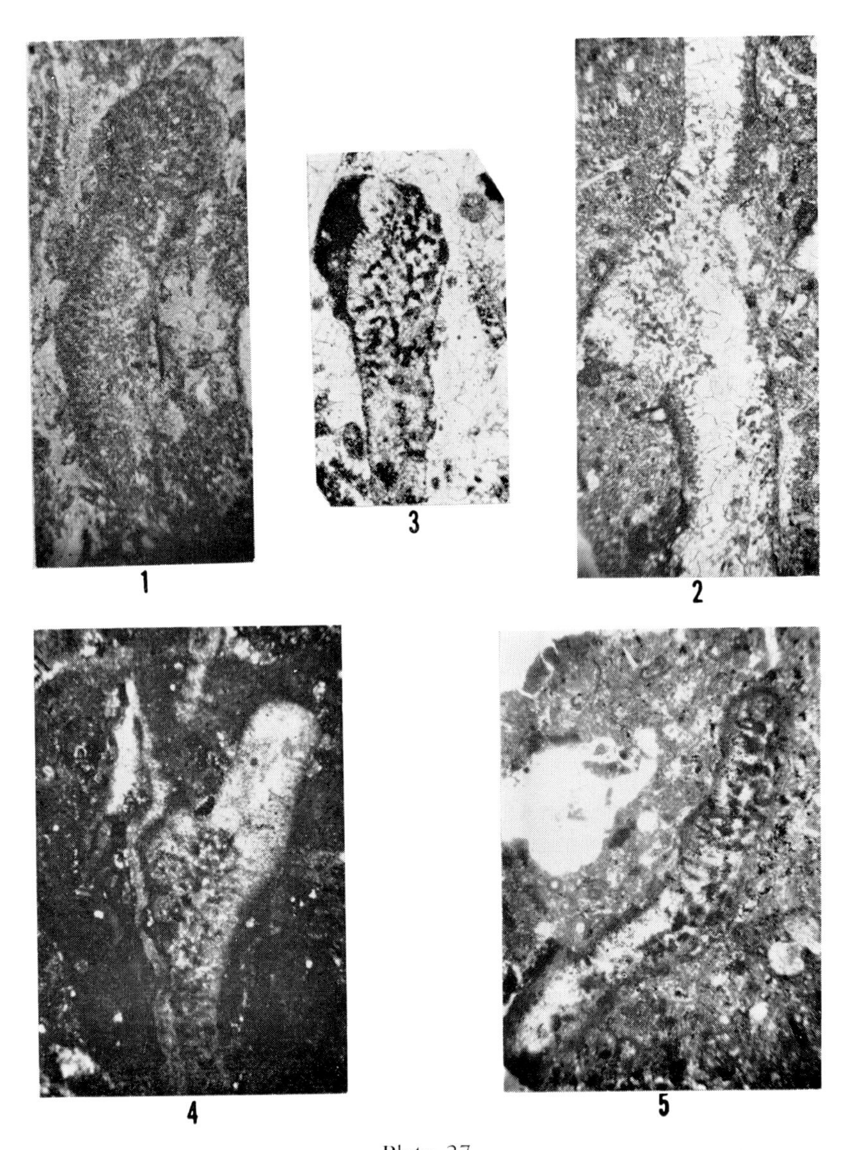

Plate 27
Genus *Anchicodium*

Figures 1-2. *Anchicodium nodosum* Johnson (x30). Upper Pennsylvanian, Deer Creek limestone, Doniphan County, Kansas.

Figures 3-4. *Anchicodium funili* Johnson (x36). Upper Pennsylvanian Auburn shale and Wakarusa limestone, Lyon, Brown and Shawnee Counties, Kansas.

Figure 5. *Anchicodium undulatum* Johnson (x32), Upper Pennsylvanian, Lecompton limestone, Jefferson County, Kansas.

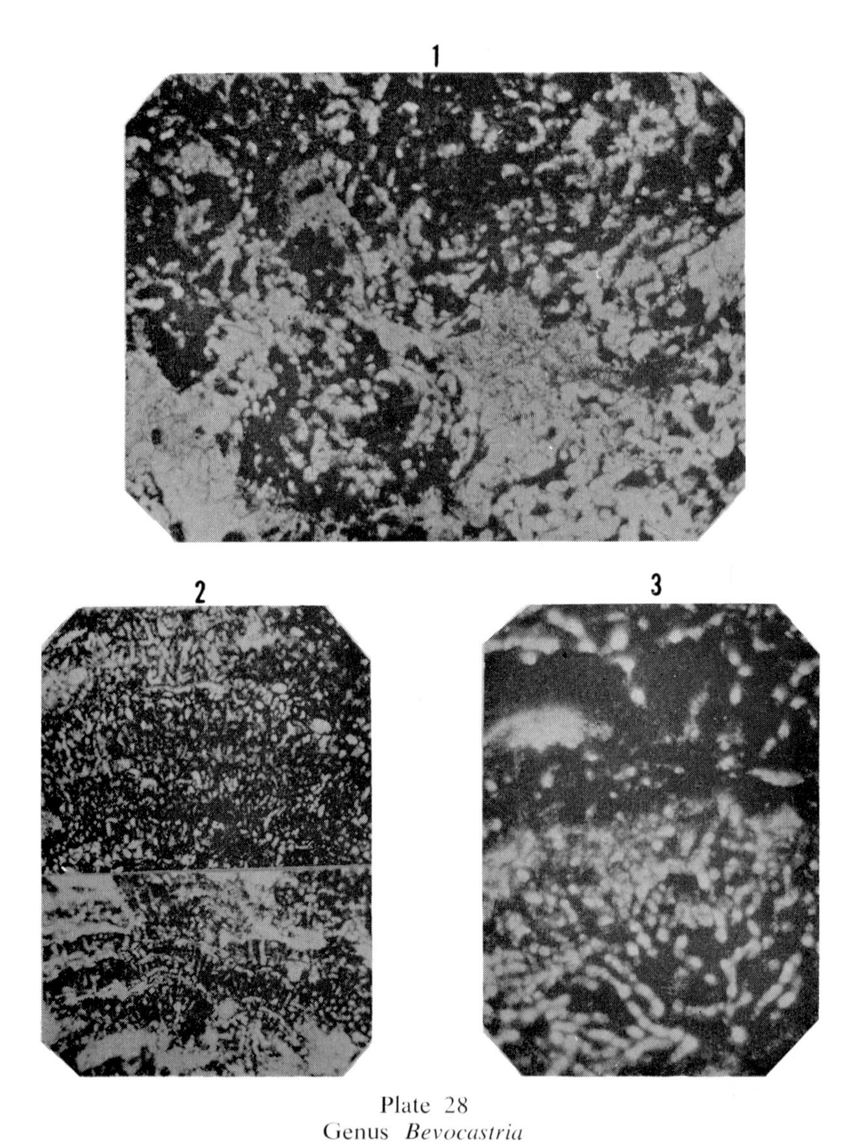

Plate 28
Genus *Bevocastria*
Figures 1-3. *Bevocastria conglobata* Garwood, Mississippian, Christianbury, Cumberland, England (from Garwood, 1931, pl. 12). 1 and 3. Details (x40). 2. *Ortonella* and *Bevocastria* mixed (x25).

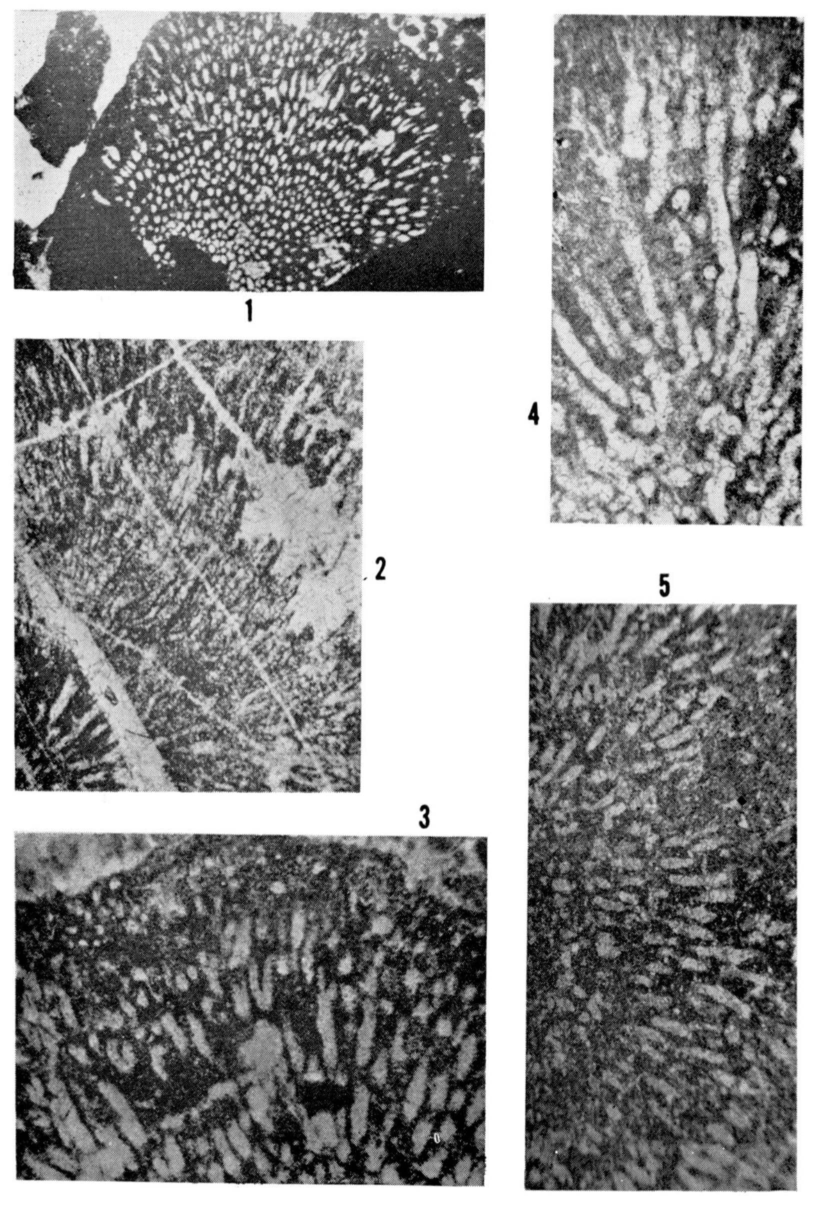

Plate 29
Genus *Cayeuxia*

Figures 1, 3. *Cayeuxia moldavica* Frollo (x16), Jurassic, Hungary.
Figure 2. *Cayeuxia piae* Frollo. (x16), Jurassic, Hungary.
Figure 4. *Cayeuxia americana* Johnson (x40), Jurassic Van Zant County, Texas.
Figure 5. *Cayeuxia* species ? Lower Cretaceous, Guatemala.

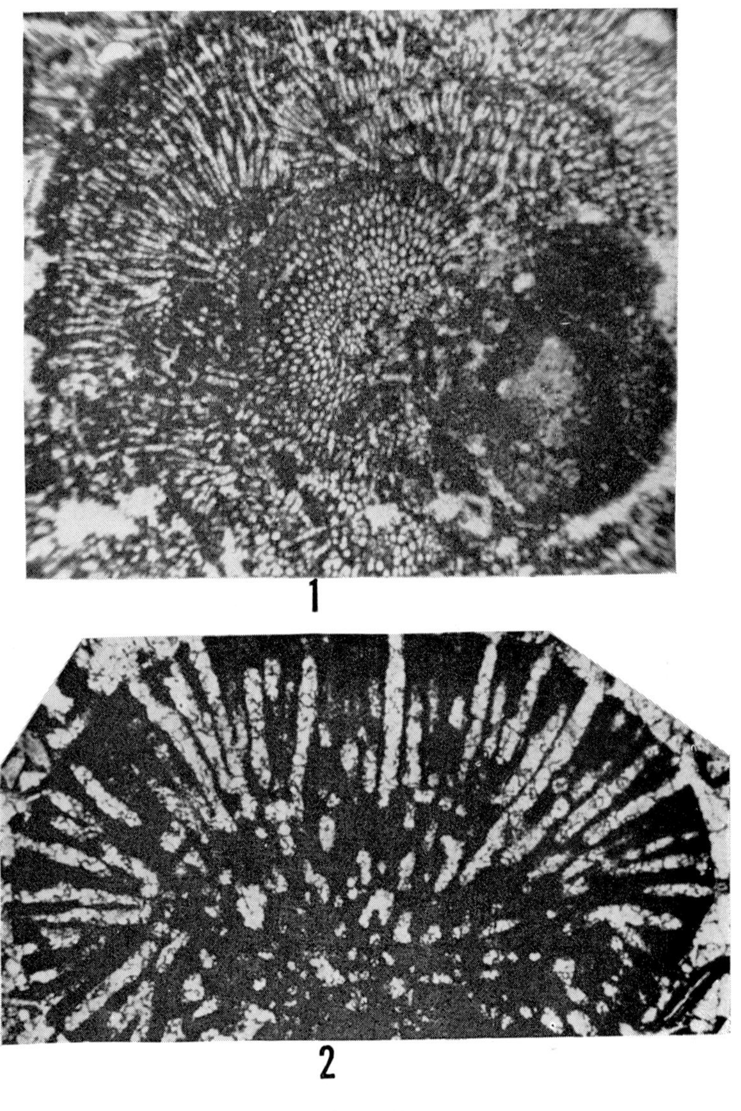

Plate 30

Genus *Garwoodia (Mitcheldeania)*

Figure 1. Section of a nearly spherical colony (x25) of *Garwoodia gregaria* (Nicholson). Bristol Region, England.

Figure 2. *Garwoodia gregaria* (Nicholson) (x45). Detail showing the unusual type of branching which is characteristic of this genus. (After Pia, 1937.)

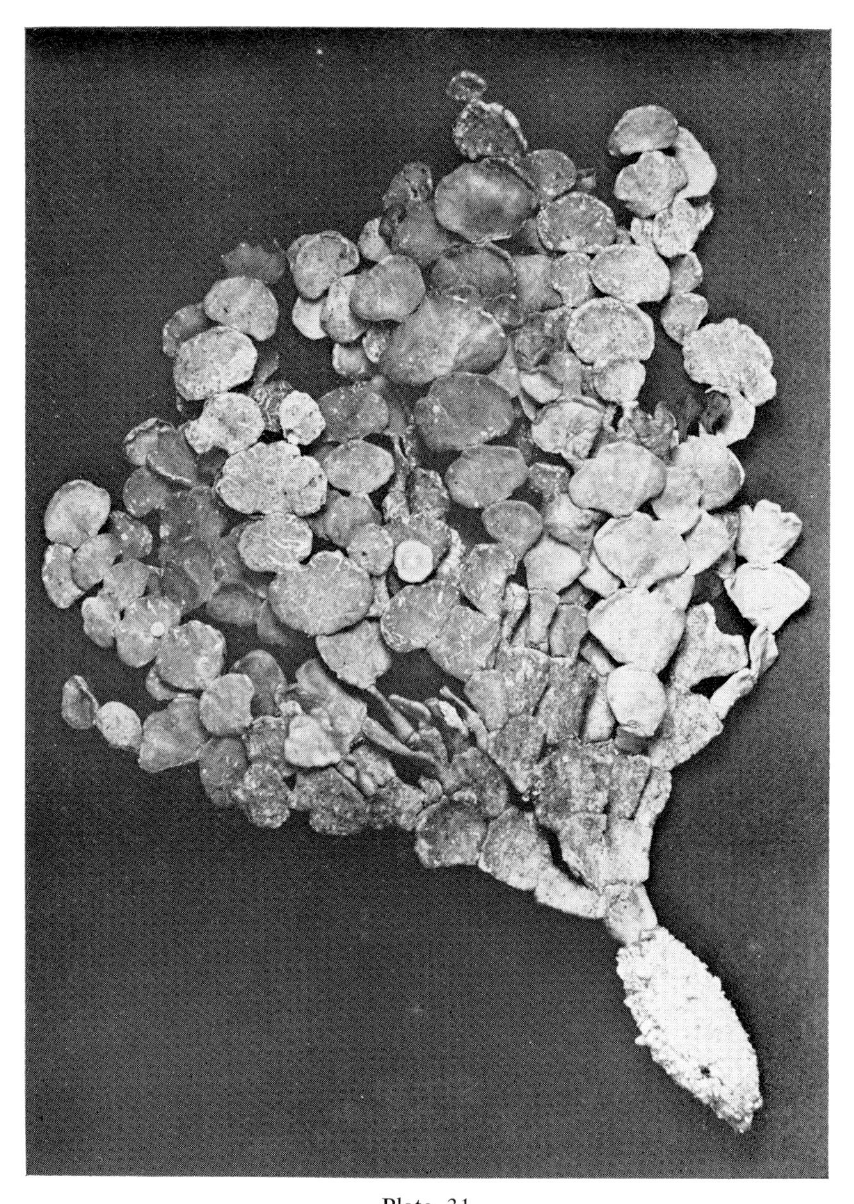

Plate 31
Genus *Halimeda*
(Published by permission, U. S. Geological Survey)
Figure 1. A Recent *Halimeda* plant from the Palau Islands (x1).

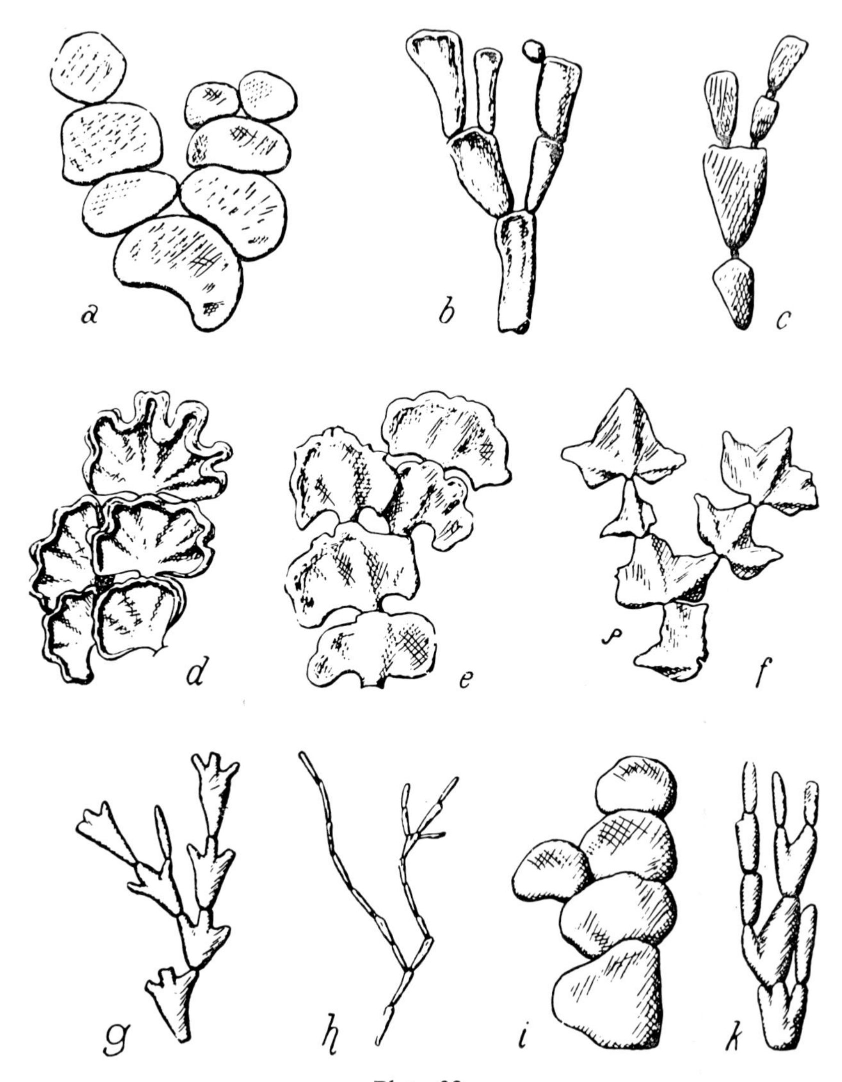

Plate 32
Genus *Halimeda*
Varied leaf forms after Barton.

a. *Halimeda tuna Lmx.* (x6.5) b. *Halimeda tuna forma albertsi* Peccona (x8.5). c. *Halimeda cuneata* (x7.5). d. *Halimeda cuneata forma undulata* Barton (x4.5). e. *Halimeda opuntia* Lmx. (x2). f. *Halimeda opuntia forma hederacea* Barton (x1). g. *Halimeda opuntia forma elongata* Barton (x8.5). h. *Halimeda gracilis* Harvey *forma laxa* Barton (x7.5). i. *Halimeda macroloba* Descaisne (x6.5). k. *Halimeda incrassata* Lmx. *forma monilis* (x8.5).

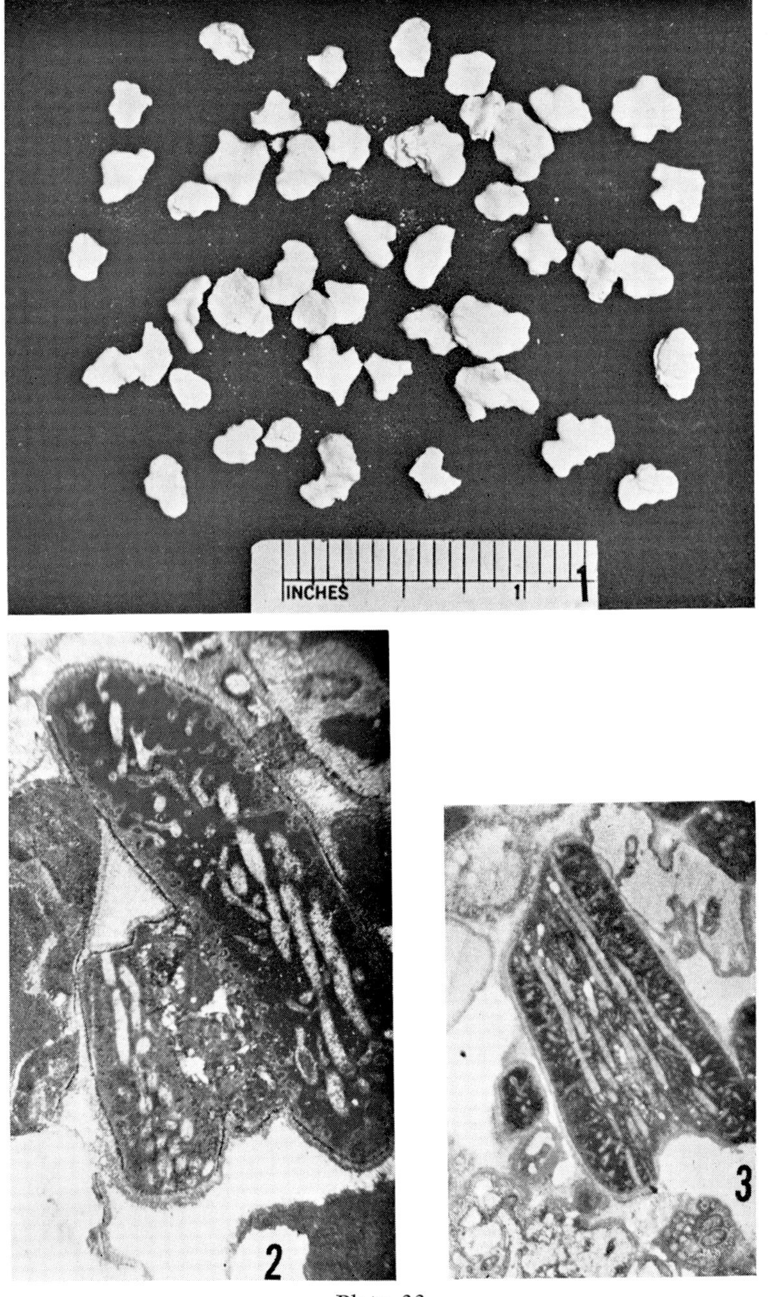

Plate 33
Genus *Halimeda*
Figure 1. *Halimeda* fragments. Figure 2. Sections of *Halimeda* (x25). Figure 3. Sections of *Halimeda* (x15). All three from Bikini Atoll.

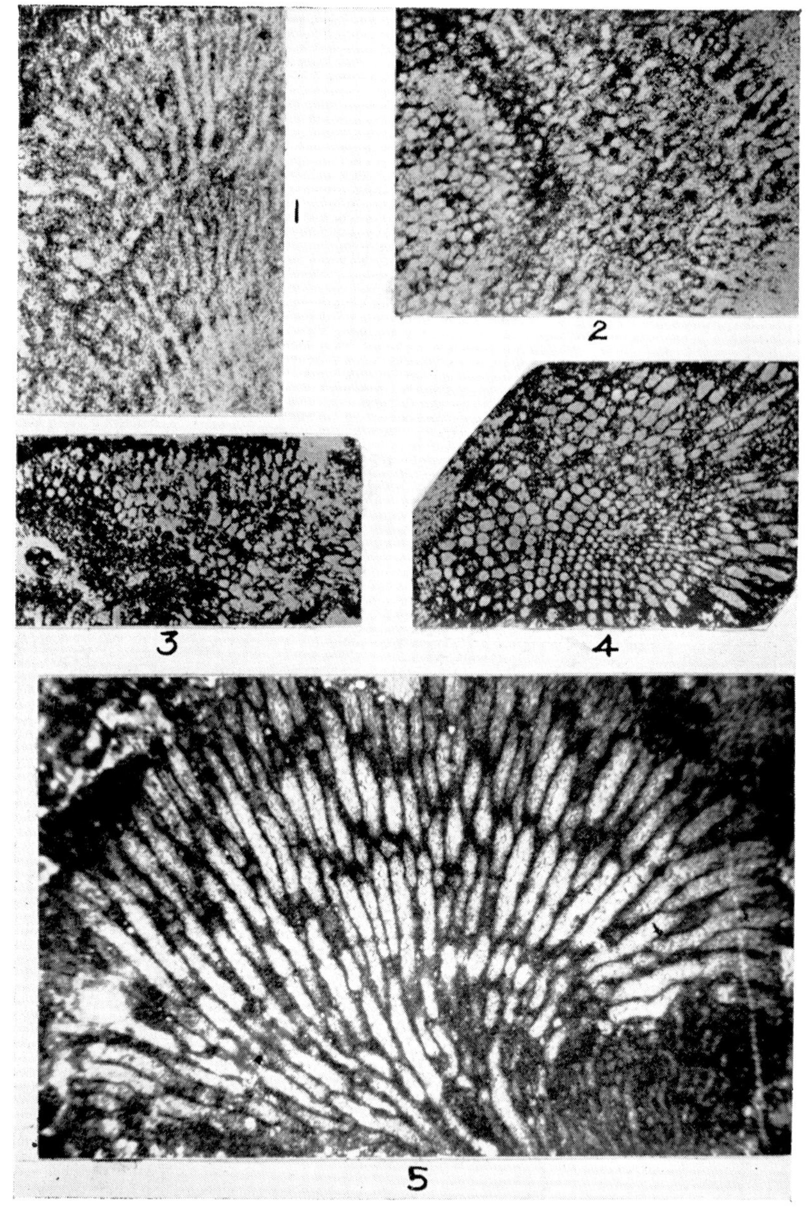

Plate 34

Genus *Hedstroemia*

Figures 1-2. *Hedstroemia aequalis* Høeg (x50). Norway. (After Høeg.)
Figures 3-4. *Hedstroemia nidarosiense* Høeg (x25). Norway. (After Høeg.)
Figure 5. *Hedstroemia* sp. ? (x40). Oslo Fiord, Norway. (Johnson.)

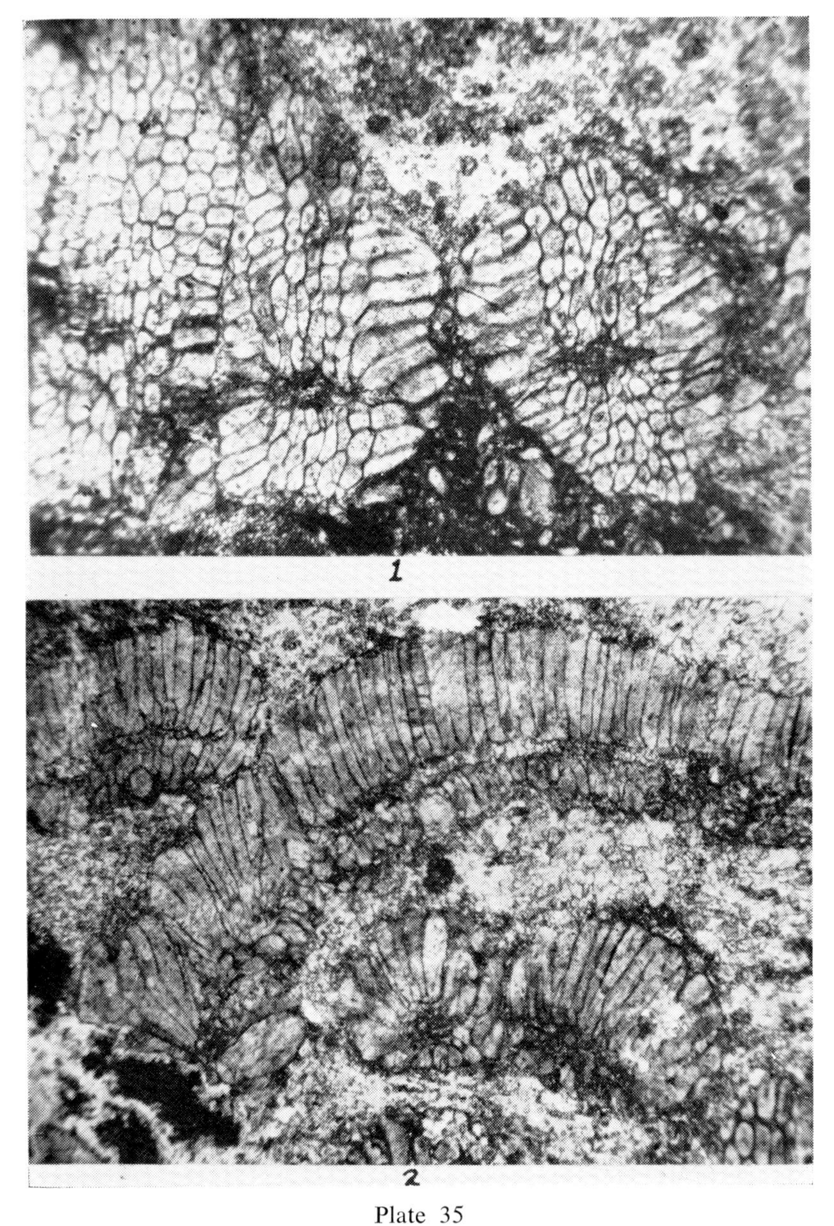

Plate 35
Genus *Microcodium*
Figures 1-2. Sections (x40), Eocene, France.

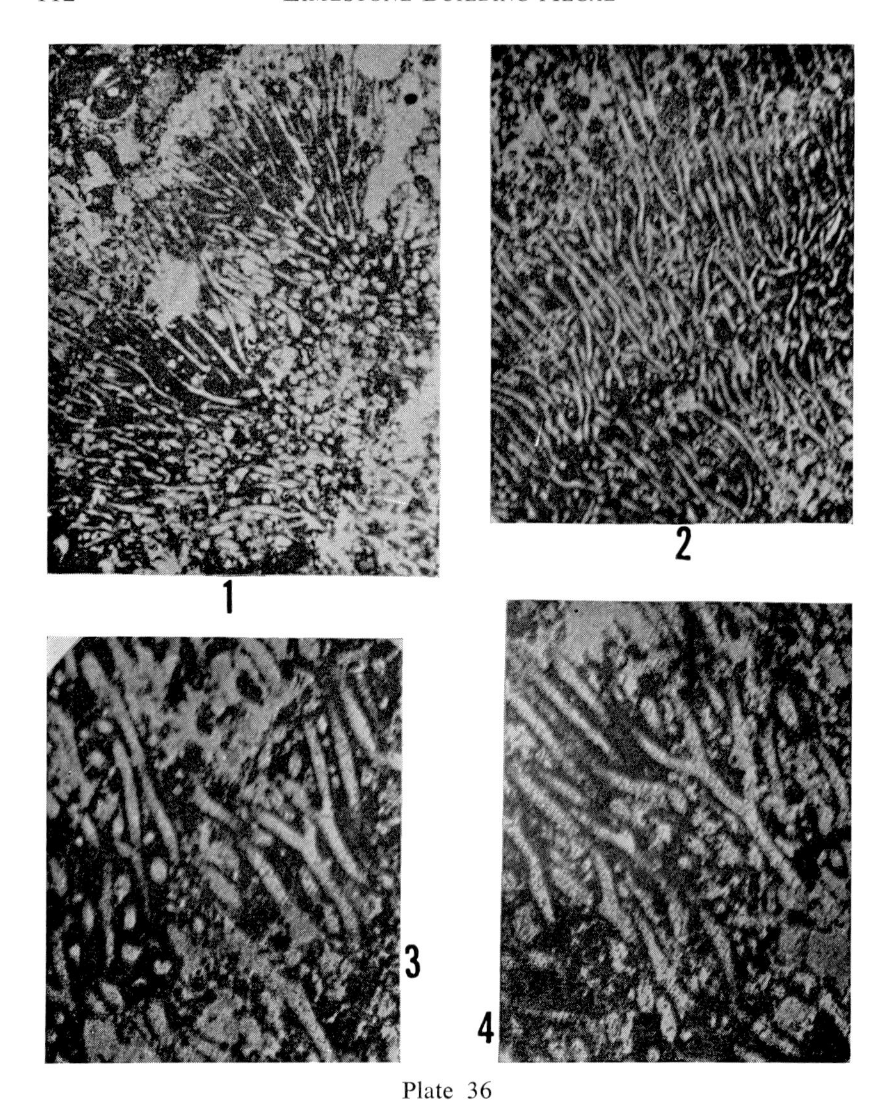

Plate 36
Genus *Ortonella*

Figures 1-4. *Ortonella furcata* Garwood, England. (From Garwood, 1914 and 1931.) 1. Section of a large growth (x30). 2. Section through a large thick growth (x20). 2-4. Details (x50) showing character of the branching.

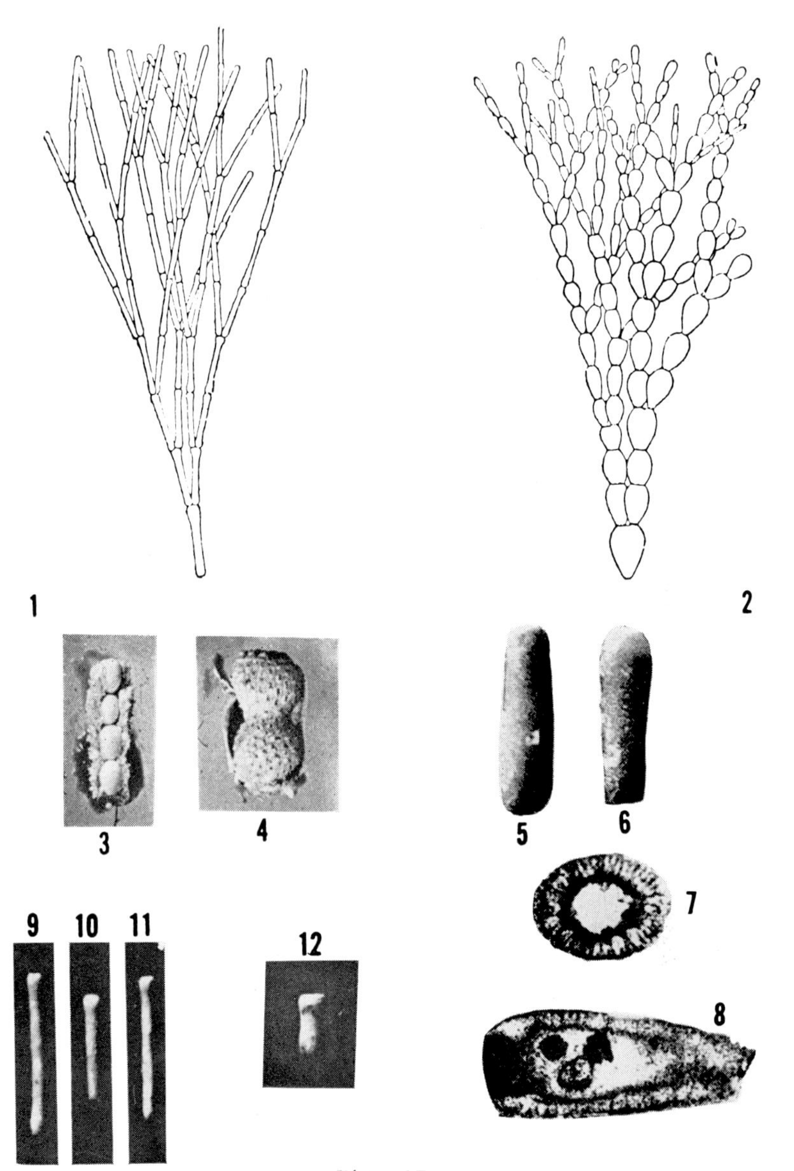

Plate 37
Genus *Ovulites*

Figure 1. Restoration of branch of *Ovulites elongatus* LK (x4) after Munier-Chalmas, 1879.
Figure 2. Restoration of branch of *Ovulites margaritula* LK (x2) after Munier-Chalmas, 1879.
Figures 3-4. *Ovulites margaritula* LK (x5), from L. and J. Morellet, 1939. Eocene, Paris basin. 3. Four segments held together by an encrusting bryozoan. 4. Two segments coated with a bryozoan.
Figures 5-8. *Ovulites morelleti* Elliott, Paleocene, northern Iraq (from Elliott, 1955). 5-6. Two segments (x28). 7. Transverse section (x50). 8. Oblique-vertical section (x50).
Figures 9-11. *Ovulites elongatus* LK segments (x15). Eocene, Paris basin (from L. and J. Morellet, 1939).
Figure 12. *Ovulites oehlerti* Munier-Chalmas (x15), Eocene, Paris basin (from L. and J. Morellet, 1939).

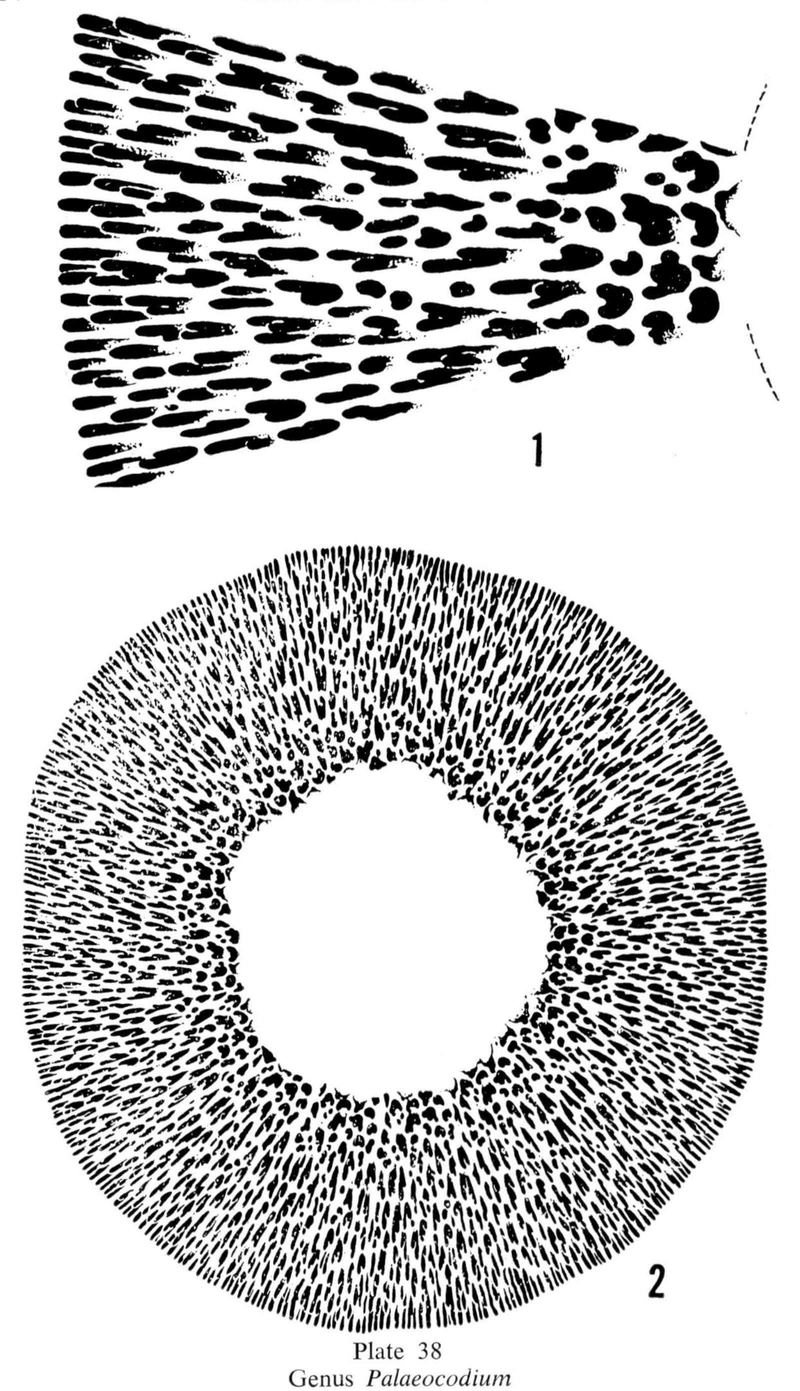

Plate 38
Genus *Palaeocodium*

Figure 1. Diagrammatic sketch of a section (x9) giving details of arrangement of the branching filaments.

Figure 2. Sketch (x3) of a transverse section through a thallus.

Plate 39
Genus *Palaeoporella*
Figure 1. *Palaeoporella* sp. ? (x1.5). Upper Ordovician (4-d). Frognøya, Ringerike, Norway. Specimen in Paleontologisk Museum, Oslo, Norway.

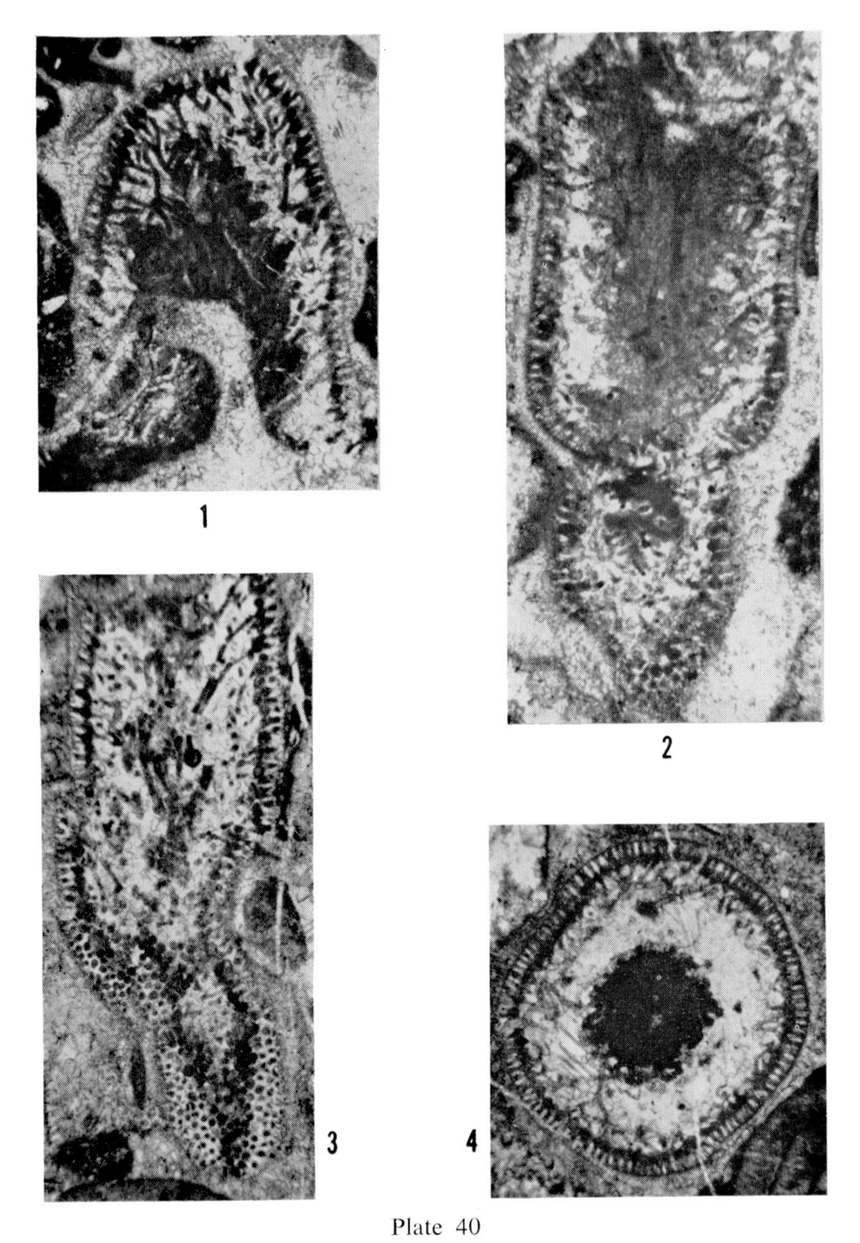

Plate 40
Genus *Succodium*
Figures 1-4. *Succodium multipilularum* Konishi (x40). Upper Permian, Shikoko, Japan.

Order DASYCLADALES Pascher, 1931
Family DASYCLADACEAE Kützing, orth. mut. Stizenberger, 1860

General Description

The thallus of a dasycladacean alga consists of a central stem from which primary branches develop. Among the earlier more primitive genera, these primary branches are not regularly arranged around the central stem. However, in most of the genera, the characteristic structure is to have the primary branches arranged in regular whorls which radiate out from the central stem like the spokes of a wheel. In a few genera, the primary branches radiate out more or less in all directions from the central stem, forming spherical or egg-shaped knobs. Commonly, the primary branches bear terminal tufts of secondary branches, which, in turn, may give rise to tufts of tertiary branches. In a few genera, quaternary branches may develop. Most of the genera of Dasycladaceae form cylindrical plants, although some are known which are club shaped, spherical, ovoid, or segmented. The plant is attached to the substratum by simple rhizoids or hold-fasts. The sporangia are spherical, egg shaped, or cylindrical, and are developed 1) within the central stem, 2) upon the central stem, 3) on the primary branches, or 4) among the secondary branches. Types 2 and 3 are those most commonly observed in fossils.

The thallus normally is more or less calcified. Calcification appears to follow one of two patterns. In the first type of calcification, fine particles of calcium carbonate are precipitated as a crust around the central stem and primary branches, and this crust may eventually become thick enough to envelop secondary or tertiary branches. Plants calcified in this manner make good fossils. In the second type of calcification, the calcium carbonate merely forms a thin crust which envelops the outer portions of the plant. Normally, this external calcification leaves the central stem and primary branches uncovered, and they are not preserved in the fossils.

The fossils consist of calcareous molds of the stem and branches. The molds usually appear as small cylinders or spheres, or break into disc-shaped objects. Commonly, the fossils are about the size of a large fusulinid. How much structure is shown in the fossil depends on the type of calcification. If it was of the first type and if the crust was thick enough, the fossil may show the central stem and branches as cavities which may or may not be filled with secondary calcite. The fossils may break into discs or bead-like fragments. Fossil dasyclads show a great range in size, from smaller than an average pin head to more than 18 inches in length.

The family Dasycladaceae has had a long geologic record, extending from the Cambrian to the Recent. It appears to have reached the zenith of its development during the Late Triassic and Jurassic, when some of its members developed the most complex structures known in the family.

During those periods, the family was represented by the greatest number of genera and species, and some of the genera attained very large size. Since Jurassic times, the Dasycladaceae appear to have slowly declined in both variety and numbers, and some genera have shown a simplification of structure.

Fossil dasycladacean remains puzzled early paleontologists who frequently confused them with animals, particularly foraminifera. Munier-Chalmas (1877) clearly demonstrated their algal nature. Thanks to the work of Pia in Austria and L. and J. Morellet (1913-1951) in France, the Dasycladaceae are now probably the best known group of fossil algae.

Recent Dasycladaceae

Among the living Dasycladaceae, *Dasycladus* appears to have the most simple structure. The vegetable body of *Dasycladus* consists of a large central stem attached to the bottom by rhizoids. The body forms a sort of trunk, more or less encrusted by calcium carbonate, in which a large number of alternating whorls of from ten to fifteen branches are densely inserted. These primary branches trifurcate several times forming branches of the second, third, and fourth orders. Each ramification is unicellular, separated from its support by a partition. The upper (outer) whorls of the plant are fertile. The sporangia are globular, isolated, and surrounded by branches of the second order. The sporangia appear to be at the ends of the primary branches but, in reality, are slightly lateral.

The structure of other Recent Dasycladaceae is very similar to that of *Dasycladus*. The genus *Batophora* is not as strongly encrusted with calcium carbonate. The ramification of the branches is less dense, and the sporangia are sub-terminal at the ends of the primary and secondary branches.

The modern *Neomerites* are strongly encrusted with lime. The ramification of the branches does not progress beyond the second order. The sporangia are more or less globular, inserted on the primary branches (in the genus *Bornetella*) or in sub-terminal positions at the extremities of the primary branches (as in *Neomeris* and *Cymopolia*). The thallus of *Cymopolia* is further ramified in a dichotomous manner and is articulated. Whorls of sterile branches alternate with whorls of fertile branches.

Among the *Acetabularia,* the fertile and sterile whorls succeed each other in varying numbers, and among the common species, only a single fertile whorl develops. These whorls give the alga an aspect very similar to that of a cap mushroom or a small extended parasol. The sporangia are very elongated.

All these modern forms obviously represent the end products of a long evolutionary history. The complex ramification, the distribution of the whorls, and the structure of the sporangia do not in any way indicate primitive structures.

Fossil Dasycladaceae

Three tribes of Dasycladaceae, the Dasyporelleae, the Cyclocrineae, and the Primicorallineae, are known only from the Paleozoic Era. The main stems of these plants are simple. They branch rarely with a dichronous habit *(Anthracoporella)* or have branches inserted abundantly but not in regular whorls. The branches themselves are habitually simple. Except among the Primicorallineae and *Anthracoporella,* the branches are more or less covered by a calcareous muff which sometimes develops upon the extremities of the outer branches, forming a crust of calcareous plates as in *Cyclocrinus* and *Macroporella.*

The sporangia of the early forms are not known. It is supposed that they developed within the main stem.

Cambrian Dasycladaceae are not well known. Russian geologists have described several genera, but, unfortunately, the descriptions are meager and the illustrations do not show nearly as much of the structures as one would wish.

We have a better picture of the Ordovician representatives of the family. Among the most interesting of these are the genera *Rhabdoporella, Cyclocrinus,* and *Primicorallina.* These probably represent the primitive types from which all later Dasycladaceae developed. Not one of these very ancient types has survived to the present day.

Dasycladaceae with truly whorled branches appear about the Middle Silurian. Of particular interest because of the multiplicity of tendencies which they manifest is the tribe of the Diploporaria from the Late Permian and Triassic. Among these forms, one sees the beginning of the structures which characterize Recent genera.

The genus *Macroporella* can be thought of as a Mesozoic *Rhabdoporella* with branches in clearly defined rows.

The classification given in Table XI is based on the works of Pia (1927) and Emberger (1944) with some additions by the author.

The Recent Dasycladaceae have all the characteristics of a typical relic family as shown by the following data summarized by Emberger (1944, p. 73).

1. Of the sixteen known tribes, twelve are entirely fossil and only one, the Dasycladees, is composed exclusively of modern types. Of the best known 58 genera, 48 are known only from fossils. Of the ten living genera, six are not known in the fossil state. The other four genera have both living and fossil representatives.

2. None of the living genera are represented by many species. *Dasycladus, Batophora, Chlorocladus, Chalmasia,* and *Acicularia* have only one species each, *Halcoryne* has two, *Cymopolia* three, *Bornetella* four, *Neomeris* six. The genus *Acetabularia* has the most representatives with 15 living species.

3. The modern genera from which we do not know any fossil species have the following peculiar geographic distribution. *Dasycladus clavaeformis* (Rothpletz) is found in the Mediterranean and the West Indies. *Batophora oerstedi* is localized in the West Indies and Florida. *Chlorocladus australasiaticus* exists only in the warm Australian waters. The genera *Bornetella* and *Halicoryne* are localized in the oceans around

Malaysia and Australia. *Halicoryne* is also known from Okinawa. *Chalmasia antillana* is known from the West Indies.

These peculiar geographical distributions might suggest a recent origin for those genera. However, the following genera which include both fossil and recent members have a similar erratic distribution and this fact suggests that they are relics of an older, more widely distributed flora. The genus *Neomeris* exists throughout the West Indies and in the islands of Sonde. The three species of *Cymopolia* live today, one in Malaysia, another in the Gulf of Mexico, and the third in the subtropical and warm-temperate areas of the western Atlantic.

The genus *Acicularia* has only one living species which lives in the inter-tropical zone along the east coast of the American continent, although more than a dozen fossil species have been found in other regions.

The geographical distribution of the genus *Acetabularia* is interesting, particularly when one considers that of the various subdivisions of the genus. The section *Acetabulum* has only one species, *A. mediterranea,* which is localized in the Mediterranean and the adjoining areas of the Atlantic. The section *Acetabuloides* is found in the East Asiatic and Australian waters, and is also found in the West Indies and along the coast of Florida. The section *Polyphysa* is known from the Indo-Australian region and from the West Indies.

The genus *Halicoryne* is represented by two species which inhabit the southeastern area of the warm Pacific and the region around Okinawa.

Fossil Dasycladaceae are common in many areas and locally were sufficiently abundant to be limestone builders. Such limestones are well known from the Ordovician of Norway (pls. 42 and 46), the Silurian of the Baltic region, the Permian of the Mediterranean region, Japan and the southwestern United States (plate 46), the Triassic of the Tyrol, and in the Cretaceous and Tertiary at numerous localities.

Classification of Fossil Dasycladaceae

The classification used in this paper is based on the system proposed by Pia (1920) and used by Rezak (1959). It is based on the following characteristics:

1. General form of thallus (cylindrical, club-shaped, etc.)
2. Type of thallus (segmented, unsegmented, branched, etc.)
3. Form of primary branches (rays) (simple, branched, cylindrical, tapering, etc.)
4. Disposition of rays (in whorls, irregular, in tufts)
5. Shape of central stem (stipe)
6. Sporangia (position and shape)
7. Dimensional data (measurements and ratios)

The determination of species requires a specialist, as the separation is often subtle and is based on a number of variable features.

The family is subdivided into tribes and genera as shown in the following table (Table XI). Only the better known genera are shown.

Several other tribes and quite a number of additional genera have been described, but are not included in this table.

TABLE XI
CLASSIFICATION OF THE DASYCLADACEAE

Tribe	Genus	Period	Characteristic structures of tribe
1—Dasyporelleae (through the Paleozoic)	*Rhabdoporella* Stolley *Vermiporella* Stolley *Anthracoporella* Pia	Silurian Ordovician to Permian Pennsylvanian-Permian	Central stem simple or bifurcating (in *Vermiporella*). Branches irregularly arranged, simple, rarely branching (*Anthracoporella*), not closely packed, barely extending beyond the calcareous sheath. Sporangia unknown, probably in central stem.
2—Cyclocrineae (exclusively Paleozoic)	*Coelosphaeridium* Roemer *Cyclocrinus* Eichwald *Mastopora* Eichwald *Apidium* Stolley *Mizzia* Schubert *Epimastopora* Pia	Ordovician Ordovician Ordovician Ordovician Permian Pennsylvanian-Permian	Spherical or oval, simple or articulated (*Mizzia*). Branches irregularly arranged. Terminal tufts of secondary branches. Calcified crust forms a shell or crust. Sporangia unknown.
3—Primicoral-lineae (Lower Paleozoic)	*Primicorallina* White *Callithamniopsis* White	Ordovician Ordovician	Long, relatively slender, central stem, branches few and irregularly arranged, ramified into secondary and tertiary branches, articulated. Sporangia unknown.
4—Verticillo-poreae (Silurian)	*Verticillipora* Rezak *Phragmoporella* Rezak	Silurian Silurian	Cylindrical or beaded unbranched thalli. Central stem cylindrical. Branches in regular whorls. No secondary branches. Sporangia unknown, probably in central stem.
5—Diploporeae (Triassic to Eocene)	*Macroporella* Pia *Stichoporella* Pia *Thaumatoporella* Pia *Gyroporella* Gümbel *Oligoporella* Pia *Actinoporella* Alth	Permian-Triassic Triassic Cretaceous Permian-Triassic Triassic Jurassic	Stem and branches not ramified. Branches usually in clusters of 3 or 6 and often, but not always, in regular whorls (verticille). Generally the thallus is cylindrical or club-shaped. Sporangia in the central stem or in the branches. Sometimes the branches are differentiated into sterile and fertile.

TABLE XI (Continued)

	Physoporella Steinmann	Triassic	
	Clavaphysoporella Endo	Permian	
	Diplopora Schafhäutl	Permian? Triassic	
	Uragiella Pia	Jurassic	
	Muniera Deecke	Cretaceous	
	Salpingoporella Pia	Cretaceous	
	Clypeina Michelin	Eocene	
6—Teutlo-porelleae (Triassic)	*Teutloporella* Pia	Triassic	Thallus consists of a long central stem with very numerous thin branches which often are arranged in regular whorls (verticille). Sporangia unknown, probably within the central stem.
7—Linoporelleae (Jurassic)	*Linoporella* Steinmann	Jurassic	Slender ramified branches regularly arranged in whorls. Sporangia not in branches, probably in the central stem.
	Myrmekioporella Pia	Jurassic	
8—Triploporel-leae (Triassic to Eocene)	*Triploporella* Steinmann	Jurassic-Cretaceous	Thallus club-shaped to spherical. Branches numerous, regularly arranged in whorls. Numerous long secondary branches. Sporangia abundant, among the primary branches.
	Goniolina D'Orbigny	Jurassic	
	Sestrosphaera Pia	Triassic-Jurassic	
	Brockella Morellet	Paleocene-Eocene	
9—Uterieae (Lower Tertiary)	*Uteria* Michelin	Paleocene-Eocene	Not well known. Structure close to the Triploporelleae. According to Morellet the position of the sporangia and the definite development of sterile and fertile whorls would place this tribe between the Neomereae-Bornetelleae and the Acetabularieae.
	Brockella Morellet	Paleocene-Eocene	

TABLE XI (Continued)

Tribe	Genus	Age	Characteristics
10—Thyrso-porelleae (Cretaceous-Eocene)	*Trinocladus* Raineri	Cretaceous	Branches numerous, in regular whorls. All branches, even those of the second, third, and fourth orders thicken and probably contain sporangia.
	Thyrsoporella Gümbel	Eocene	
	Belzungia Morellet	Paleocene-Eocene	
11—Petrasculeae (Jurassic)	*Petrascula* Gümbel	Jurassic	Branches in regular whorls. Primary branches are dimorphous. Possibly a sexual differentiation.
12—Coniporeae (Jurassic)	*Palaeodasycladus* Pia	Jurassic	Branches in regular whorls with secondary and tertiary branches. Extremities of the tertiary branches form a dense brush whose calcified envelope forms a crust or carapace as in *Bornetella.* Sporangia known but slightly.
	Palaeocladus Pia	Jurassic	
	Conipora d'Archiac	Jurassic	
13—Dactylo-poreae (Early Tertiary)	*Dactylopora* Lamouroux	Eocene	Simple branches in whorls. Sporangia individualized. Close to the modern Dasycladaceae such as *Neomeris.*
	Digitella Morellet	Eocene	
	Zittelina Munier-Chalmas	Eocene	
	Montiella Morellet	Paleocene	
14—Dasycladeae (Recent)	*Dasycladus* Ag.	Recent	Branches in whorls developing into secondary and tertiary branches. Individualized sporangia sub-terminal inserted at the ends of the primary or secondary branches.
	Batophora Ag.	Recent	
	Chlorocladus Sonder	Recent	
15—Neomereae (Cretaceous to Recent)	*Meminella* Morellet	Eocene	Branches in whorls. Only primary branches develop. Sporangia terminal on primary branches. (The Morellets place the genera *Parkerella* and *Jodetella* in a separate tribe, the Parkerellideae, which they consider to be intermediate between the modern Dasycladaceae and the Neomereae.)
	Lemoinella Morellet	Eocene	
	Neomeris D'Archiac	Cretaceous-Present	
	Cymopolia Lamouroux	Cretaceous-Present	
	Larvaria Defrance	Cretaceous-Oligocene	
	Karreria Munier-Chalmas	Cretaceous-Miocene	
	Jodetella Morellet	Paleocene	
	Parkerella Morellet	Paleocene	
	Bornetella Munier-Chalmas	Recent	

TABLE XI (Continued)

	Halicoryne Harvey	Recent	
16—Acetabularieae (Paleocene to Recent)	*Chalmasia* Solms-Laubach	Recent	Thallus with structure reduced to a central stem bearing only one whorl of branches (*Acetabularia*) or to relatively few whorls. Sporangia in the branches.
	Acetabularia Lamouroux	Oligocene-Present	
	Acicularia d'Archiac	Cretaceous-Eocene	
	Orioporella Munier-Chalmas	Paleocene	

Descriptions of Genera

Only some of the more important and better known genera are discussed here. The genera are arranged according to tribes as shown in Table XI.

For descriptions of genera not included and of species the reader is referred to the list of references given at the end of the section on the Dasycladaceae.

Tribe 1—Dasyporelleae
Genus *Rhabdoporella* Stolley, 1893
Plate 41, figures 4-5.

Description—

Thallus small in size, calcareous, cylindrical, straight or slightly curved. Thallus consists of a thin wall (cortex) and a large central stalk. Branches (only primary ones exist) are fine, simple, unbifurcated. They arise perpendicularly and radially from the central stalk, extend through the cortex, and are spaced more or less evenly.

Remarks—

Morphologically, this genus represents one of the most primitive Dasycladaceae. Locally, as for instance in Sweden (Hadding, 1933), the genus occurs so abundantly as to be a limestone builder.

Generic range—

Basal Ordovician (Ellenberger) (Cloud and Barnes, 1948) to Silurian.

Geographic distribution—

Llano region of Texas, the Baltic region.

Genus *Vermiporella* Stolley, 1893
Plate 41, figures 2-3.

Description—

Thallus cylindrical, dichotomously ramified, and sometimes strongly undulated, bent, or sinuous, building a complex network as much as a few centimeters in diameter. Branches (no secondary branches exist) of uniform width or slightly expanded toward the outside, arising perpendicular to the central stalk. Wall thickness varied, generally thin. The central stalk is considerably larger in diameter than the branches. Gametangia possibly inside of central stalk.

Remarks—

One of the most primitive of the dasyclads.

Generic range—

Ordovician, Silurian, Pennsylvanian, and Permian (?).

Geographic distribution—

The Baltic region, central and southern Europe, Japan, Washington, Oregon, British Columbia, Texas.

Genus *Anthracoporella* Pia, 1920

Plate 47, figures 1-3; plate 127.

Description—

Thallus branches dichotomously, not segmented. Thallus is composed of a moderately narrow, cylindrical, central stem. Primary branches numerous, closely packed, short, nearly cylindrical, thickening slightly toward the outside. Secondary branches in tufts of two (as a result of dichotomous branching of some of the primary branches). Primary branches not regularly arranged. Sporangia unknown, probably in the central stem.

Remarks—

Some members of this genus were several inches long. The fossils usually appear as cylindrical fragments, covered with numerous pores. In some cases, the fossils are covered by small, dermal plates.

Generic range—

Pennsylvanian and Permian.

Geographic distribution—

Texas, New Mexico, Japan, Italy.

Tribe 2—Cyclocrineae

Genus *Coelosphaeridium* Roemer, 1883, emend. Stolley, 1896.

Plate 42, figure 1; plate 117, figure 1.

Description—

Thallus spherical, with a spherical central stem on a long slender stem. Primary branches radiate out from the central stem and end in a wide bowl-like extension, which forms a terminal plate. These plates are calcified and form a dermal layer.

Remarks—

Coelosphaeridium closely resembles *Cyclocrinus*. It differs in having shorter and thicker primary branches with smaller, deeper bowl-shaped dermal plates.

Generic range—

Ordovician.

Geographic distribution—

Baltic region, Holland, Great Britain, India.

Genus *Cyclocrinus* Eichwald (1840), emend. Stolley, 1896
Plate 41, figure 6; plate 43, figures 1-2.

Description—

Thallus composed of a club-shaped central stem from which radiate numerous primary branches. These primary branches consist of a long slender shaft that terminates abruptly in a bowl-shaped cortical "cell" with a flattened top. These "cortical cells" represent calcified tufts of secondary branches. The "cortical cells" merge to form an outer calcareous cortical membrane which makes the outer layer of the fossils. The tops of the "cortical cells" appear as hexagonal plates, perforated by numerous regularly-spaced openings.

Remarks—

In shape and general structure *Cyclocrinus* greatly resembles *Coelosphaeridium* to which it is closely related. It differs mainly from *Coelosphaeridium* in having a longer "drumstick-shaped" central stem and longer and more slender primary branches with wider bowl-shaped tips. Normally, *Cyclocrinus* is more weakly calcified. Originally, *Cyclocrinus* was described as a cystoid because of the shape and because of its external "plates." The internal structure, however, proves its algal nature.

Generic range—

Ordovician.

Geographic distribution—

Norway, the Baltic region, northeastern North America.

Genus *Mastopora* Eichwald, 1840 (1860)
Plate 44, figures 1-5; plate 45, figures 1-3.

Description—

Thallus spheroidal to ovoid or club-shaped to cylindrical, perhaps 5 cm or more in diameter (as large as an apple). Calcification generally occurs only around the cortical cells. Primary branches apparently numerous and irregularly spaced. Cortical cells (secondary cells) prismatic or funnel-shaped, with a gently concave small circular bottom, which is the tubular end of the lateral branches. Immediately below the cortex is a zone containing a more or less continuous ring of roughly circular areas which are interpreted as gametangia.

Remarks—

"The genus *Mastopora,* originally instituted by Eichwald in 1840, though not figured until 1860, was recognized as an alga by Stolley (1896), who gave a very full and clear account of it . . . and showed that Salter's problematical *Nidulites* (1851) was identical with *Mastopora* . . . (nevertheless) the identification of *Nidulites* with *Mastopora* and its recognition as a calcareous alga have been largely overlooked by paleontologists. Nicholson, indeed (1889), with material of Eichwald's *Mastopora concava* (type species) in his hands, had realized the identity of the two genera, though he was under the impression that Salter's name was the earlier, and he had even commented on the resemblance of the specimens to the Dasycladaceae" (Currie and Edwards, 1943.) Pia (1927) placed the genus under his sub-tribe Mastoporineae of the tribe Cyclocrineae along with *Apidium* Stolley and *Epimastopora* Pia, 1922.

Generic range—

Ordovician to Silurian.

Geographic distribution—

Baltic region, Great Britain, Appalachian region, northern Mississippi Valley.

Genus *Mizzia* Schubert, 1907

Plate 46, figures 1-3; plate 48, figure 1; plate 128, figures 1-2.

Description—

Thallus segmented, consisting of several spherical or elongated members growing on a common stem, suggesting a string of beads. Central stem thickens within each segment and narrows in the constrictions between segments. Primary branches thicken gradually, then abruptly toward the exterior. Primary branches are arranged in concentric rows around the stem. Outer surface commonly covered by calcareous plates. No secondary branches. Sporangia unknown, probably produced within the central stem.

Remarks—

The fossils commonly consist of the individual segments although occasionally pieces containing several connected segments may be found.

Mizzia may be distinguished from other genera of Dasycladaceae by its beaded structure, by the spherical form of the segments (in most species), and by the simple structure of the branches.

The genus is a typical Permian genus reaching its greatest development during the late Middle and Upper Permian. It was very widespread geographically and locally occurred so abundantly as to be a limestone builder.

Generic range—

Permian, especially Middle and Upper Permian.

Geographic distribution—

Essentially world wide in warmer seas.

Genus *Epimastopora* Pia, 1922

Plate 48, figures 2-4.

Description—

Dasycladacean algae of unusually large size known only from fragments. Thallus probably cylindrical. Central stem possibly thick. Numerous, relatively long, primary branches occur in fairly regular, closely spaced whorls, with successive whorls alternating in position, so the branches appear to be arranged in diagonal spiraled rows around the primary stem. Sporangia unknown. Calcification appears to form a crust or rind around the tips and outer portions of the primary branches.

Remarks—

Pia gave a meager and inadequate description of the genus accompanied by a very hypothetical drawing reconstructing the plant. He visualized it as probably spherical. Later students (Wood, 1943; Johnson, 1946) have interpreted the plant as having a long, more or less cylindrical thallus. To date it is known only from small fragments which locally (in Kansas and Japan) may be very numerous.

Generic range—

Upper Pennsylvanian to Lower Permian.

Geographic distribution—

Kansas, Japan, southern Europe.

Tribe 3—Primicorallineae
Genus *Primicorallina* Whitfield, 1894
Plate 41, figure 1.

Description—

Thallus consists of a long slender central stem bearing widely spaced, slender primary branches. These branches bear tufts of four secondary branches which in turn bear four tertiary branches. Primary and secondary branches commonly thicken outward. The tertiary branches terminate in dull points. Calcification slight. Sporangia unknown.

Remarks—

The name suggests that Whitfield thought this form represented an articulated coralline algae. Pia considered it to be a dasyclad on the basis of Whitfield's description and drawings. The type specimen has recently been found in the U. S. National Museum. This specimen shows the primary branches are not as regularly arranged as shown in Whitfield's drawings.

Generic range—

Middle Ordovician (Trenton limestone).

Geographic distribution—

New York.

Tribe 4—Verticilloporeae
Genus *Verticillopora* Rezak, 1959
Plate 49, figures 1-8; plate 50, figures 1-5, 7.

Description—

Thallus cylindrical. Central stem cylindrical or pentagonal in cross section, segmented, with one whorl of branches per segment. Branches regularly arranged in whorls. Branches cylindrical to oval in cross section, expanding at distal ends to form a strong cortex.

Remarks—

Suggests somewhat *Rhabdoporella* and *Vermiporella,* but differs from them in having the branches arranged in whorls and in being segmented.

Generic range—

Middle Silurian.

Geographic distribution—

California, Nevada, Utah.

Genus *Phragmoporella* Rezak, 1959
Plate 50, figure 6.

Description—

Thallus beaded. Central stem cylindrical, segmented, with one whorl of branches per segment. Branches regularly arranged in whorls. Branches are cylindrical in cross section and expanded at the end to form a cortex.

Remarks—

Resembles the Permian genus *Mizzia* except for attaining a much larger size and for having the branches in definite whorls.

Generic range—

Silurian.

Geographic distribution—

Utah.

Tribe 5—Diploporeae
Genus *Diplopora* Schafäutl, 1863
Plate 51, figures 1-4; plate 52, figures 1-4.

Description—

Thallus cylindrical to club-shaped. Central stem cylindrical to club-shaped. Its relative thickness varies considerably among different species from slender (about 1/7 diameter of thallus) to thick (about 1/2 diameter of thallus). Branches in bundles, commonly of 3 or 6, rarely of 4. Branches vary considerably in shape and length in different species. Normally, they are thick at the base and thin toward the end, in some cases terminating in a long hair-like growth. Sporangia are in the central stem.

Remarks—

This is a highly variable genus containing a large number of species.

Generic range—

Permian and Triassic, with the greatest development during the Middle Triassic.

Geographic distribution—

West Texas, New Mexico, Iraq, Austria, Japan, Italy, Yugoslavia.

Genus *Macroporella* Pia, 1912
Plate 54, figures 1-2.

Description—

Thallus cylindrical with a moderately thick central stem. Strong primary branches thicken toward the outer end. These branches are arranged in irregular whorls. No secondary branches. Sporangia probably in the central stem.

Remarks—

This genus makes its greatest development during the Triassic, but it appears to have been fairly common and widespread during the Permian.

Generic range—

Pensylvanian to Jurassic.

Geographic distribution—

Texas, New Mexico, Japan, central and southern Europe.

Genus *Oligoporella* Pia, 1912
Plate 54, figures 3-5.

Description—

Thallus cylindrical and unbranched. Central stem cylindrical and moderately thick. Primary branches, in clusters of three, arranged in whorls. In general, the whorls are fairly widely spaced from one another. Primary branches long, normally thick at base or close to base, then tapering toward the exterior, in some species ending in a hair-like tip. Branches grow perpendicular (or nearly so) to the central stem. Secondary branches slender to hair-like, in tufts of three. These bear tufts of short, slender, tertiary branches. Sporangia probably in the central stem, possibly in the swollen bases of the primary branches.

Remarks—

Structurally, this genus appears to be between *Macroporella* and *Teutloporella*. Well preserved specimens can be easily distinguished, but poorly preserved fragments are hard to identify.

Generic range—

Permian to Triassic, with greatest development during the Triassic.

Geographic distribution—

Southern Europe, Near East and Japan.

Genus *Clavaphysoporella* Endo, 1958
Plate 55, figures 1-2.

Description—

Thallus cylindrical, relatively straight, consisting of fine annulations. Central stem relatively thick, straight, cylindrical. Primary branches thick, nearly lozenge-shaped, sometimes slightly thicker just above the base, with rounded ends. Branches in closely packed bundles of three. Commonly, the branches are inclined upward. Sporangia unknown, probably developed in the central stem.

Remarks—

Endo, 1958, erected this genus with *Physoporella minutula* Gümbel as the type species. He split it from the older genus *Physoporella* Steinmann, 1903, and included those forms having inclined lozenge-shaped (rounded cylindrical) primary branches, leaving those with rounded conical branches growing nearly perpendicular to the stem in the genus *Physoporella*.

It appears that *Physoporella* is essentially a Triassic genus which began in the Permian, while most of the described species of *Clavaphysoporella* are Permian, with only a few from the Triassic.

Generic range—

Permian and Triassic.

Geographical distribution—

Japan, central and southern Europe.

Genus *Actinoporella* Gümbel, 1882.
Plate 53, figures 1-7; plate 55, figures 3-5.

Description—

Thallus cylindrical with thick, slightly curved, primary branches arranged in whorls. Commonly, a prominent calcareous tube envelops each branch. No secondary branches. Sporangia in lower (fertile) primary branches.

Remarks—

The calcified fossils commonly break into discs, each containing one whorl of branches. Pia separated *Actinoporella* from *Oligoporella* because the latter does not separate into discs.

Generic range—

Jurassic and Early Cretaceous.

Geographic distribution—

Central and southern Europe, the Middle East.

Genus *Munieria* Deecke (1883)
Plate 56, figures 1-7.

Description—

Thallus cylindrical. Central stem slender, pinching and swelling slightly, thickest at whorls of primary branches, thinnest midway between the whorls. Primary branches shaped like batons, growing perpendicular to the central stem. These branches are regularly arranged in widely spaced whorls. No secondary branches. Sporangia unknown, probably in the central stem. Calcification covers the central stem and branches individually with a rather thin sheath which commonly thickens to form a crown around the outer part of the branch.

Remarks—

This genus, with its slender pinching and swelling central stem, widely spaced whorls, and characteristic calcification, is quite distinctive.

Generic range—

Upper Jurassic to Lower Cretaceous.

Geographic distribution—

Switzerland, Germany, France, the Middle East.

Genus *Clypeina* (Michelin), 1845
Plate 57, figures 1-3.

Description—

Thallus cylindrical. Central axis moderately thick, cylindrical, pinching and swelling in some species. Whorls of primary branches regularly spaced. Primary branches curved so that each whorl is shaped like a bowl or a funnel. Sporangia in the lower ends of the branches. As a result of the structure and calcification, the plants break, or at least tend to break, into bowl-shaped segments (corresponding to a whorl of branches) which with a little wear become discs with irregular edges. These are the fossils commonly seen in sections.

Generic range—

Jurassic to Eocene (possibly early Oligocene).

Geographic distribution—

France, Belgium, Great Britain, the Mediterranean region, the Near East, Florida.

Tribe 8—Triploporelleae
Genus *Triploporella* Steinmann, 1880
Plate 58, figures 1-2.

Description—

Thallus club-shaped with upper end subspherical. Primary branches numerous, thick, cylindrical with rounded ends, arranged in regular whorls. Tufts of long slender secondary branches. Sporangia in thickened, fertile, primary branches. Calcification normally extends from the central stem up to the basal portion of the secondary branches.

Remarks—

This is one of the most spectacular and structurally complex genera of the Dasycladaceae.

Generic range—

Cretaceous.

Geographic distribution—

Mexico, Guatemala, central Europe, the Mediterranean region, the Middle East.

Genus *Goniolina* d'Orbigny, 1850
Plate 59, figure 1.

Description—

Thallus spherical to ovoid. Central stem thick or club-shaped. Thick primary branches arranged in regular whorls. Short, thick secondary branches in tufts. Calcification covers the tips of the secondary branches with a spectacular cortical layer of geometrically arranged plates. Sporangia in the primary branches.

Generic range—

Upper Jurassic.

Geographic distribution—

France and Switzerland.

Genus *Cylindroporella* Johnson, 1954
Plate 60, figures 1-8.

Description—

Thallus segmented, consisting of cylindrical segments with rounded ends. Each segment contains a relatively narrow central stem from which develop numerous whorls of six primary branches, alternating with sporangia. Tufts of short secondary branches. Sporangia large, on short primary branches. In successive whorls, the sporangia and infertile branches alternate in position giving the appearance of diagonal rows of sporangia in vertical sections.

Remarks—

Species are separated on the basis of dimensions of internal features.

Generic range—

Upper Jurassic and Lower Cretaceous.

Geographic distribution—

Texas, Iraq.

Tribe 9—Uterieae
Genus *Uteria* Michelin, 1845
Plate 61, figure 3.

Description—

Thallus barrel-shaped with a rounded upper end. Central stem cylindrical, moderately thick. Primary branches arranged in regular whorls. The primary branches have a distinctive feature—a whorl of sterile branches alternating with several whorls of fertile branches. Calcification also follows an unusual pattern. The central stem, the whorls of sterile primary branches, and the outer surface of the plant are calcified by a thin coating of calcium carbonate, while the area occupied by the fertile branches is not calcified. Thus, we have no information regarding the position of the sporangia. Inasmuch as there are many more pores in the outer wall than there are primary branches, it is assumed that the fertile branches probably bore tufts of secondary branches, but the pattern of branching is unknown.

Remarks—

Because of its unique combination of structural features, the exact systematic position of the genus has not been established. Structurally, it appears to lie between the tribes of the Acetabulareae and the Bornetelleae.

Generic range—

Eocene.

Geographic distribution—

Paris Basin of France.

Tribe 10—Thyrosoporelleae
Genus *Trinocladus* Raineri, 1922
Plate 61, figures 1, 2 and 4-10.

Description—

Thallus cylindrical or slender and club-shaped. Central stem cylindrical, moderately large. Primary, secondary, and tertiary branches present. Primary branches in regular whorls. Commonly, in the lower whorls, the primary branches alone are present; a little higher, primary and secondary branches occur; while in the upper portion, primary, secondary, and tertiary branches develop. The primary branches are thick and probably contained the sporangia.

Remarks—

The genera *Trinocladus* and *Thyrsoporella* are closely related and are very similar structurally. They can be separated by the character of the branching. Both genera have primary, secondary, and tertiary branches, but *Trinocladus* has them arranged in the manner described above.

Generic range—

Upper Cretaceous and Paleocene.

Geographic distribution—

Paris Basin, the Mediterranean region, India.

Tribe 11—Petrasculeae
Genus *Petrascula* Gümbel, 1873
Plate 62, figures 1-4; plate 63, figures 1-3.

Description—

Thallus club-shaped with an enlarged club-shaped central stem. Central stem relatively slender at base becoming enlarged above. Primary branches arranged in regular whorls. The primary branches are perpendicular to the central stem near its base but become slightly inclined above. Bifurcations give rise to secondary and tertiary branches. Sporangia are in fertile primary branches in the central and upper parts of the plant.

Remarks—

Structurally, *Petrascula* is one of the most complex of the dasyclads. Pia (1920) suggested that probably there were externally different male and female gametes in separate gametangia on each plant. Apparently, *Petrascula* represented an end point on a highly specialized side line of the Dasycladaceae.

Some plants attained a length of at least six inches.

Generic range—

Upper Jurassic.

Geographic distribution—
Florida, Yugoslavia, Switzerland.

Tribe 12—Coniporeae
Genus *Palaeocladus* Pia, 1920
Plate 64; plate 65, figures 1-5.

Description—
Thallus long, more or less club-shaped. Central stem long, relatively slender, nearly cylindrical. Well developed primary, secondary, and tertiary branches. Primary branches inclined to the central stem, arranged in regular whorls. Secondary and tertiary branches in clusters (probably 4 to 6) except at base of plant. Sporangia not known with certainty.

Generic range—
Lower and Middle Jurassic.

Geographic distribution—
Italy, Yugoslavia, Greece, India.

Tribe 13—Dactyloporeae
Genus *Dactylopora* Lamarck, 1816
Plate 66, figures 1-4.

Description—
Thallus finger-shaped, non-segmented. Central stem large, cylindrical. Numerous, closely spaced whorls of primary branches. Each whorl consists of numerous, straight, relatively slender branches. Sporangia numerous, small, spherical, clustered around the sides of the primary branches, especially toward the outer end.

Remarks—
This is a distinctive genus of relatively simple structure. It is probably the most abundant form in the middle Eocene of the Paris Basin.

Generic range—
Middle Eocene.

Geographic distribution—
Paris Basin of France, England.

Tribe 15 — Neomereae
Genus *Neomeris* Lamouroux, 1816
Plate 66, figures 5-11.

Description—
Plant consists of a central stem from which arise very regular whorls of primary branches. Each primary branch ends in a tuft of secondary branches, each of which, in turn, ends in a terminal hair. Sporangia spherical, ovoid, or pyriform, growing at the ends of specially developed secondary branches.

Calcification light or absent around the central stem and most of the primary branches; strong around the sporangia. Normally, the calcification envelops the secondary branches, often extending beyond the outer ends of the primary branches.

Remarks—

Among Recent species, the terminal hairs drop off early in the life of the plant. They have not been observed in fossils.

The genus *Neomeris* was founded on Recent species, several of which are known and all of which are easily grouped generically. However, when we come to fossil forms, especially those from the Early Tertiary, we find considerable confusion and a wide difference of opinion as to what should be included in the genus. (Compare Pia, 1927, and Morellet and Morellet, 1922 and 1939).

Section 1. *Neomeris (Decaisnella)*

Description—

Calcification is such that the remains of the plant easily separate into ring-like segments, although, rarely, undivided cylindrical masses are found.

Remarks—

This section includes Recent forms and some fossil species.

Geologic range—

Cretaceous ? to Recent.

Geographic distribution—

France, Belgium, England, Florida, Middle East, North Africa, India, Caribbean and Mediterranean regions.

Section 2. *Neomeris (Vaginopora)*

Description—

Calcification stronger than in Section 1, so that the rings are fused together, and the fossils appear as hollow cylinders with numerous rings of pores.

Remarks—

All known members of this group are fossil forms.

Geologic range—

Eocene, Oligocene, Miocene (?).

Geographic distribution—

France, Belgium, England, Mediterranean region, Middle East, India.

Genus *Cymopolia* Lamouroux, 1816

Plate 67, figures 1-13; plate 68, figures 1-3.

Description—

Thallus segmented. Segments cylindrical or cushion-shaped. Each segment consists of a thick, cylindrical or nearly cylindrical, central stem, from which develop regular whorls of long, straight, slender, primary branches. Primary branches develop at regular intervals. Tufts of secondary branches grow from tips of the primary branches.

Sporangia spherical to rounded-cylindrical, on a short secondary branch, located within a cluster of secondary branches.

Remarks—

This is one of the best known and most widely distributed dasycladacean genera.

Generic range—

Cretaceous to Recent.

Geographic distribution—

Recent—most warm seas. Fossil: Paris Basin, England, Mediterranean region, Middle East, India, Saipan, Guatemala, Florida.

Genus *Larvaria* Defrance, 1822
Plate 69, figures 2-9.

Description—

Thallus forms a tall cylinder, which consists of a cylindrical central stem surrounded by closely spaced, regularly arranged, whorls of branches. Each primary branch ends in a cluster of one fertile and two sterile branches. The diameter of the central stem commonly is about one-fifth that of the thallus. The two sterile secondary branches are usually in the same plane. The fertile branch is considerably above them. Sporangia ovoid on modified secondary branches. Calcification weak around central stem, normally absent around primary branches, and strong around secondary branches and sporangia. The fossil thalli commonly break into rings, each containing a whorl of branches and their secondary appendages.

Remarks—

This genus closely resembles *Meminella,* from which it differs in having ovoid instead of spherical sporangia, and in having the calcification weak or absent around the primary branches instead of strong as in *Meminella.*

Generic range—

Late Cretaceous to early Oligocene, but greatest development during the middle Eocene.

Geographic distribution—

Paris Basin, England, Florida.

Tribe 16 — Acetabularieae
Genus *Acetabularia* Lamouroux, 1812
Plate 69, figure 1.

Description—

Thallus develops through vegetative and fertile stages. Fertile thallus consists of a slender central stem bearing at its apex one or more whorls of gametangial rays or discs. The gametangia are free or fused along their lateral margins. They contain numerous spherical cysts (sporangia) at maturity. Characteristic protuberances known as the corona superior and corona inferior often occur on the inner portions of the upper and lower surfaces close to the point of attachment to the central stem.

Remarks—

The gametangial rays represent highly modified primary branches. This umbrella-shaped plant is one of the most spectacular of the Recent Dasycladaceae. The fossils commonly represent fragments of discs, rays, or pieces of rays, although, rarely, entire discs are found.

Generic range—

Tertiary, Pleistocene, and Recent.

Geographic distribtuion—

Mediterranean and Caribbean regions, Japan, East Indies, north Australian waters.

Genus *Acicularia* d'Archiac, 1843.
Plate 70, figures 1-10.

Description—

The plant consists of a slender stem from which arise whorls of regularly arranged primary branches. These branches are of two types, fertile and sterile, and they are segregated in separate whorls. Normally, the whorls of sterile branches occur on the lower part of the central stem, and the fertile whorls are above. In other cases, the sterile branches develop first, then drop off when the fertile whorls (or whorl) appear.

Sporangia circular or elongated, always lateral, borne on specially developed primary branches.

Generally, the central stem, the bases of the sterile branches, and the walls of the sporangia are but feebly calcified. However, the fertile whorls are strongly calcified. The sporangia of a branch are surrounded by calcium carbonate forming a spicule in which the spores are firmly embedded. In numerous cases, these spicules fuse laterally to a daisy-like or umbrella-shaped disc.

Remarks—

The fossils found usually consist of the calcified sporangial spicules or segments of them.

This genus is a large and varied assemblage. It is divided into two sections.

Section 1. *Acicularia* (sensu stricto)

Description—

Spicules somewhat rounded or swollen with depressed calcified areas between.

Remarks—

Includes Recent and some fossil species.

Geologic range—

Upper Jurassic to Recent.

Geographic distribution—
Nearly worldwide, in warm marine waters.

Section 2. *Briardina*

Description—
Spicules commonly flat, or nearly so, separated by strongly calcified partitions.

Remarks—
Includes only extinct species.

Geologic range—
Eocene to Miocene.

Geographic distribtuion—
Paris Basin.

REFERENCES

Cloud, P. E., Jr., and Barnes, V. E., 1948, The Ellenberger group of central Texas: Texas Univ. Bur. Econ. Geol. Pub. 4621, 473 p., 45 pls.

Currie, E. D., and Edwards, W. N., 1943, Dasycladaceous algae from the Girvan area: Geol. Soc. London Quart. Jour., v. 98, pt. 3-4, p. 235-240, 1 fig., 1 pl.

Egerod, L. E., 1952, Analysis of the Siphonous Chlorophycophyta: Univ. California Pub. in Botany, v. 25, no. 5, p. 325-454, pls. 29-42, 23 text-figs.

Emberger, L., 1944, Les plantes fossiles dans leur rapports avec les vegetaux vivants: Paris, Masson et Cie, p. 66-78 (statement on Dasycladaceae).

Endo, Riuji, 1951, Stratigraphical and paleontological studies of the Later Paleozoic calcareous algae in Japan I: Trans. Proc. Palaeont. Soc. Japan, N.S., no. 4, p. 121-129, pls. 10-11.

————, 1952, Stratigraphical and paleontological studies of the Later Paleozoic calcareous algae in Japan II: Trans. Proc. Palaeont. Soc. Japan, N.S., no. 5, p. 139-144, pl. 12.

————, 1952, Stratigraphical and paleontological studies of the Later Paleozoic calcareous algae in Japan III: The Science Reports of the Saitama Univ., v. 1, no. 1, p. 23-28, pl. 1.

————, 1952, Stratigraphical and paleontological studies of the Later Paleozoic calcareous algae in Japan IV: Trans. Proc. Palaeont. Soc. Japan, N.S., no. 8, p. 241-248, pl. 23.

————, 1953, Stratigraphical and paleontological studies of the Later Paleozoic calcareous algae in Japan V: Japanese Jour. Geol. and Geog., v. 23, p. 117-126, pls. 11-12.

————, 1953, Stratigraphical and paleontological studies of the Later Paleozoic calcareous algae in Japan VI: The Science Reports of the Saitama Univ., ser. B., v. 1, no. 2, p. 97-104, pl. 9.

————, 1954, Stratigraphical and paleontological studies of the Later Paleozoic calcareous algae in Japan VIII: The Science Reports of the Saitama Univ., ser. B., v. 1, no. 3, p. 209-216, pl. 18.

————, 1954, Stratigraphical and paleontological studies of the Later Paleozoic calcareous algae in Japan IX: The Science Reports of the Saitama Univ., ser. B., v. 1, no. 3, p. 217-221, pl. 19.

————, 1956, Stratigraphical and paleontological studies of the Later Paleozoic calcareous algae in Japan X: The Science Reports of the Saitama Univ., ser. B., v. 2, no. 2, p. 221-248, pls. 22-31.

————, 1957, Stratigraphical and paleontological studies of the Later Paleozoic calcareous algae in Japan XI: The Science Reports of the Saitama Univ., ser. B., v. 2, no. 3, p. 279-305, pls. 37-44.

————, 1958, Stratigraphical and paleontological studies of the Later Paleozoic calcareous algae in Japan XIII: Trans. Proc. Palaeont. Soc. Japan, N.S., no. 31, p. 265-269, pl. 39.

———, 1959, Stratigraphical and paleontological studies of the Later Paleozoic calcareous algae in Japan XIV: The Science Reports of the Saitama Univ., ser. B., v. 3, no. 2, p. 177-207, pls. 30-42.

———, and Horiguchi, M., 1957, Stratigraphical and paleontological studies of the Later Paleozoic calcareous algae in Japan XII: Japanese Jour. Geol. and Geog., v. 28, no. 4, p. 169-177, pls. 13-15.

———, and Kanuma, M., 1954, Stratigraphical and paleontological studies of the Late Paleozoic calcareous algae in Japan VII: The Science Reports of the Saitama Univ., ser. B., v. 1, no. 3, p. 177-207, pls. 13-17.

Hadding, A., 1933, The pre-Quaternary sedimentary rocks of Sweden. Part V: Kyngl. Fysiografiska Sällskapets Handl. Lund N.F., v. 44, no. 4, Medd. L.G.M.E., p. 75.

Johnson, J. Harlan, 1942, Permian lime-secreting algae from the Guadalupe Mountains, New Mexico: Geol. Soc. America Bull., v. 53, no. 2, p. 195-226, 7 pls., 5 figs.

———, 1946, Lime-secreting algae from the Pennsylvanian and Permian of Kansas: Geol. Soc. America Bull., v. 57, p. 1087-1120, 10 pls., 5 figs.

———, 1951, Permian calcareous algae from the Apache Mountains, Texas: Jour. Paleont., v. 25, no. 1, p. 21-30, pls. 6-10.

———, 1960, Jurassic algae from the subsurface of the Gulf Coast: Jour. Paleont., v. 35, p. 147-151, pls. 31-32.

———, and Dorr, M. E., 1942, The Permian algal genus *Mizzia:* Jour. Paleont., v. 16, no. 1, p. 63-67, pls. 9-12.

———, and Konishi, Kenji, 1956, Studies of Mississippian algae: Colorado School of Mines Quarterly, v. 51, no. 4, 132 p., illus.

———, 1858, Studies of Devonian algae: Colorado School of Mines Quarterly, v. 53, no. 2, 114 p., illus.

———, 1959, Studies of Silurian (Gotlandian) algae: Colorado School of Mines Quarterly, v. 54, no. 1, p. 1-82, 131-162, illus.

Maslov, V. P., 1956, Fossil algae in the U.S.S.R.: Acad. Sci. U.S.S.R. Proc. Inst. Geol. Sci., no. 160, p. 54-62, 176-182 (on Dasycladaceae).

Morellet, L., and Morellet, J., 1913, Les Dasycladacees du Tertiaire Parisien: Soc. Geol. France Mem. 47, v. 21, no. 1, p. 1-43, pls. 1-3.

———, 1922, Nouvelle contribution a l'etude des Dasycladacees Tertiaires: Soc. Geol. France Mem. 58, v. 25, no. 2, p. 1-35, pls. 1-2.

———, 1939, Tertiary siphoneous algae in the W. K. Parker collection: British Mus. Nat. History, London, 55 p., 6 pls.

Munier-Chalmas, E., 1877, Observations sur les algues calcaires appartenant au groupe des siphonees verticillees (Dasycladees Harv.) et confondues avec les foraminiferes: Acad. Sci. Paris, Comptes Rendus, v. 2, no. 85, p. 814-817.

Pia, Julius, 1920, Die Siphoneae Verticillatae vom Karbon bis zur Kreide: Abhandl. Zool.-Botan. Gesellschaft in Wien, Vienna, v. 11, pt.2, p. 1-263, 8 pls., 25 figs.

———, 1926, Pflanzen als Gesteinsbildner: Berlin, Gebruder Borntraeger, 355 p., 166 figs.

———, 1927, Thallophyta *in* Hirmer, M., Handbuch der Paläobotanik: Berlin and Munich, p. 1-136.

Rezak, Richard, 1959, New Silurian Dasycladaceae from southwestern United States: Colorado School of Mines Quarterly, v. 54, no. 1, p. 115-129, 4 pls.

Stolley, E., 1896, Über gesteinsbildende algen und die Mitwirkung solcher bie der Bildung der Skandinavisch-baltischen Silurablagerungen: Naturwiss, Wochenschr., v. 11, p. 173.

———, 1896, Untersuchungen über *Coelosphaeridium, Cyclocrinus, Mastopora,* und verwandte genera des silur: Anthropologie und Geologie Schleswig-Holstein Archv., v. 1, p. 177.

Wood, Alan, 1943, The algal nature of the genus *Koninckopora* Lee, its occurrence in Canada and western Europe: Geol. Soc. London Quart. Jour., v. 98, pt. 3-4, no. 391-392, p. 205-221, 3 pls. 3 figs.

Yabe, H., and Toyama, S., 1949, New Dasycladaceae from the Jurassic Torinosu limestone of the Sakawa Basin II: Proc. Japan Acad., v. 25, no. 7, p. 40-44, 2 text-figs.

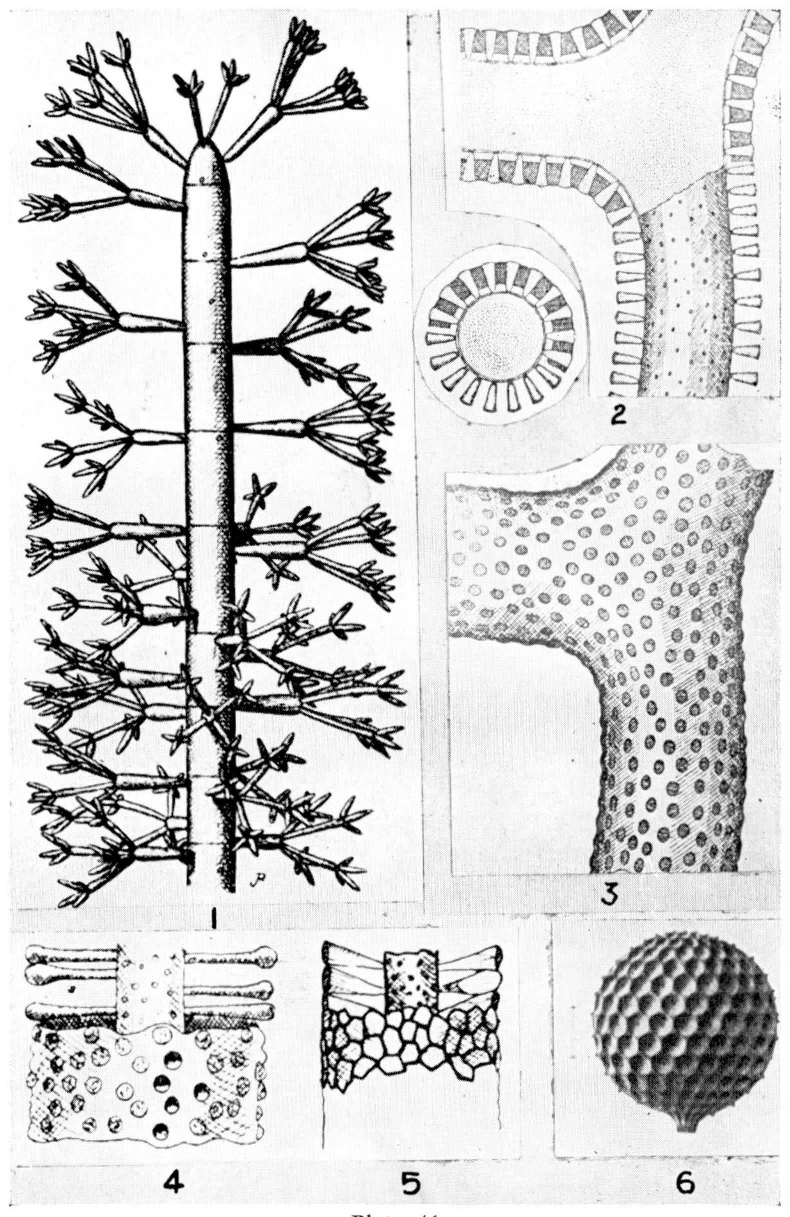

Plate 41
Dasycladaceae
Genera *Primicorallina, Vermiporella, Rhabdoporella,* and *Cyclocrinus*

Figure 1. *Primicorallina trentonensis* Whitfield. Diagrammatic sketch showing structure (x10). After Ruedemann.

Figures 2-3. *Vermiporella.* Diagrammatic sketches showing structure. After Pia, 1920.

Figures 4-5. *Rhabdoporella pachyderma* Rothpletz (x180). After Pia.

Figure 6. *Cyclocrinus pyriformis* Stolley (x1). After Stolley, 1896. External view.

Plate 42
Genus *Coelosphaeridium*
Coelosphaeridium cyclocrinophilum Roemer (x1). Middle Ordovician (4-b). *Coelosphaeridium* beds, Fangberget, Veldre, Mjosa District, Norway. Specimen in Paleontologisk Museum, Oslo, Norway.

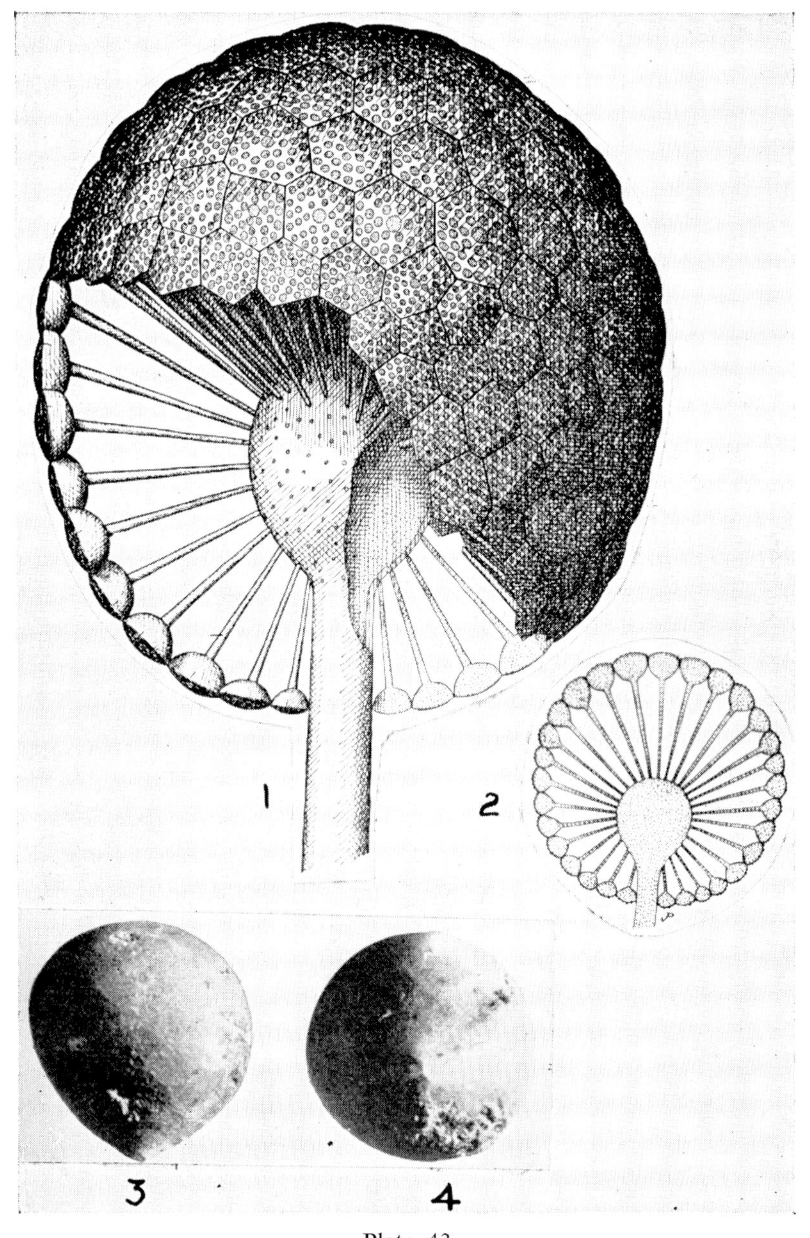

Plate 43

Genera *Cyclocrinus* and *Apidium*

Figure 1. *Cyclocrinus porosus* Stolley (x14). Diagram showing structure and outer appearance. After Stolley.

Figure 2. *Cyclocrinus porosus* Stolley. Structural diagram. After Pia, 1932.

Figures 3-4. *Apidium rotundum* Høeg (x5). Ordovician, Trondheim district, Norway. After Høeg, 1932.

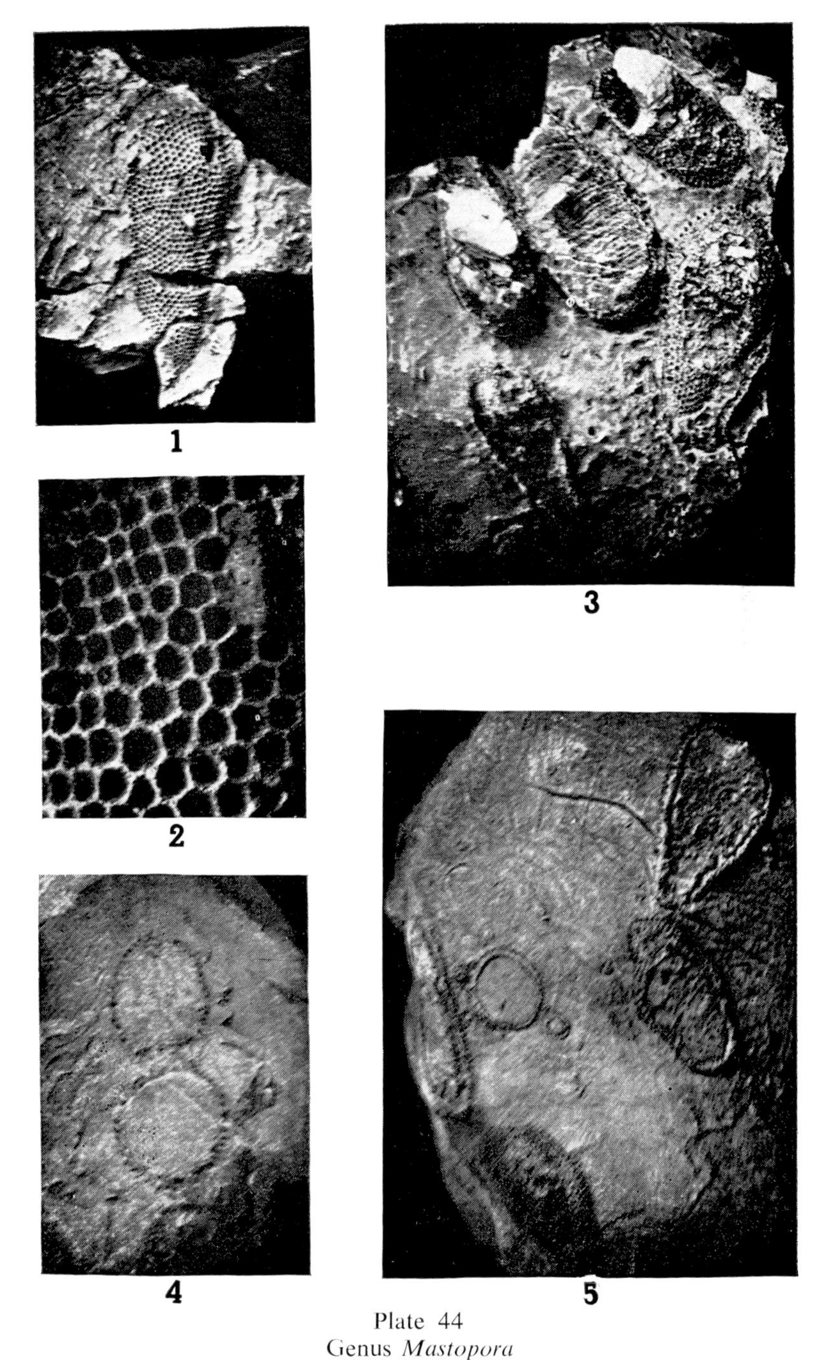

Plate 44

Genus *Mastopora*

Figures 1-4. *Mastopora pyriformis* Bassler (after Bassler). 1. A nearly complete example partially embedded in limestone. The pyriform shape of the organism is shown. 2. Surface of same showing the usual aspect of the cups (x4). 3. A group of examples on a limestone slab with their interior filled with crystalline calcite. 4. Fractured piece of limestone with cross sections of this organism.

Figure 5. Weathered rock surface with *Nidulites* in various positions, Ordovician (Mohawkian). *Nidulites* bed of Chambersburg limestone. Strasburg, Virginia.

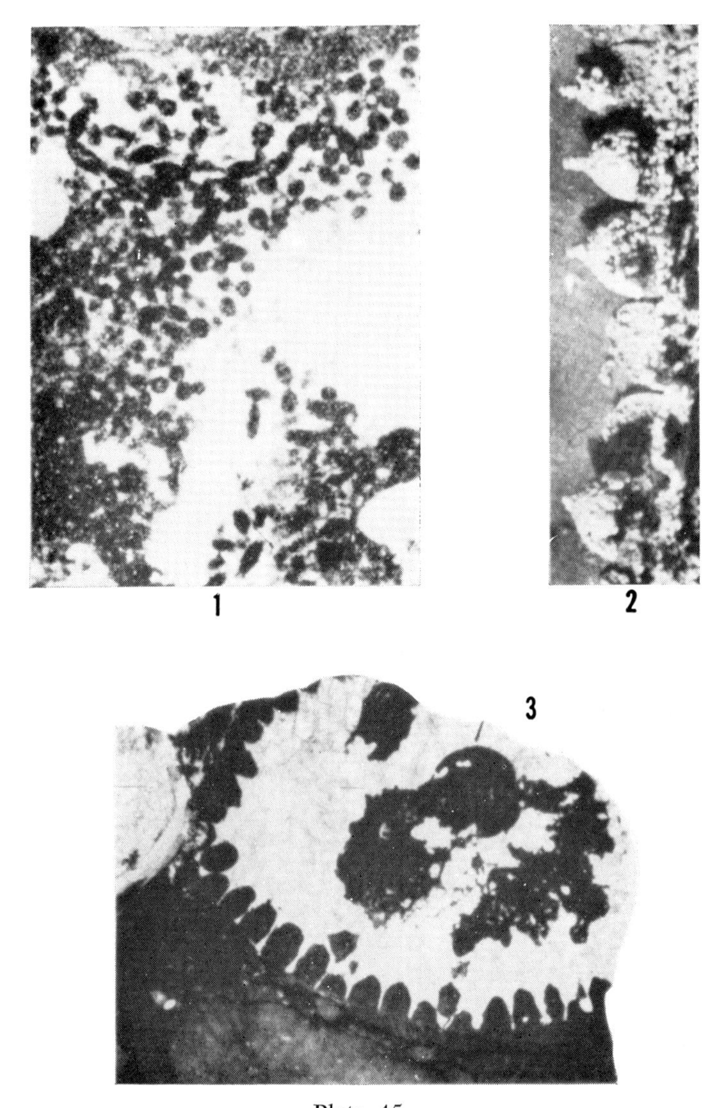

Plate 45
Genus *Mastopora*

Figures 1-3. *Mastopora pyriformis* (Bassler). Ordovician, Chambersburg limestone, Virginia. 1. Section (x5) showing probable sporangia (?). External molds of cortical cups (x11). 3. Section (x5) showing cortical cups. From Osgood and Fischer, 1960.

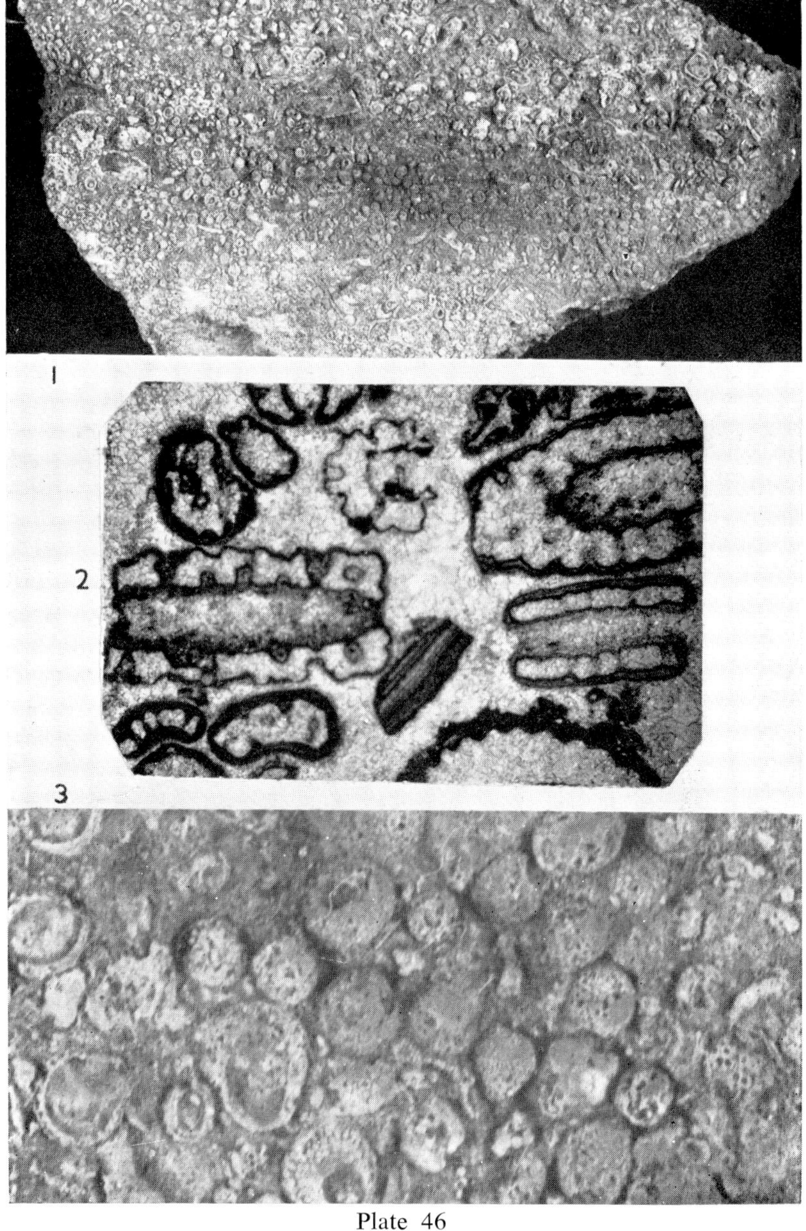

Plate 46

Genus *Mizzia*

Figure 1. *Mizzia* limestone (x1), Permian, Apache Mountains, Texas.

Figure 2. Section *Mizzia* limestone (x35). Permian, Guadalupe Mountains, New Mexico.

Figure 3. *Mizzia* limestone. Detail of weathered surface (x11). Permian, Apache Mountains, Texas.

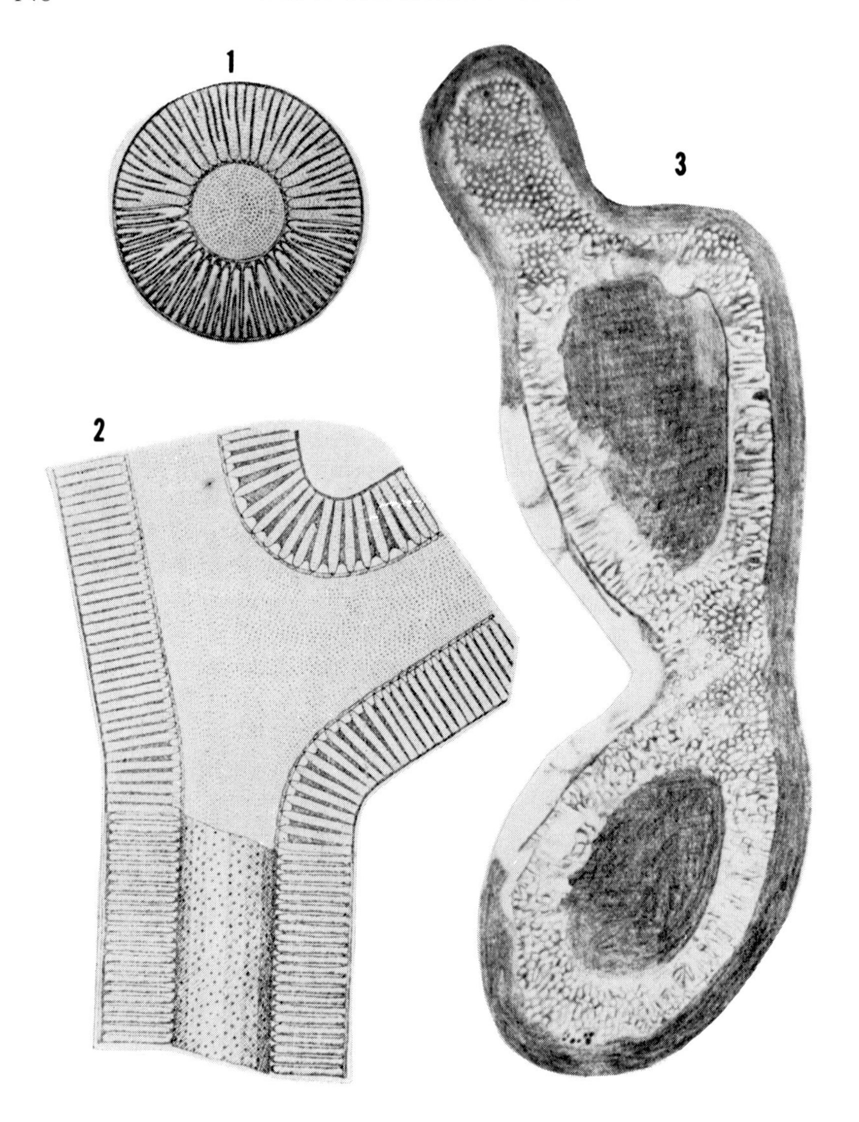

Plate 47
Genus *Anthracoporella*
Figures 1-2. Reconstructions of *Anthracoporella spectabilis* Pia (from Pia, 1920).
Figure 3. Longitudinal section of a slightly curved piece of *Anthracoporella spectabilis* Pia (x13) (from Pia, 1920).

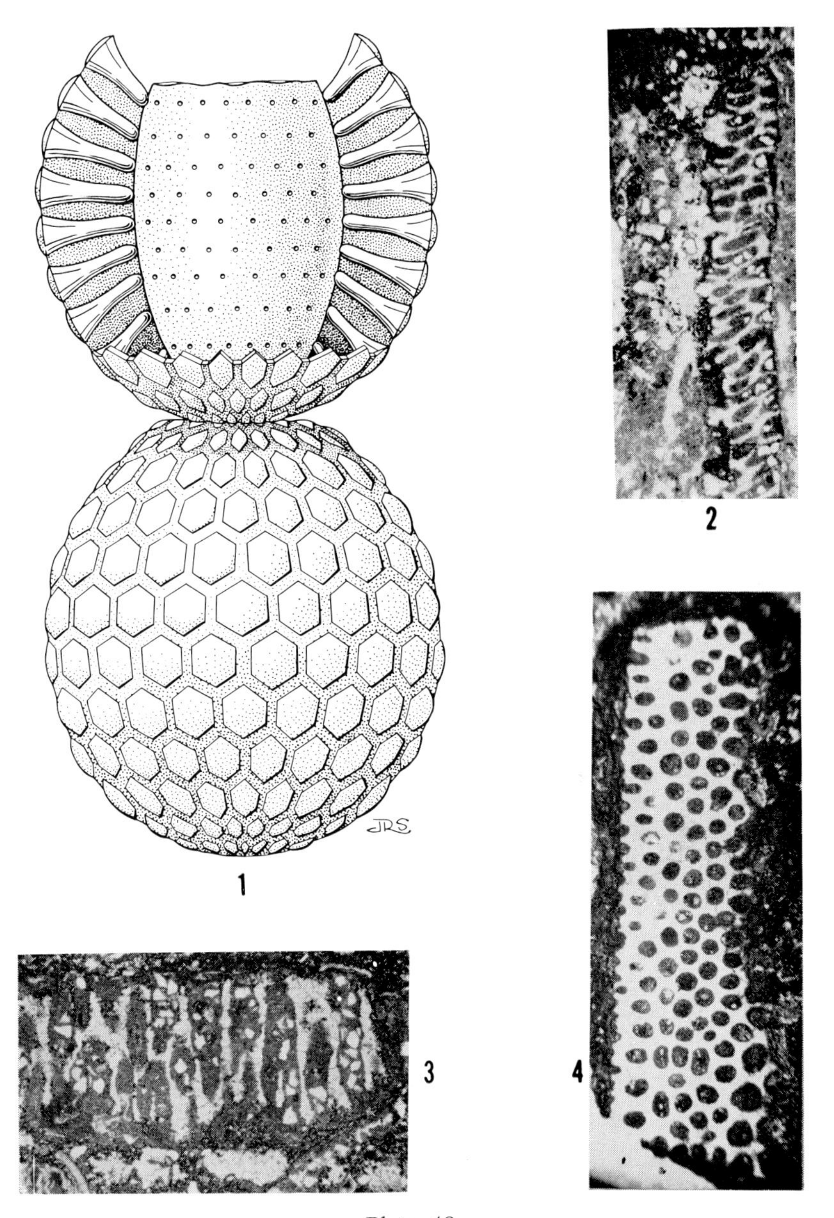

Plate 48
Genera *Mizzia* and *Epimastopora*

Figure 1. Reconstruction of *Mizzia velebitina* by Rezak, 1958.
Figures 2-3. *Epimastopora kansasensis* Johnson (x30). Longitudinal sections, Pennsylvanian Lecompton limestone, Elk County, Kansas.
Figure 4. *Epimastopora regularis* Johnson (x25). Tangential section, Lower Permian, Riley County, Kansas.

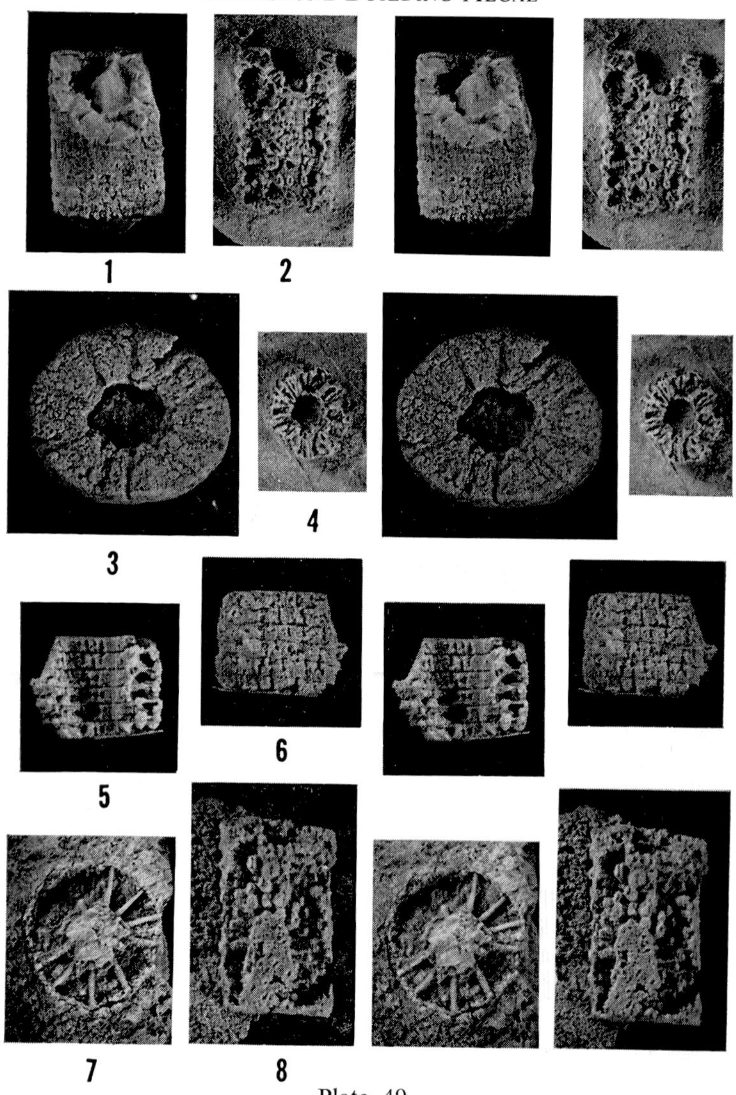

Plate 49
Genus *Verticillopora*
Verticillopora annulata n. sp. (x1)

Figures 1-8. *Verticillopora annulata* Rezak (x1). 1. U.S.G.S. Algae No. a660, lateral view of rather poorly preserved specimen showing rays and undulant surface of thallus. 2. U.S.G.S. Algae No. a648, lateral view of a deformed specimen showing bases of rays on segmented stipe. 3. U.S.G.S. Algae No. a657, surface normal to axis of thallus. Note pentagonal cross section of stipe. 4. U.S.G.S. Algae No. a649, surface similar to 3 above. 5 and 6. U.S.G.S. Algae No. a664, inner and outer views of a thallus fragment. Note oval shaped cross section of rays. 7. U.S.G.S. Algae No. a666, well preserved rays and segment walls. 8. U.S.G.S. Algae No. a667, thallus fragment with part of cortex removed to show numerous rays.

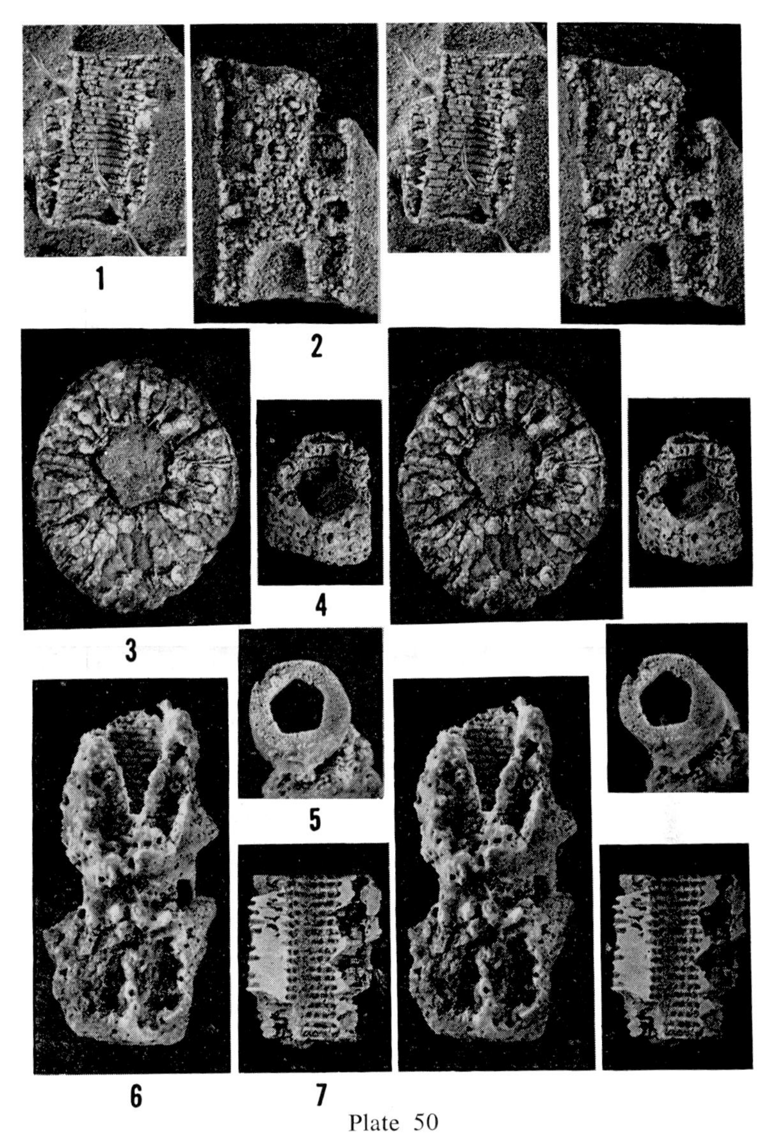

Plate 50

Genera *Verticillopora* and *Phragmoporella*

Figures 1-5, 7. *Verticillopora annulata* Rezak.

Figure 6. *Phragmoporella monilis* Rezak.

1. U.S.G.S. Algae No. a658, internal view of a rather thoroughly recrystallized and poorly replaced specimen. 2. U.S.G.S. Algae No. a659, internal view showing ray bases. Poor arrangement is due to recrystallization and incomplete replacement by silica. 3. U.S.G.S. Algae No. a652, surface almost normal to axis of thallus. Note pentagonal cross section of stipe and enlargement of some rays through secondary deposits of lime.

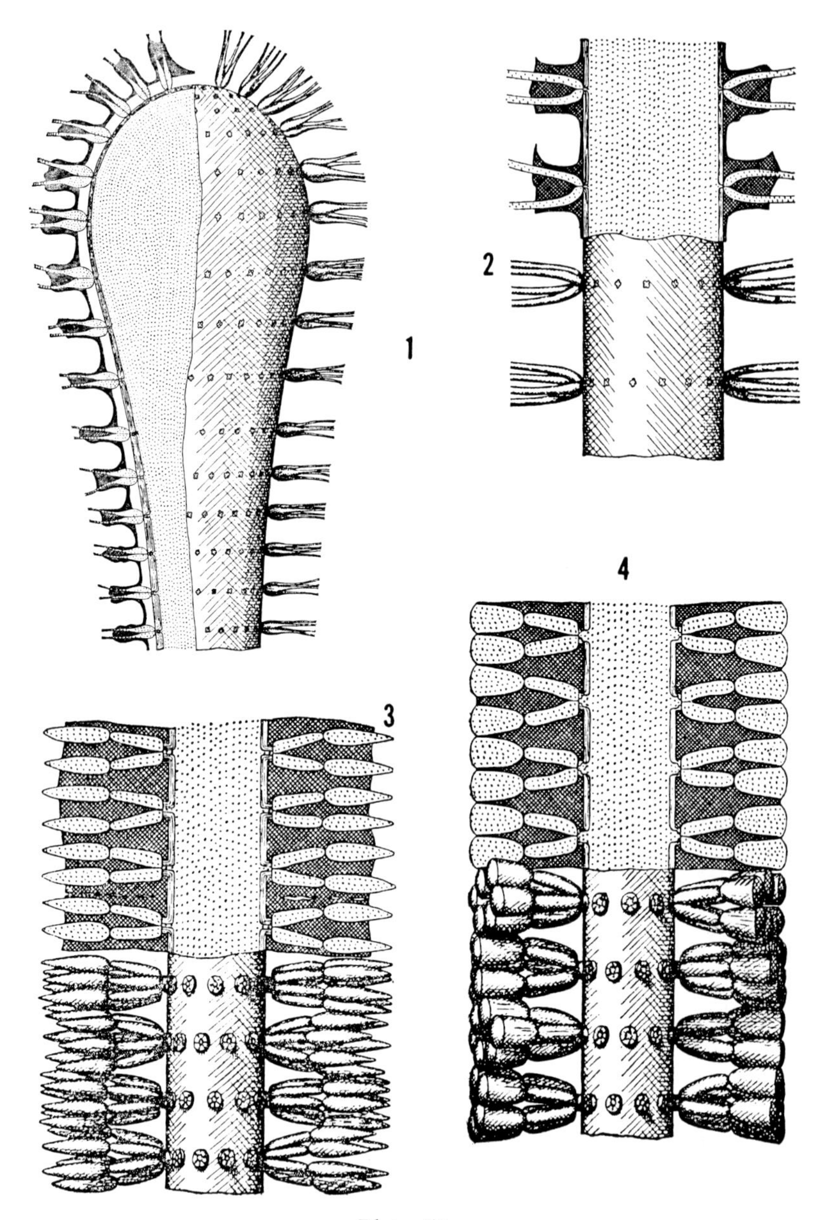

Plate 51

Genus *Diplopora*

Figure 1. Reconstruction of *Diplopora clavaeformis* Pia
Figure 2. Reconstruction of *Diplopora uniserialis* Pia
Figure 3. Reconstruction of *Diplopora hexaster* Pia
Figure 4. Reconstruction of *Diplopora helvetica* Pia
All of Triassic age, from Pia, 1920

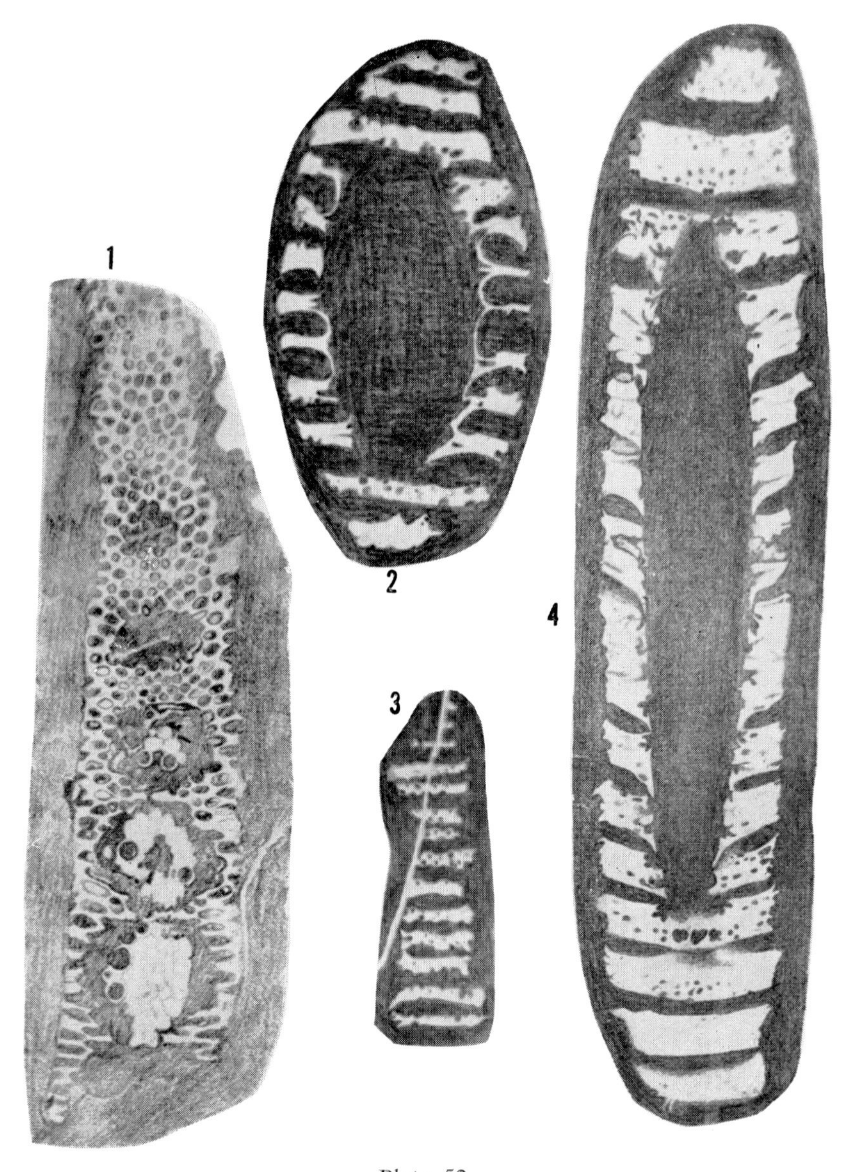

Plate 52
Genus *Diplopora*
Figure 1. *Diplopora phaneraspora* Pia. 1. Tangential section (x10).
Figures 2, 3. *Diplopora annulatissima* Pia. Tangential sections (x12).
Figure 4. *Diplopora annulata* Schfhäutl (x8).
(All Triassic, from Pia, 1920.)

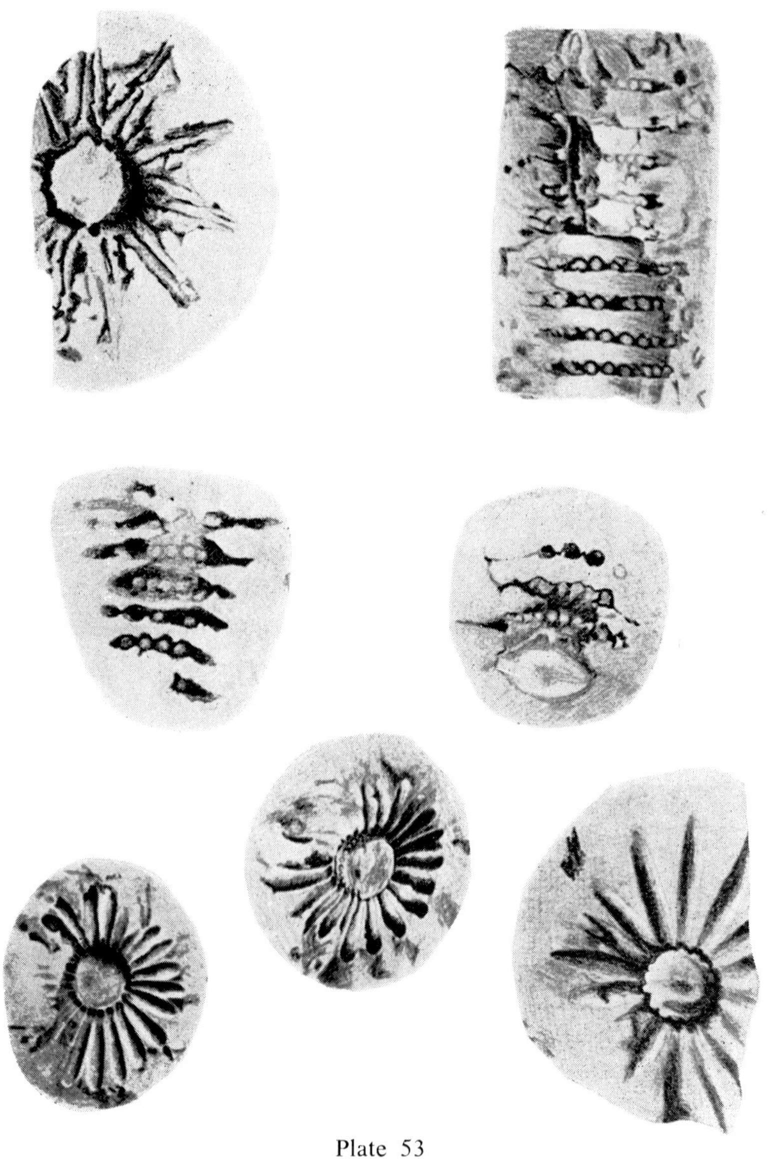

Plate 53
Genus *Actinoporella*
Figures 1-6. *Actinoporella podolica* Alth. Polished sections at various angles (about x12). Jurassic, Austria.
Figure 7. *Actinoporella sulcata* Alth. Cross section (x14). Jurassic, Austria. (From Pia, 1920.)

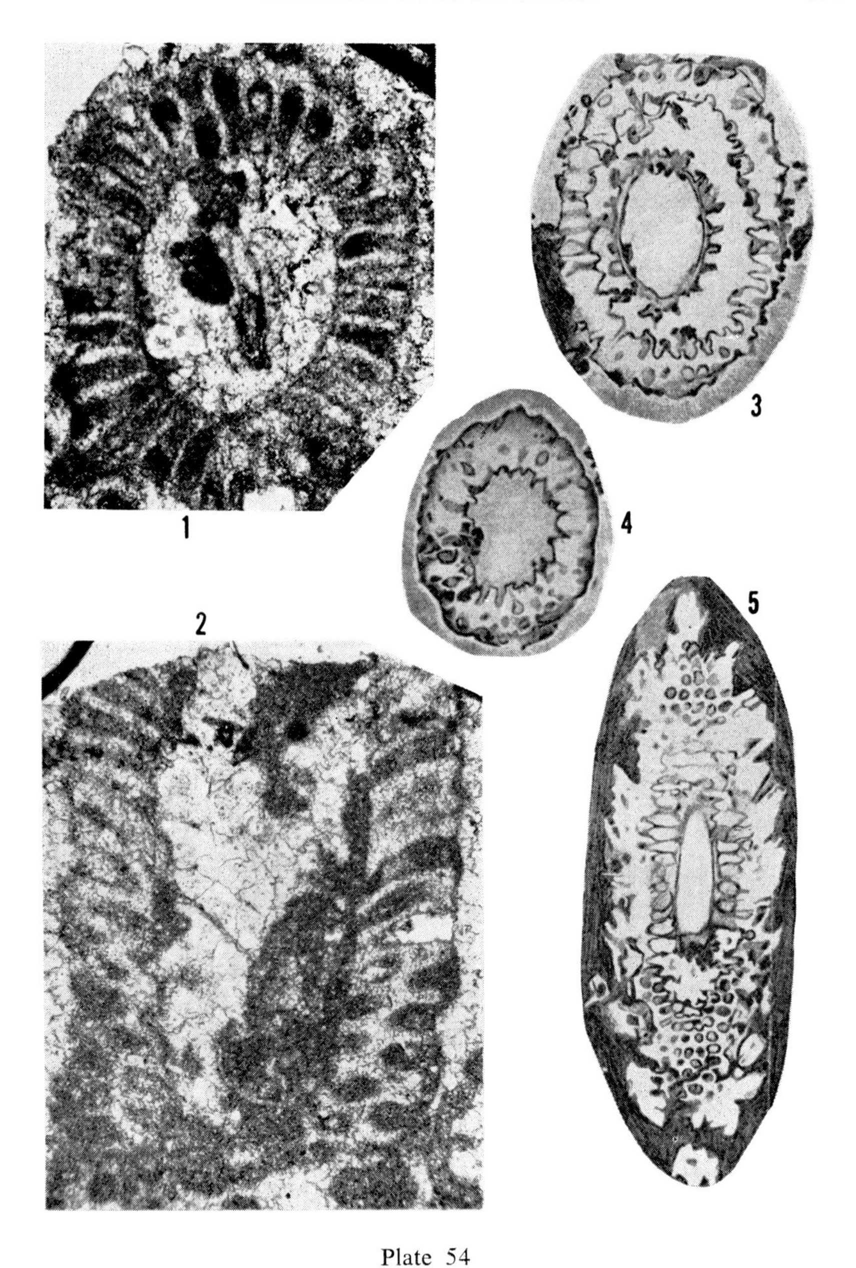

Plate 54
Genera *Macroporella* and *Oligoporella*
Figures 1, 2. *Macroporella* Jurassic, Switzerland (from Carozzi).
Figures 3, 4, 5. *Oligoporella duplicata* Pia (x13) Triassic (from Pia, 1920).

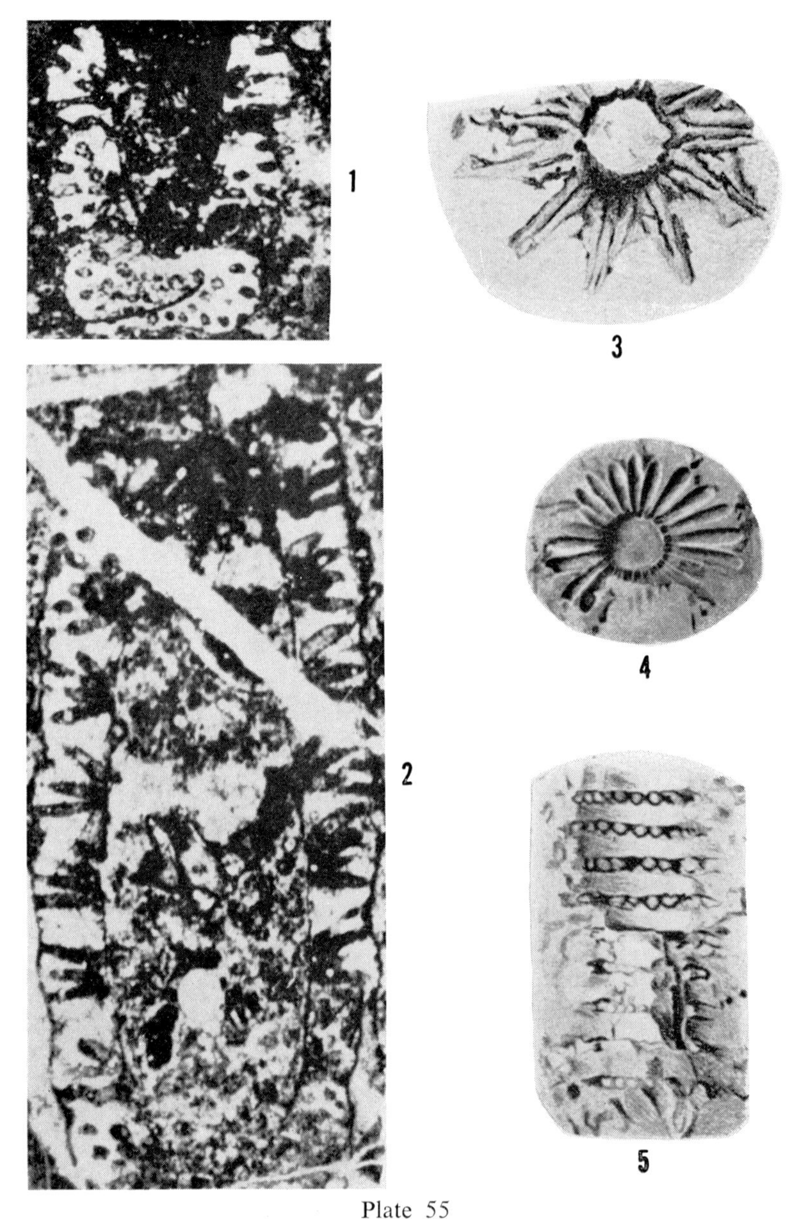

Plate 55

Genera *Clavaphysoporella* and *Actinoporella*

Figures 1, 2. *Clavaphysoporella faceta* Endo. (x25) Permian, Figu district, Honshu, Japan. 1. A longitudinal section showing the clusters of primary branches. 2. An oblique longitudinal section showing the base.
Figure 3. *Actinoporella sulcata* Alth. (x14) cross section Jurassic.
Figures 4, 5. *Actinoporella podica* Alth. Jurassic. 4. A cross section (x10), 5. a longitudinal section (x10).

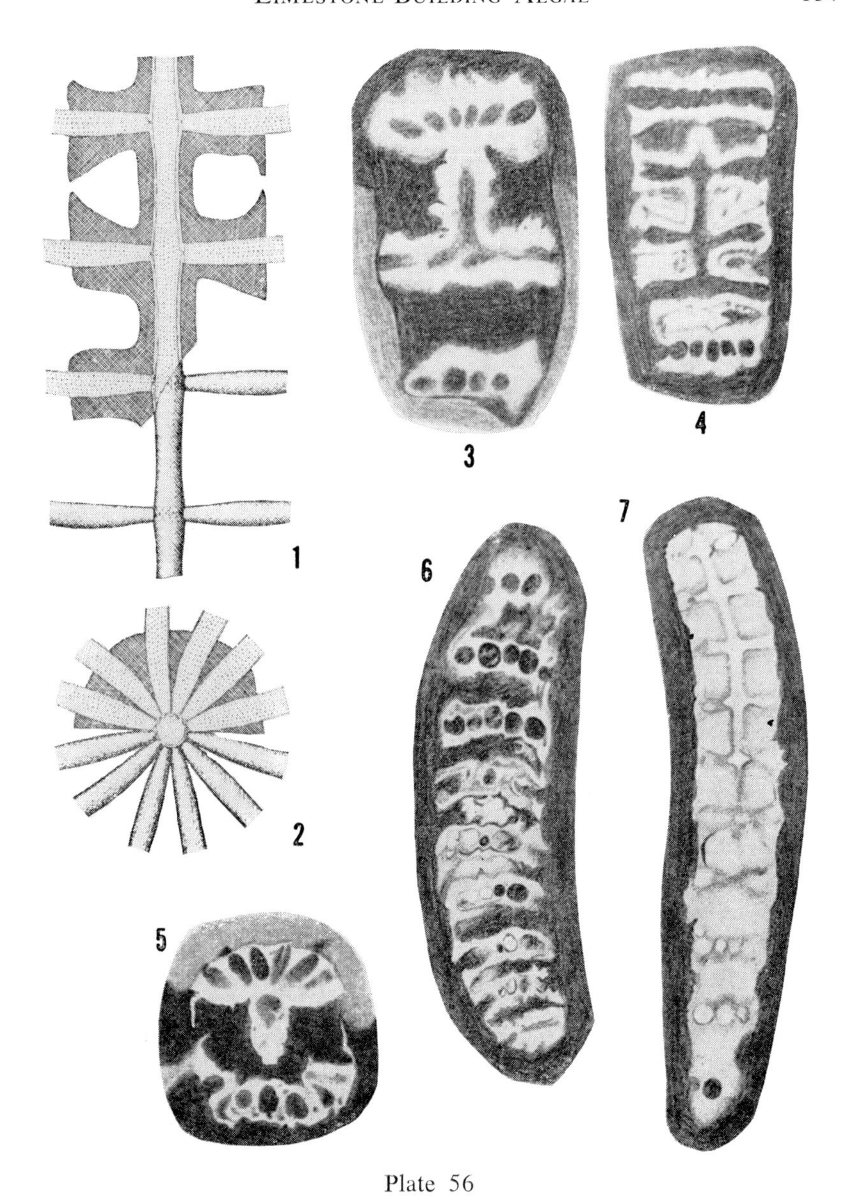

Plate 56
Genus *Munieria*
Figures 1-7. *Munieria baconica* Deecke. 1, 2, Diagrammatic reconstructions. 3-7, Drawings of polished sections (x23) Upper Jurassic. (From Pia, 1920.)

Plate 57
Genus *Clypeina*

Figures 1-2. *Clypeina lucasi* Emberger (washed specimens x15). Jurassic of Algeria (from Emberger, 1955).
Figure 3. *Clypeina jurassica* Farre. (x20) Jurassic of Algeria (from J. Morellet, 1950).

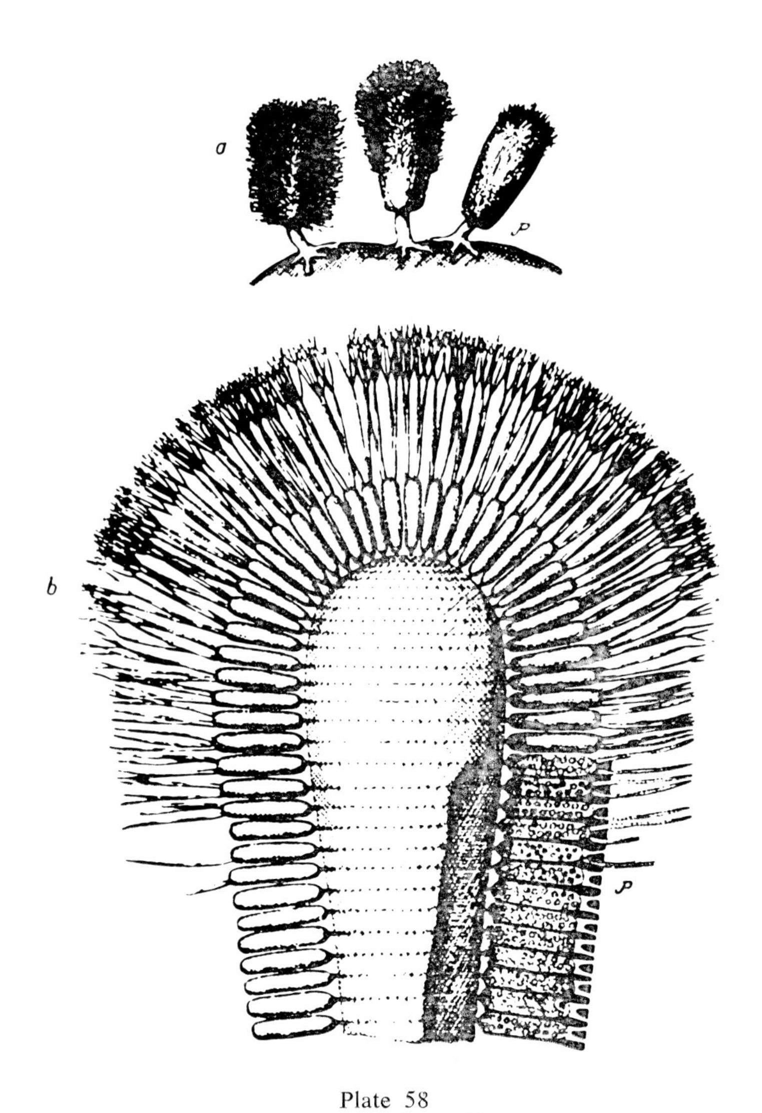

Plate 58
Genus *Triploporella*
Figure 1. Reconstruction of *Triploporella* (x8) after Steinmann and Pia.

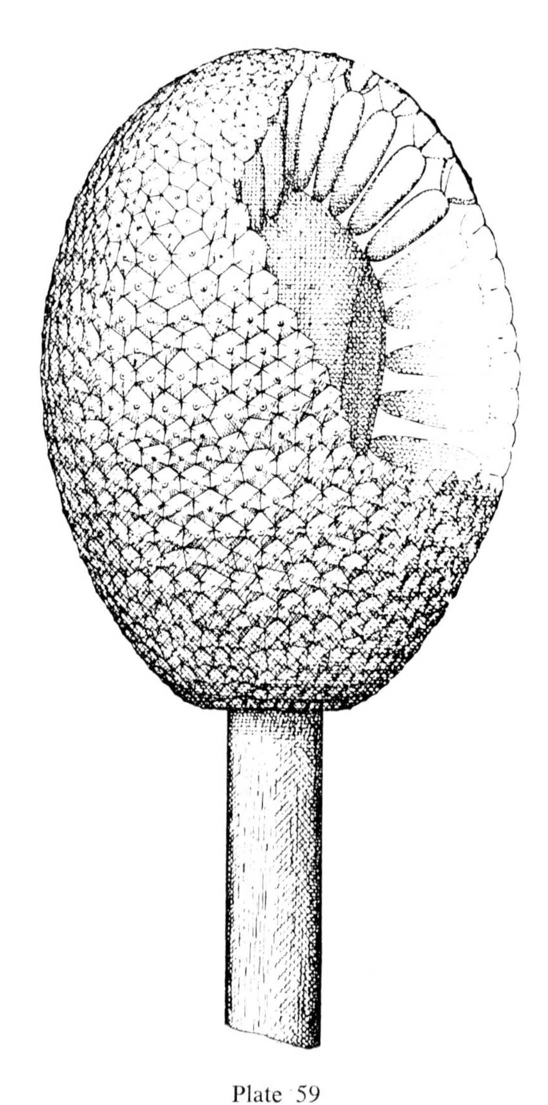

Plate 59
Genus *Goniolina*
Figure 1. Reconstruction of *Goniolina geometrica* Roemer. Jurassic. (From Pia, 1920).

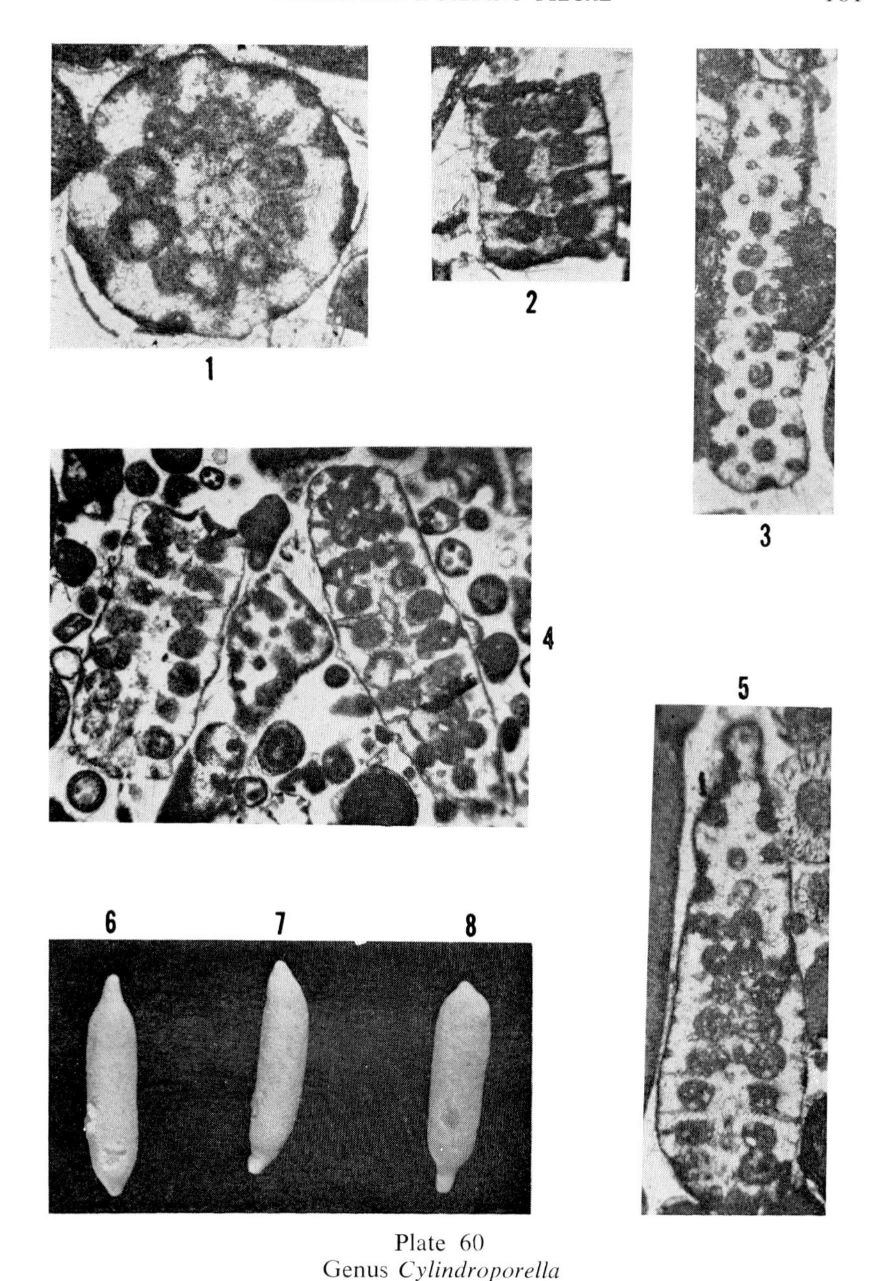

Plate 60
Genus *Cylindroporella*
Figures 1-5. *Cylindroporella texana* Johnson. Jurassic, Texas. 1. Cross section (x50). Nearly vertical sections (x25).
Figures 6-8. *Cylindroporella barnesii* Johnson, Lower Cretaceous, Texas. External views of segments (x10).

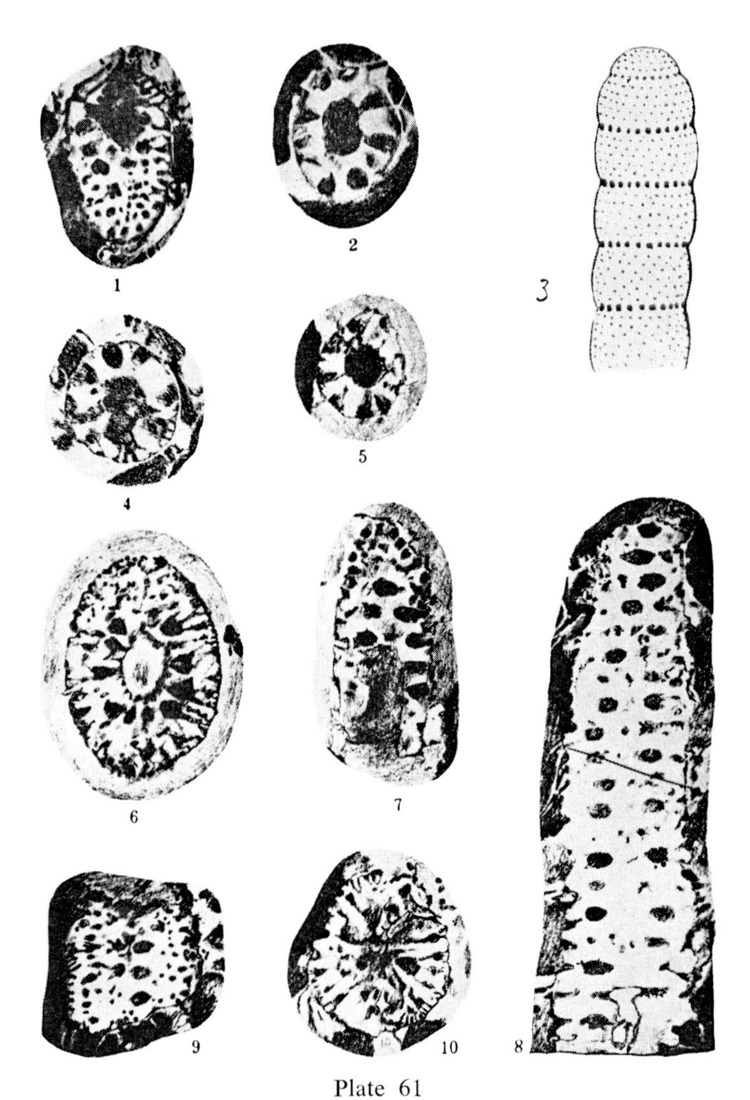

Plate 61
Genera *Trinocladus* and *Uteria*
Figures 1, 2, 4-10. *Trinocladus tripolitanus* Raineri (x30). Cretaceous, Tripoli. (From Pia, 1936.)
Figure 3. *Uteria encrinella* Michelin. Sketch of exterior (from Morellet and Morellet).

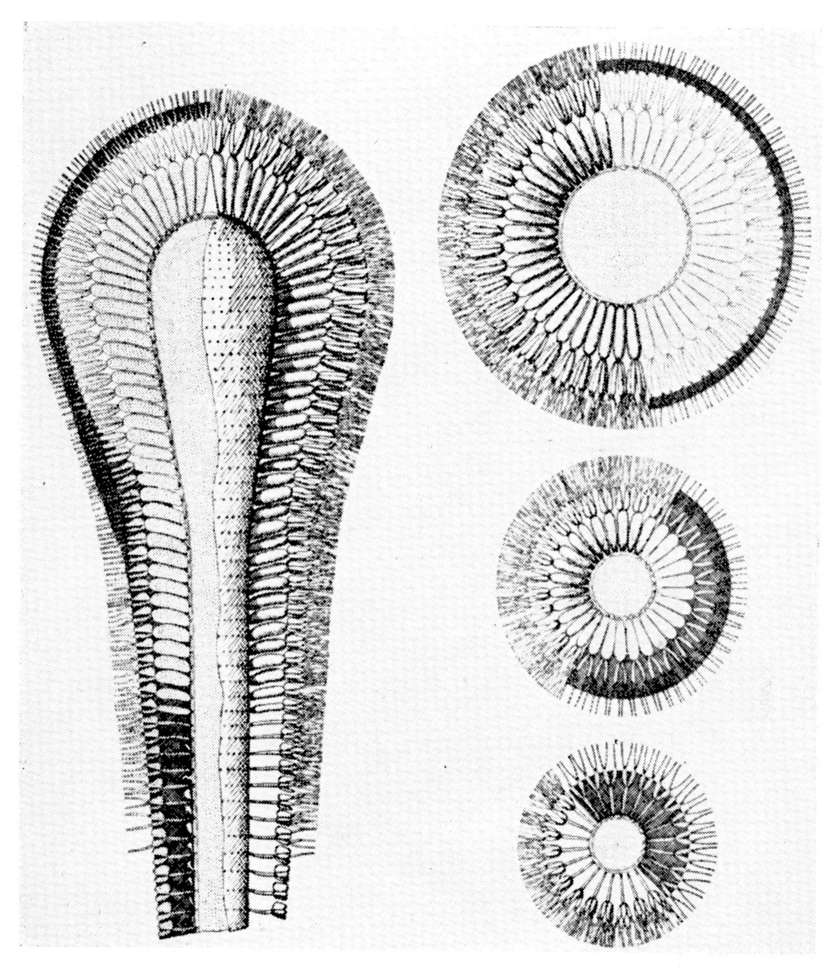

Plate 62
Genus *Petrascula*
Pia's reconstruction of *Petrascula bursiformis* from the Jurassic of southern Europe. (From Pia, 1920.)

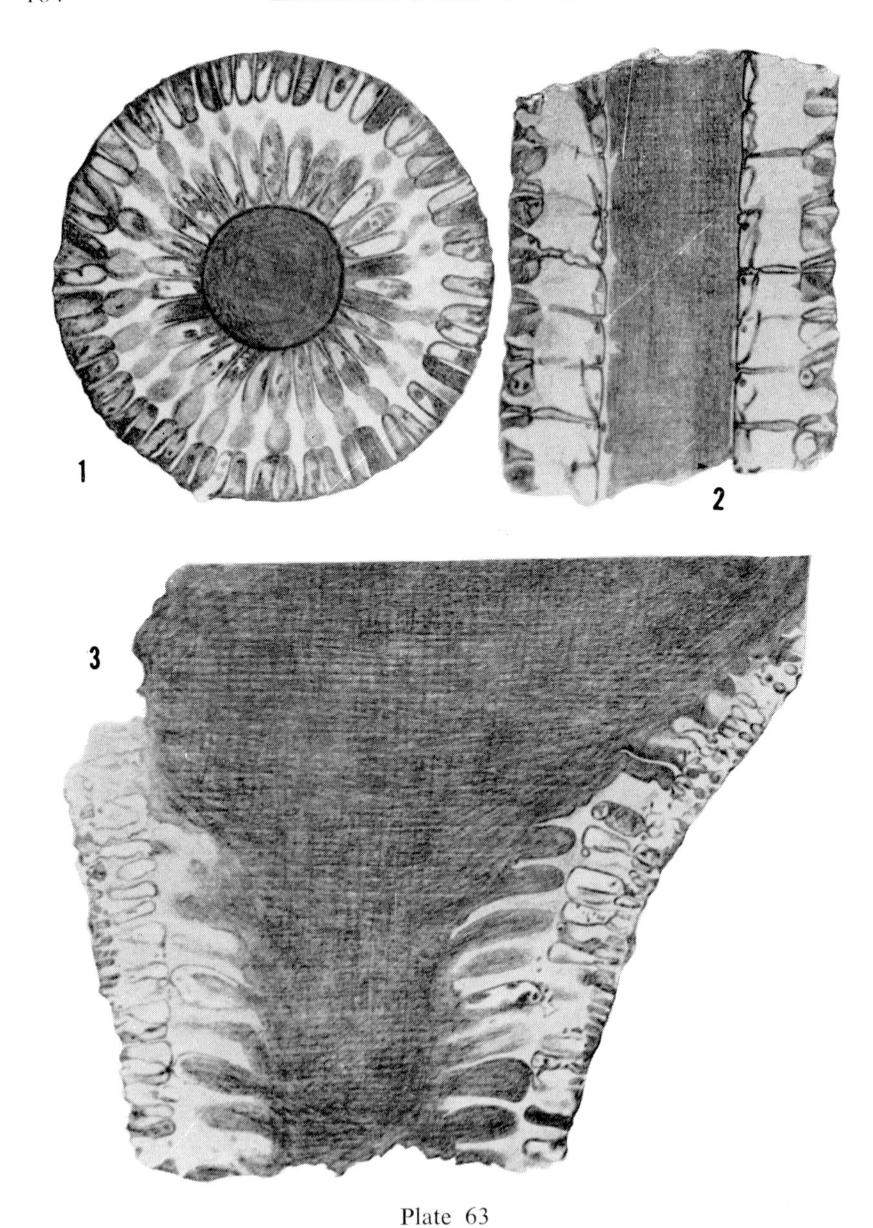

Plate 63
Genus *Petrascula*
Figures 1-3. *Petrascula bursiformis* (Etallon), (1-x15, 2-x17, 3-x12). Upper Jurassic of Austria. (From Pia, 1920.)

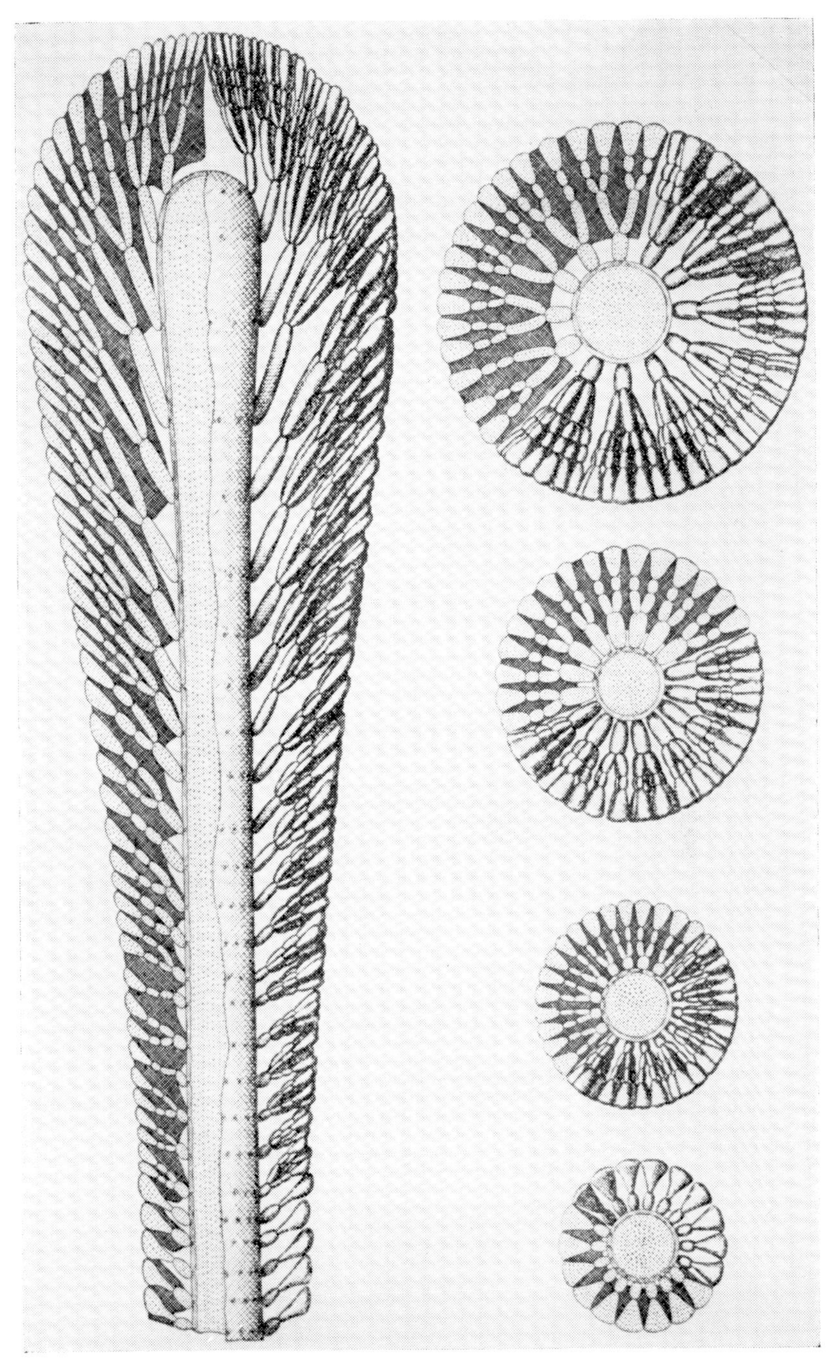

Plate 64
Genus *Palaeocladus*
Reconstruction of *Palaeocladus mediterraneous* Pia. Jurassic, southern Europe (after Pia, 1920).

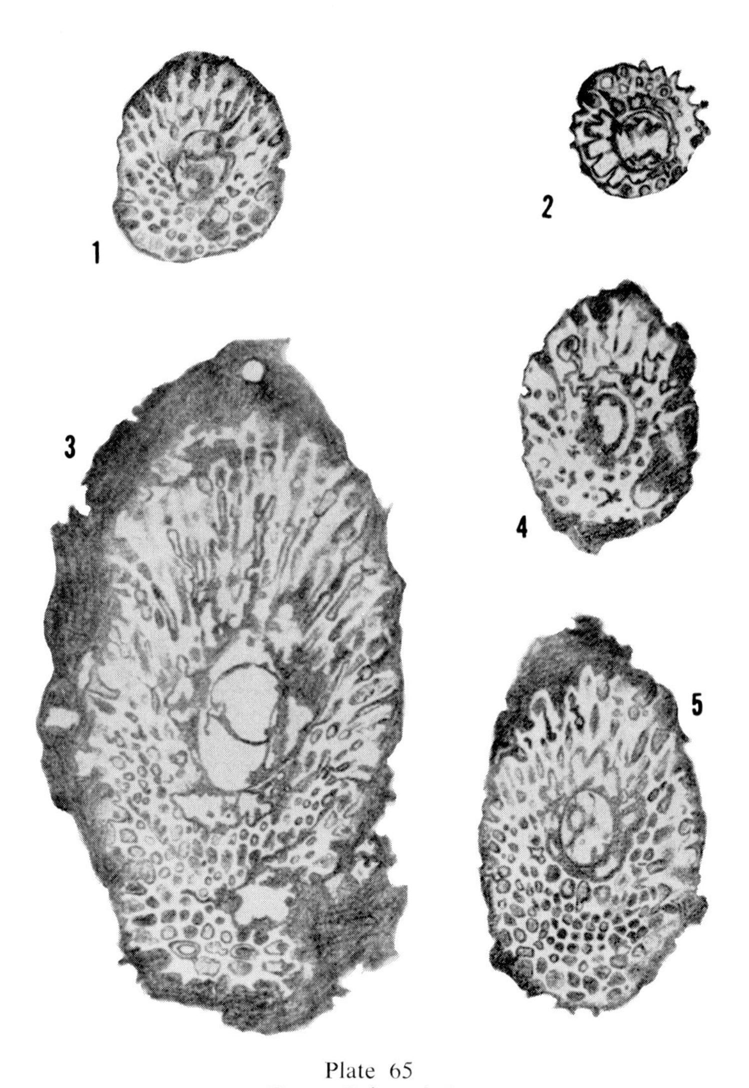

Plate 65
Genus *Palaeocladus*
Figures 1-5. *Palaeocladus mediterraneous* Pia (x20). Jurassic. Austria. A series of oblique sections.

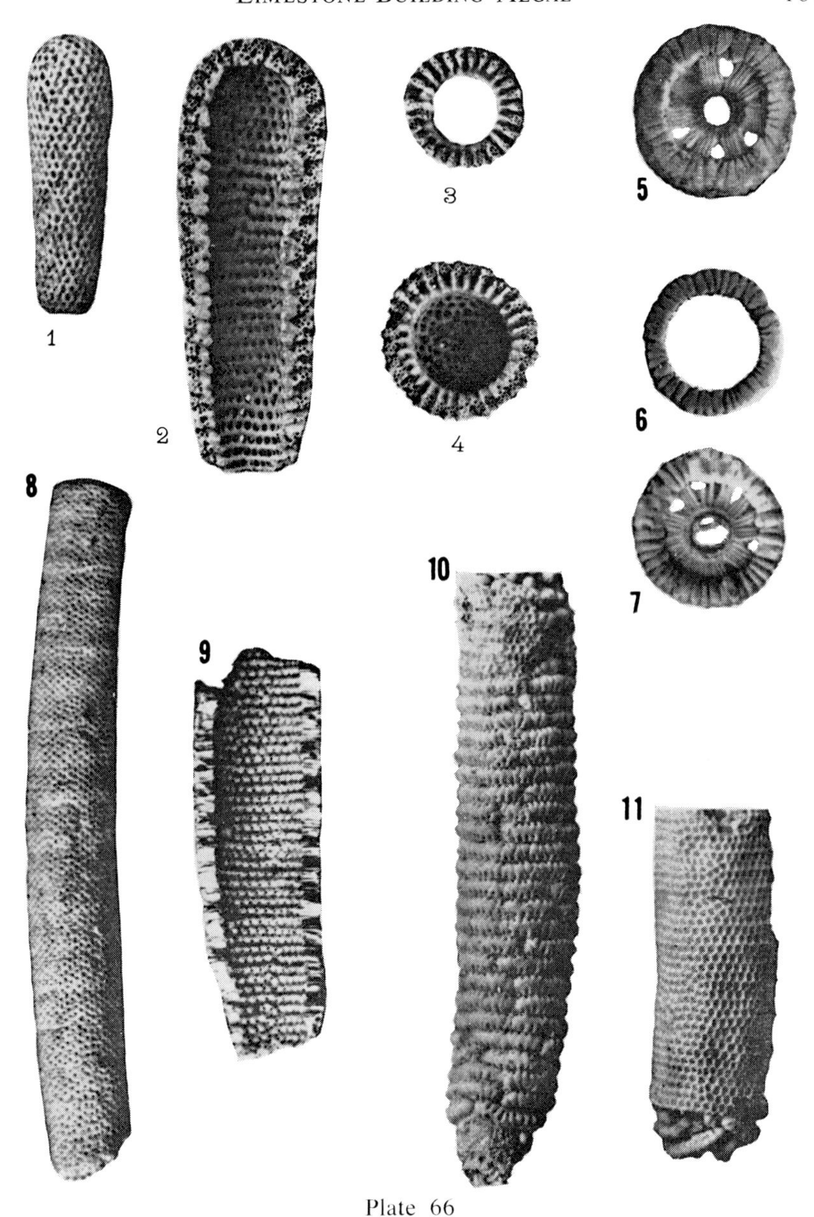

Plate 66
Genera *Dactylopora* and *Neomeris*

Figures 1-4. *Dactylopora cylindracea* Lmx. (1-x3; 2, 3, 4-x5). Eocene, Paris Basin, France.
Figures 5-7. *Neomeris (Decaisnella) annulata* Dickie, (x12) sporangial rings. Recent, Bermuda Islands.
Figures 8-9. *Neomeris (Vaginopora) herouvalensis* Munier-Chalmas. (x10).
Figures 10-11. *Neomeris (Descaisnella) annulata* Dickie (x10). Recent, Bermuda Islands.

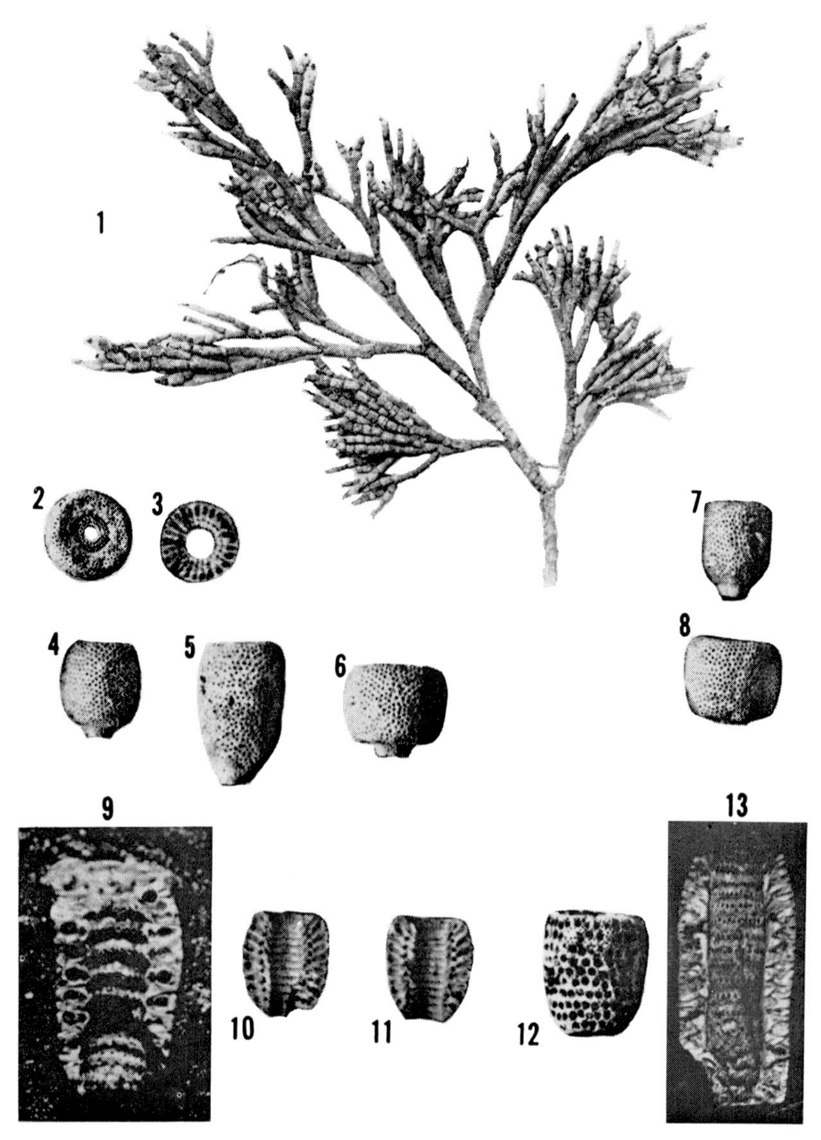

Plate 67
Genus *Cymopolia*
(From Morellett and Morellett 1913, 1922, 1939)

Figure 1. *Cymopolia barbata* (L.) Harvey. (x1) Recent. Near Havana, Cuba.
Figures 2-8 and 10-12. *Cymopolia (Karreria) zitteli* Morellet and Morellet (x10), segments. Eocene, Paris Basin, France.
Figures 9 and 13. *Cymopolia edwardsi* Morellet and Morellet. Vertical sections (x15), Eocene. Paris Basin, France.

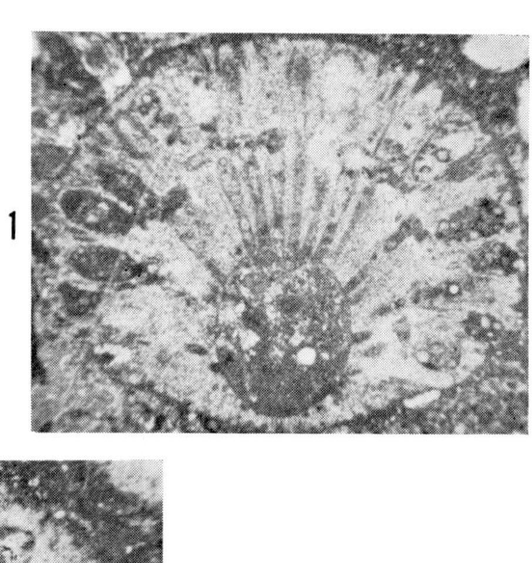

1

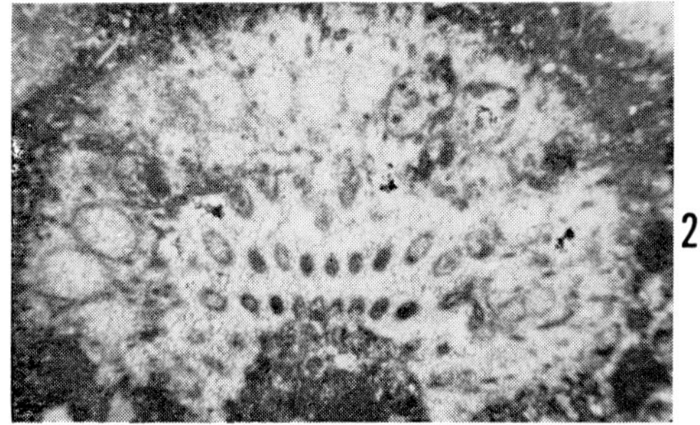

2

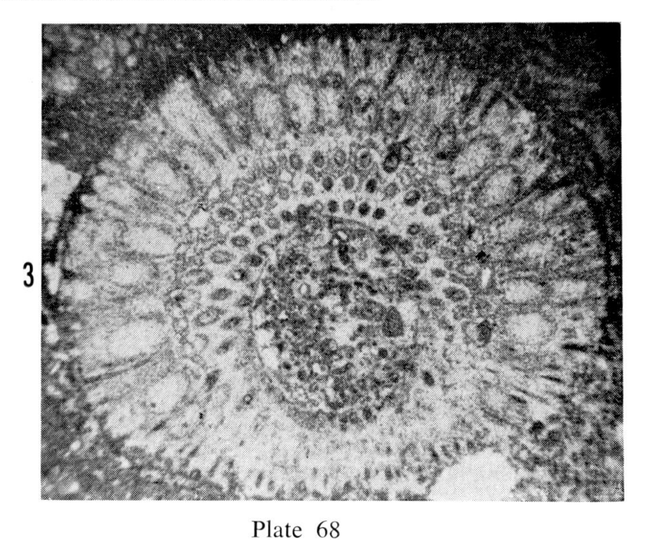

3

Plate 68
Genus *Cymopolia*
Figures 1-3. *Cymopolia pacifica* Johnson. Eocene of Saipan. (Sections x25.)

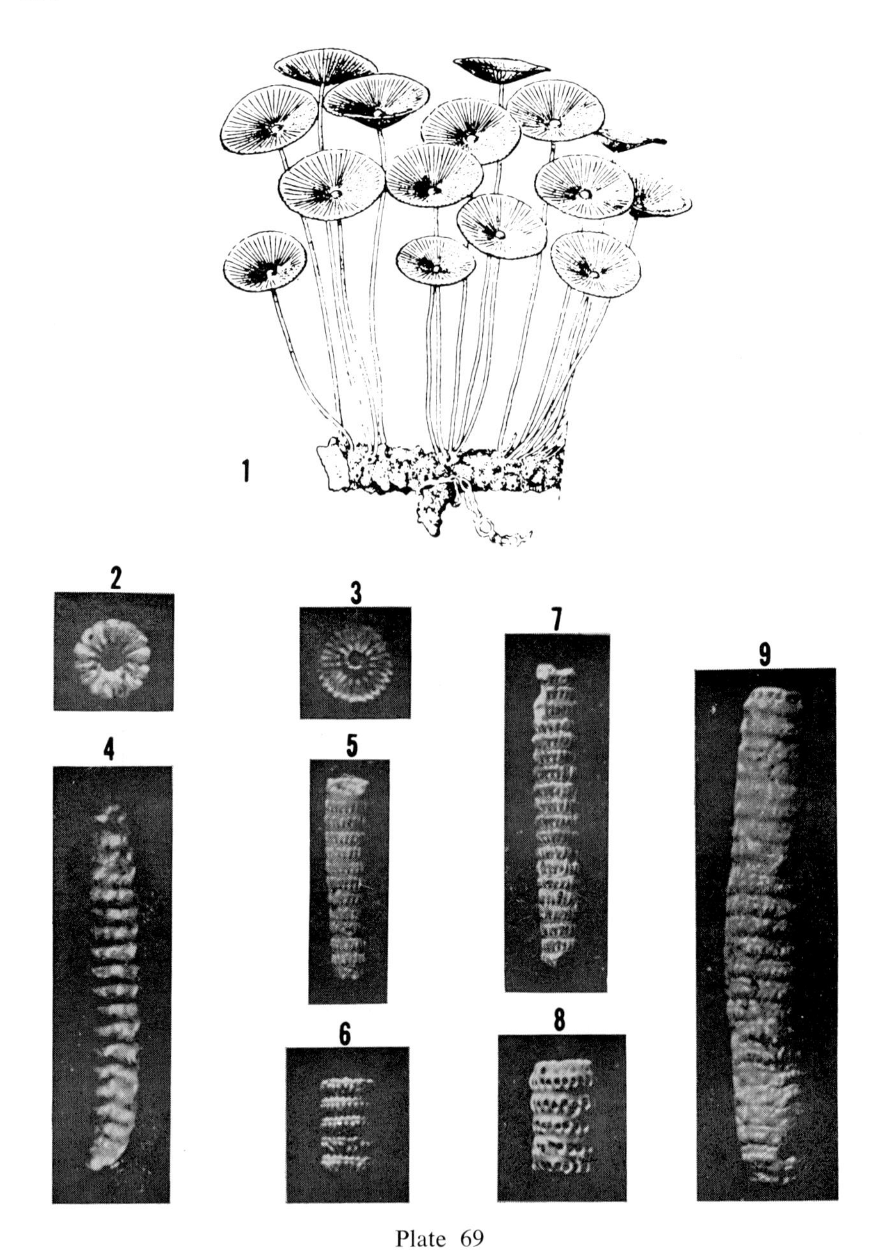

Plate 69

Genera *Acetabularia* and *Larvaria*

Figure 1. *Acetabularia* Recent, Okinawa (after Yamada).

Figures 2-9. *Larvaria* several species. Washed specimens (x15). Eocene of Paris Basin, France. (From Morellet and Morellet, 1939.)

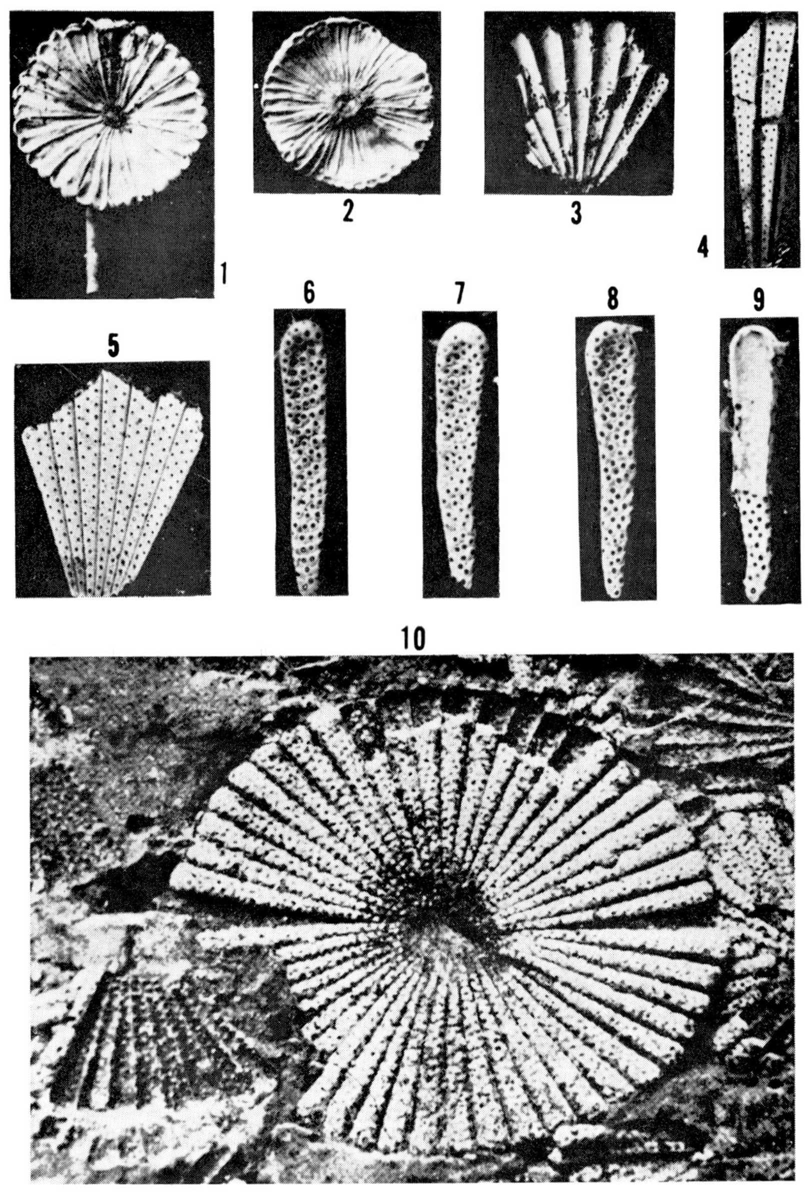

Plate 70
Acicularia

Figures 1-4 and 6-9. *Acicularia schenckii* Solms. Recent, Bermuda Islands. 1-2 Fertile discs (x5), 3-4 clusters of spicules from broken discs (x10). 6-9 Individual spicules (x10).

Figure 5. Fragment of a fertile disc (x10), Oligocene, Paris Basin.

Figure 10. *Acicularia persica* Morellet, Upper Miocene of Iran. A nearly complete fertile disc (x10). (From Morellet and Morellet, 1922 and 1939b.)

Phylum CHAROPHYCOPHYTA Papenfuss, 1946
(= Charophyta)

The Charophycophyta are a group of highly developed non-marine algae. They appear in the fossil record during the Lower Devonian, with all the characteristic features of the group already developed. Charophytes differ appreciably from all other algae in several respects, but they appear to resemble the green algae most closely. As a result of their unusual and characteristic features, there have been considerable differences of opinion as to their systematic position. Groves and Bullock-Webster (1920-1924) considered the charophytes as a separate division (phylum) of plants; Wood (1952) revised the Recent forms and put them with the green algae (Chlorophycophyceae); Fritsch (1935, 1944) classed them as an order of green algae; Smith (1950) called them a class equivalent in rank to the green algae; Moore (1954), followed later by Papenfuss (1955), raised all the classes to the rank of phyla.

As observed growing in ponds, the Recent charophytes look very much like bushy higher plants. One of the distinctive features of the Charophycophyta is the development of highly specialized reproductive organs, especially the large female oogonia. Commonly, these oogonia or portions of them, particularly the spiraled gyrogonites, become strongly calcified and are preserved as fossils.

Charophytes may be extremely useful in dating non-marine strata. Frequently, they are the only fossils found in such beds. Sometimes they are so abundant as to form lenses of fresh-water limestone (plate 130, figure 1).

Recently, Peck (1957) published a review of North American Mesozoic Charophyta. Much of the data used in the present chapter, including the descriptions of Mesozoic genera, was taken from his report.

Key to genera of North American Mesozoic Charophyta
(Peck, 1957)

I. Gyrogonites enclosed in utricles (Clavatoraceae).
  A. Utricles composed of essentially vertical units.
    1. Cortex of 12 continuous tubes, spines small and in clusters .... *Clavator.*
    2. Cortex composed of 12 units; 6 descend and 6 ascend from adjoining nodes and interfinger in the center of the internode; spines long and tapering .... *Echinochara.*
  B. Utricles with 3-rayed symmetry .... *Atopochara.*
  C. Surface units irregularly spiral, antheridia (?) divided

into upper and lower hemispheres; 4 openings to interior, 2 polar and 2 equatorial .... *Perimneste.*

II. Gyrogonites not enclosed in utricles (Characeae).
  A. Distal ends of spiral units not modified, meeting at center of summit at point or along short irregular line.
    1. Large spherical gyrogonites .... *Sphaerochara.*
    2. Gyrogonites with distinctly flattened summits .... *Obtusochara.*
    3. Apical pole (and generally basal pole) slightly protruding .... *Praechara.*
  B. Distal ends of spirals distinctly modified.
    1. Summit depressed.
      a. Spirals end in shallow summit depression—if ends are not preserved gyrogonite has large summit opening .... *Aclistochara.*
      b. Spiral ends enclose throatlike opening to gyrogonite .... *Stellatochara.*
    2. Summit not depressed.
      a. Spirals narrow at periphery of summit, then expand to close summit with swollen tips forming a rosette .... *Tectochara.*
      b. Spiral ends form pyramidal projection on summit .... *Latochara.*

III. Gyrogonites unknown, vegetative parts resembling those of the living *Chara* .... *Charaxis.*

Genus *Aclistochara* Peck, 1937, emend. Mädler, 1952
Plate 71, figures 1-11.

Description—
Gyrogonites with five sinistrally spiraled units that ascend to a truncate summit, bend onto the summit to form its outer rim, then bend down into a central depression for a short distance, finally turn sharply into the center of the summit depression and expand to fill the space. Horizontal part thin to swollen and bulbous. Abrupt downward turning and thinning produces a circular furrow around the expanded ends of spirals.

Remarks—
This genus was proposed by Peck in 1937, emended by him in 1941, and somewhat redefined and limited by Mädler in 1952. *Aclistochara* has been used with different ideas in mind by a number of authors for a number of Early Tertiary gyrogonites, with the result that there is some confusion as to which species should be included in this genus and which should be put under *Chara.* In this work, the present author follows the usage of Peck.

Generic range—
Jurassic to Oligocene, possibly to Pliocene (according to Mädler).

Geographic distribution—
Peru, Colorado, Wyoming, Idaho, South Dakota, Germany, France.

Genus *Atopochara* Peck, 1938
Plate 72, figures 1-8.

Description—
Clavatoraceae with utricles possessing three-rayed symmetry. Each ray consists of short vertical units originating at or near basal opening and ascending to or near equatorial plane; one to several small units grouped on or near equatorial plane; several sinistrally spiraled units extend from equator to or near to summit opening.

Remarks—
The construction of the utricle indicates that a few vertical cells were sufficient to cover the gyrogonite near the base. As the diameter of the gyrogonite increased, the original cells divided, increasing in number in proportion to the increased space to be covered; and, as the circumference decreased toward the summit, cells pinched out in that direction until only a few actually reached the summit opening.

Generic range—
Cretaceous (Aptian to Cenomanian-Turonian).

Geographic distribution—
Texas, Oklahoma, Rocky Mountain region, Utah.

Genus *Chara* Vaillant, 1719
Plate 74; plate 75, figures 1-12; plate 76, figures 5-13; plate 130.

Description—
Stems and branches with or without cortex. Plants may be monoecious or sexual. Oogonia consist of five sinistrally spiraled enveloping cells, and five coronula cells which do not calcify.

Remarks—
This is an important and widespread Recent genus. Fossils consist almost entirely of the calcified oogonia. Originally, nearly all fossil charophyte oogonia were attributed to this genus, but during the last 35 years, numerous groups of such remains have been placed in special genera, until at present, only those of Cenozoic age, not put in other genera, are referred to *Chara*.

Generic range—
Eocene to Recent.

Geographic distribution—
Essentially worldwide.

Genus *Clavator* Reid and Groves, emend. Harris, 1939
Plate 77, figures 1-6.

Description—
Oogonia arranged in a single row on the adaxial side of the branchlet, one on each joint. Utricles composed of essentially vertical units.

Gyrogonites small, strongly beaked, generally with both outer and inner walls of the oogonia calcified.

Remarks—

The genus *Clavator,* with two species, was described in great detail from the British Purbeck by Harris in 1939. Vegetative parts of the plant are silicified in the British Purbeckian, and both generic and specific characters are largely based on them.

In the United States, *Clavator* is confined to rocks of Aptian age, where it is represented by 4 species. *C. harrisi* Peck is widely distributed and is an excellent guide fossil.

Generic range—

Jurassic (Purbeckian) and Cretaceous (Aptian).

Geographic distribution—

Europe and North America.

Genus *Echinochara* Peck, 1957
Plate 78, figures 1-8.

Description—

Charophytes with six cortical tubes budding from each end of the nodal cells and extending as tapering dextral coils around the central tube, interfingering with the cortical tubes from adjoining nodes. Individual cortical cells short, of uniform length and, at their distal ends, giving rise to long spines that extend over and almost completely cover the next cortical cell.

Branchlets massive, stubby, bearing oogonia in whorls of six, the oogonia enclosed in thick utricles. Utricle cells originate below base of oogonia as short stubby units that repeatedly branch and resemble spines on cortical cells. Oogonia normal for family.

Remarks—

The genus is known only from the Morrison (Kimmeridgian) formation of Colorado, eastern Wyoming, and western South Dakota and from the Kimmeridgian of northwest Germany.

Generic range—

Jurassic (Kimmeridgian).

Geographic distribution—

Colorado, Wyoming, South Dakota, Germany.

Genus *Latochara* Mädler, 1955
Plate 79, figures 1-12.

Description—

Gyrogonites composed of five sinistrally spiraled units that narrow appreciably at rim of summit, level off, and turn inward, expanding in width, and then turn abruptly upward into almost vertical position to form a small pyramidal projection in center of summit. Vegetative parts unknown.

Remarks—

On representatives of *Tectochara* Grambast and Grambast, the spirals narrow at the rim of the summit and then turn inward and expand rapidly to form a rosette of five triangular terminations that close the summit. The ends of the spirals do not turn upward in the center on representatives of *Latochara*. On specimens in which the summit area is not preserved, it is difficult to determine if the gyrogonite should be referred to *Tectochara* or *Latochara*.

The gyrogonites of the genera *Clavator, Atopochara,* and *Echinochara* do have the spirals turning upward at their terminal ends to form a cylindrical neck. But gyrogonites belonging to these genera are never flattened on the summit, and the spirals turn to a vertical or almost vertical position. The projection thus formed is cylindrical not conical.

Representatives of the genus are not recorded by Mädler (1952) from the German Kimmeridgian, but they evidently are abundant in the English Purbeck. In North America the genus is abundant in the Morrison formation and has been found at only two localities in the Lower Cretaceous (Aptian).

Generic range—

Jurassic and Cretaceous.

Geographic distribution—

Europe and North America.

Genus *Perimneste* Harris, 1939
Plate 80, figures 8-16.

Description—

This genus is described by T. M. Harris (1939) as follows:

> "Stem corticated by six rows of cells, all bearing spines. Node bearing eighteen leaves (branchlets) in three whorls of six; one whorl of upward-pointing leaves on the same radius as the cortical cells, and two whorls of short leaves on the alternate radii, one pointing upward and one downward. Leaves uncorticated, bearing a few simple spine-like leaflets in small whorls.
>
> "Reproductive organs borne near the bases of the short upward-pointing leaves, each consisting of an oogonium surrounded at maturity with leaf segments bearing antheridia. Outer surface of oogonium (as well as inner parts of spiral cells) calcified. Antheridial wall calcified."

Remarks—

To date, the only species referred to this genus are *Perimneste horrida* Harris from the Purbeckian of England and *P. corrugata* Peck from the Aptian of the United States.

Generic range—

Jurassic (Purbeckian) and Cretaceous (Aptian).

Geographic distribution—

England and North America.

Genus *Praechara* Horn af Rantzien, 1954
Plate 80, figures 1-7.

Description—

Horn af Rantzien (1954) gave the following diagnosis:

"Fossil Characeae of the subfamily Chareae represented by small, medium-sized or large, terete gyrogonites of oblong or ovoid shape. Apical pole ± conically protruding. Enveloping cells 5, sinistrally spiralled, without inclination to form tubercles, turning towards one another apically at a point, along a short broken line, or leaving a small rounded opening. The convolutions do not form a grooved zone of weakly calcified cells in the apical periphery, and do not expand towards the center. Basal pole rounded or slightly protruding, the enveloping cells turned in towards the small, rounded basal pore without any change of shape. Oospore membrane thin, its structure unknown."

Remarks—

Fossil gyrogonites on which the calcareous spirals turn onto and meet in the center of the summit area without modification have always been troublesome to the systematist. These forms are similar to the lime shells of the living Charophyta, and most workers have been content to refer them to the genus *Chara*. Strong objections have been made to this procedure (Peck, 1953, p. 219), and recent articles (Mädler, 1952; Horn af Rantzien, 1954) have differentiated several genera among this group.

Generic range—

Jurassic and Cretaceous.

Geographic distribution—

Rocky Mountain region, Great Britain, Germany.

Genus *Sphaerochara* Mädler, 1952
Plate 81, figures 1-8.

Description—

Gyrogonite of the summit structure of the Characeae and of more or less spherical form without decoration. (Mädler, 1952, p. 6.)

Remarks—

Mädler (1952) proposed the genus *Sphaerochara* to contain those species that have the summit characters of the gyrogonites of the genus *Chara* and a spherical shape that prevents their assignment to *Chara*.

Generic range—

Jurassic, Cretaceous, Tertiary (Oligocene).

Geographic distribution—

Hungary, Germany, Rocky Mountain region.

Genus *Stellatochara* Horn af Rantzien, 1954
Plate 82, figures 1-11.

Description—

Gyrogonites with flat summits; entire summit area occupied by comparatively large circular to pentagonal opening that remains uniform in diameter or contracts toward interior to produce a cylindrical or a funnel-shaped throat. Five spiral units ascend to rim of opening without change in size, shape, or degree of calcification and without change in angle of ascent or deflected slightly toward vertical. Spirals end at rim of opening or turn sharply downward into an almost vertical position to line interior of throat. Vegetative parts and modification for summit closure unknown.

Remarks—

Stellatochara is represented by three species in the Triassic of Sweden, three species in the Rocky Mountains Jurassic and one species in the Rocky Mountain Aptian and Albian. Several of the species described by Mädler (1952) from the Kimmeridgian of Germany probably belong to the genus.

Generic range—

Triassic to Cretaceous.

Geographic distribution—

North America, Sweden, Germany.

Genus *Sycidium* G. Sandberger, 1849
Plate 83, figures 5a-7, 14-24.

Description—

"More or less spherical oogonia with outer shell composed of a variable number of meridional units—probably never more than twenty, commonly sixteen or less—expressed on the outer surface as vertical grooves or ridges; each subdivided transversely so as to consist of a columnar series of pits or elevations varying in number from less than eleven to more than eighteen; units of adjacent columns in lateral alignment or alternating position. Oogonium with a small basal and somewhat larger summit opening. Oospore membrane of mature oogonia usually contracted." (Peck, 1934, p. 116-117.)

Remarks—

Commonly, associated with *Trochiliscus.*

Generic range—

Devonian and Lower Mississippian.

Geographic distribution—

Missouri, Russia, Germany.

Genus *Tectochara* Grambast and Grambast, 1954
Plate 76, figures 1-4.

Description—

Gyrogonites, the width of whose spiral cells diminish perceptibly in the area of the apical zone, the joined terminal swellings forming an often prominent rosette. Apical zone easily detached by a breaking of the spires at the level of their constriction, leaving an opening, the outline of which in most cases has five curved indentations. Basal pore situated at the bottom of a funnel-shaped depression, the sides of which are frequently projecting, the base generally prominent. Length exceeding width, form generally ovoid, sometimes subglobular. (Grambast and Grambast, 1954, p. 4.)

Remarks—

The genus *Tectochara* was defined by Louis and Nicole Grambast to include Tertiary species that had formerly been referred to *Chara* and *Aclistochara*.

The genus is closely related to *Latochara*. On representatives of both genera, the spirals narrow at the rim of the summit and then expand as they turn inward. Representatives of *Latochara* do not have a conspicuous expansion at the ends of the spirals, and the ends turn upward to form a small conical projection at the center of the summit that has no resemblance to a rosette. It is, however, difficult to differentiate these genera on specimens in which the summit is not preserved.

Generic range—

Lower Cretaceous and Tertiary.

Geographic distribution—

North America and Europe.

Genus *Trochiliscus* Karpinsky, 1906
Plate 83, figures 1-5, 8-13.

Description—

"Globular oogonia composed of from 10 to 17 (possibly fewer) enveloping cell units, originating around a small basal opening, and ascending dextrally with a fairly uniform spiral twist to the opposite pole of the oogonium, where the apical ends surround the summit orifice. Each spiral unit is expressed on the outer surface by a broad gently rounded ridge, usually with a median, bounded on each side by furrows, or by a comparatively broad furrow bounded laterally by ridges. Oospore membrane in the mature oogonia usually contracted but occasionally in contact with the inner surface of the oogonia walls." (Peck, 1934, p. 104-105.)

Remarks—

Primitive Paleozoic charophytes with a large number of dextrally spiraled units.

Generic range—

Lower Devonian to Lower Mississippian.

Geographic distribution—

Russia, Germany, England, Missouri.

REFERENCES

Bell, W. A., 1922, A new genus of Characeae and new Merostomata from the Coal Measures of Nova Scotia: Royal Soc. Canada Trans., ser. 3, v. 16, sec. 4, p. 159-162.

Dollfus, G. F., and Fritel, P. H., 1920, Catalogue raisonne des characees fossiles du Bassin de Paris: Soc. Geol. France Bull., ser. 4, v. 19, p. 243-261, text-figs. 1-23.

Fritsch, F. E., 1935, The structure and reproduction of the algae: Cambridge University Press, 767 p.

Grambast, Louis, and Grambast, Nicole, 1954, Sur la position systematique de quelques Charophytes tertiaires: Rev. Gen. Botanique, v. 61, p. 1-7, 1 fig.

Groves, James, 1933, Charophyta: Fosilium Catalogus, II, Plantae, pars 19, p. 1-74.

——————, and Bullock-Webster, G. R., 1920, The British Charophyta, Volume 1, Nitelleae: The Ray Soc., p. 1-141, 25 text-figs., 20 pls.

——————, 1924, The British Charophyta, Volume 2, Chareae: The Ray Soc., p. 1-129, 31 text-figs., pls. 21-45.

Harris, T. M., 1939, British Purbeck Charophyta: British Mus. Nat. History, 83 p., 17 pls.

Horn af Rantzien, Henning, 1954, Revisions of some Pliocene charophyte gyrogonites: Bot. Notes, Sverige, no. 1, p. 1-33, 8 figs.

——————, 1956, Morphological terminology relating to female charophyte gamentangia and fructifications: Bot. Notes, Sverige, v. 109, no. 2, p. 212-259, illus.

——————, 1956, An annotated check-list of genera of fossil Charophyta: Micropaleontology, v. 2, no. 3, p. 243-256.

——————, 1958, Morphological types and organ-genera of Tertiary charophyte fructifications: Stockholm Contributions in Geol., v. 4, no. 2, p. 45-197, pls. 1-21.

——————, 1959, Recent charophyte fructifications and their relations to fossil charophyte gyrogonites: Arkiv för Botanik, Utgivet av Kungl. Svenska Vetenskapsakademien, ser. 2, v. 4, no. 7, p. 165-332, pls. 1-19.

——————, 1959, Comparative studies of some modern, Cenozoic, and Mesozoic charophyte fructifications: Stockholm Contributions in Geol., v. 5, no. 1, p. 1-17.

Karpinsky, A., 1906, Die Trochilisken: Men. Comite Geol., in Russian and German, new ser., v. 27.

Mädler, Karl, 1952, Charophyten aus dem nordwestdeutschen Kimmeridge: Geol. Jahrb., v. 67, p. 1-46, 2 figs., 8 pls.

Moore, R. C., 1954, Kingdom of organisms named Protista: Jour. Paleontology, v. 28, p. 588-598.

Papenfuss, G. F., 1955, Classication of the algae *from* A century of progress in the natural sciences: California Acad. of Sciences, San Francisco, p. 115-224.

Peck, R. E., 1934, The North American Trochiliscids, Paleozoic Charophyta: Jour. Paleontology, v. 8, p. 83-119, pls. 9-13.

——————, 1937, Morrison Charophyta from Wyoming: Jour. Paleontology, v. 11, p. 83-90, pl. 14.

——————, 1941, Lower Cretaceous Rocky Mountain nonmarine microfossils: Jour. Paleontology, v. 15, p. 285-304, pls. 42-44.

——————, 1946, Fossil Charophyta: Am. Midland Naturalist, v. 36, no. 2, p. 275-278.

——————, 1953, Fossil Charophytes: Bot. Rev., v. 19, p. 209-227.

——————, 1957, North American Mesozoic Charophyta: U. S. Geol. Survey Prof. Paper 294-A, 44 p., 8 pls.

Smith, G. M., 1950, The fresh-water algae of the United States: New York, McGraw-Hill, 2d ed.

Wood, R. D., 1952, The Characeae, 1951: Bot. Rev., v. 18, no. 5, p. 316-353.

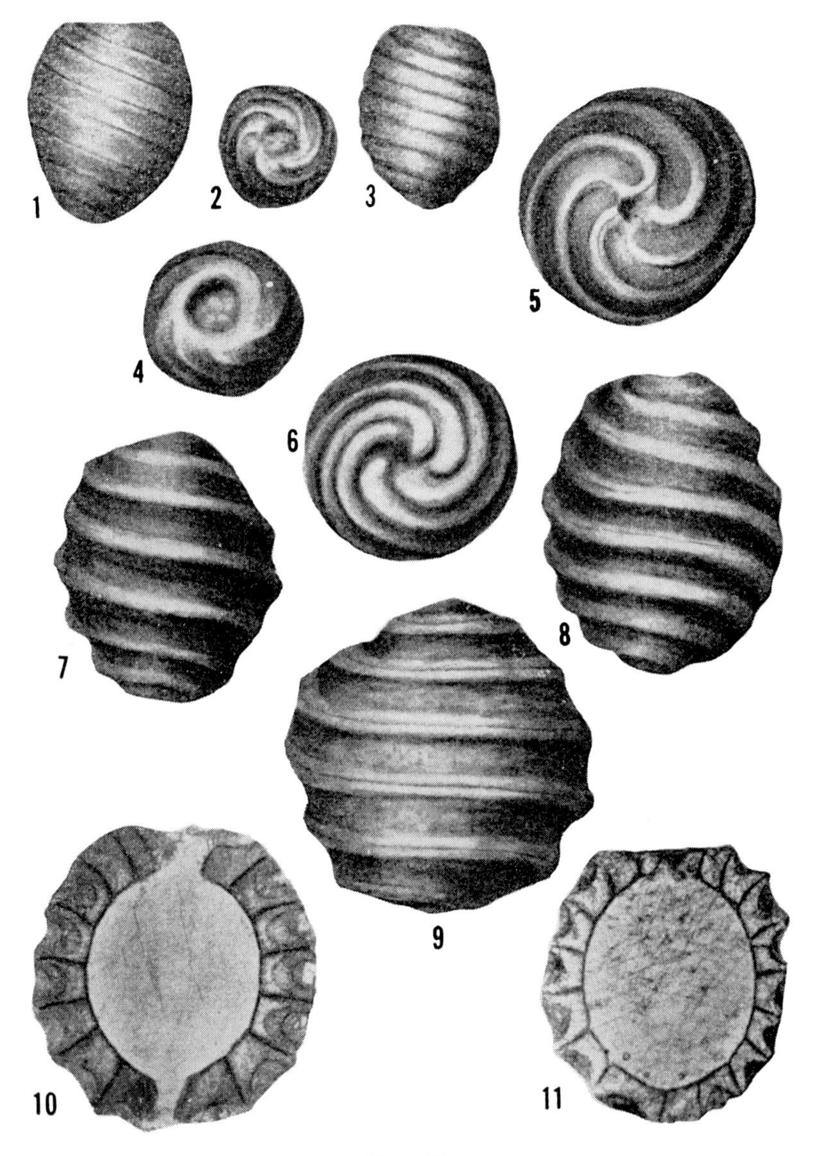

Plate 71
Genus *Aclistochara*
Figures 1-11. All from the Upper Jurassic Morrison formation of Wyoming. (From Peck, 1957, U. S. Geol. Survey Prof. Paper 294.) 1-4. *Aclistochara jonesi* Peck (x60). 5-8. *Aclistochara complanata* Peck (x60). 9-11. *Aclistochara latisulcata* Peck (x60).

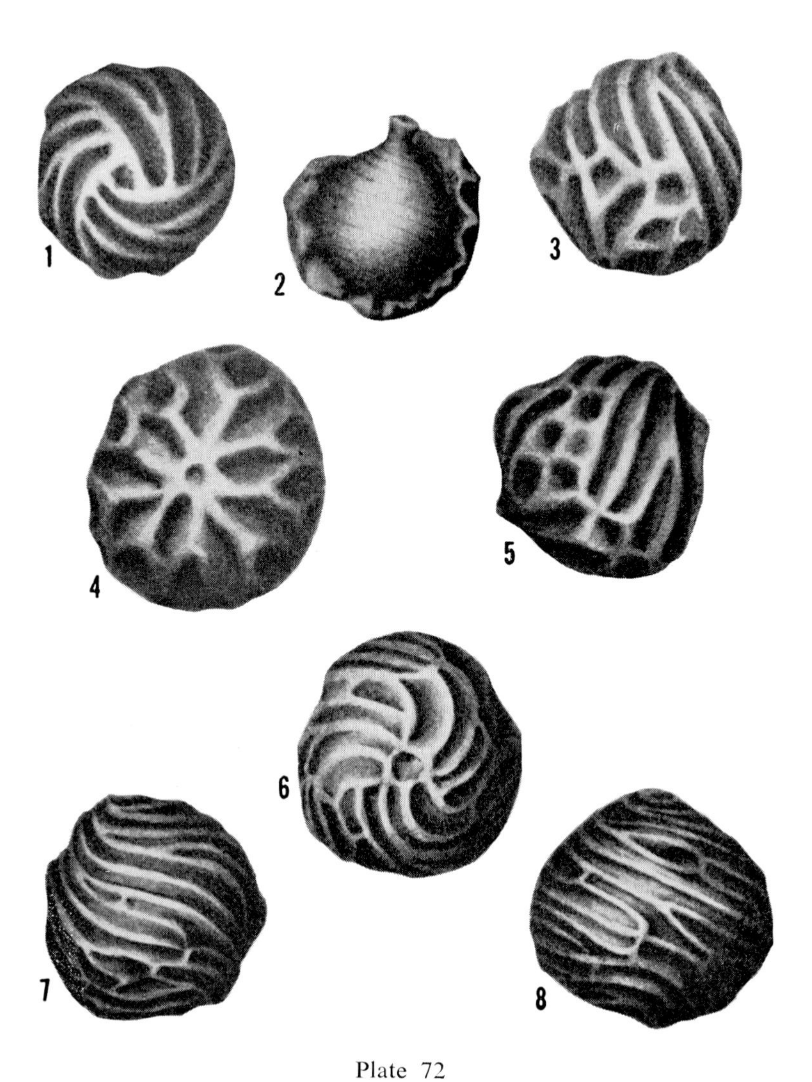

Plate 72
Genus *Atopochara*
Figures 1-5. *Atopochara trivolvis* Peck (x40). Lower Cretaceous Draney limestone, Caribou County, Idaho.
Figures 6-8. *Atopochara multivolvis* Peck (x40), Cretaceous. Tropic shale, Garfield County, Utah. (From Peck, 1957, U. S. Geol. Survey Prof. Paper 294-A.)

Plate 73
Genus *Nitella* Plant
Figure 1. The modern *Nitella hyaline* (deCand.) Agrdh (x1). After Migula.

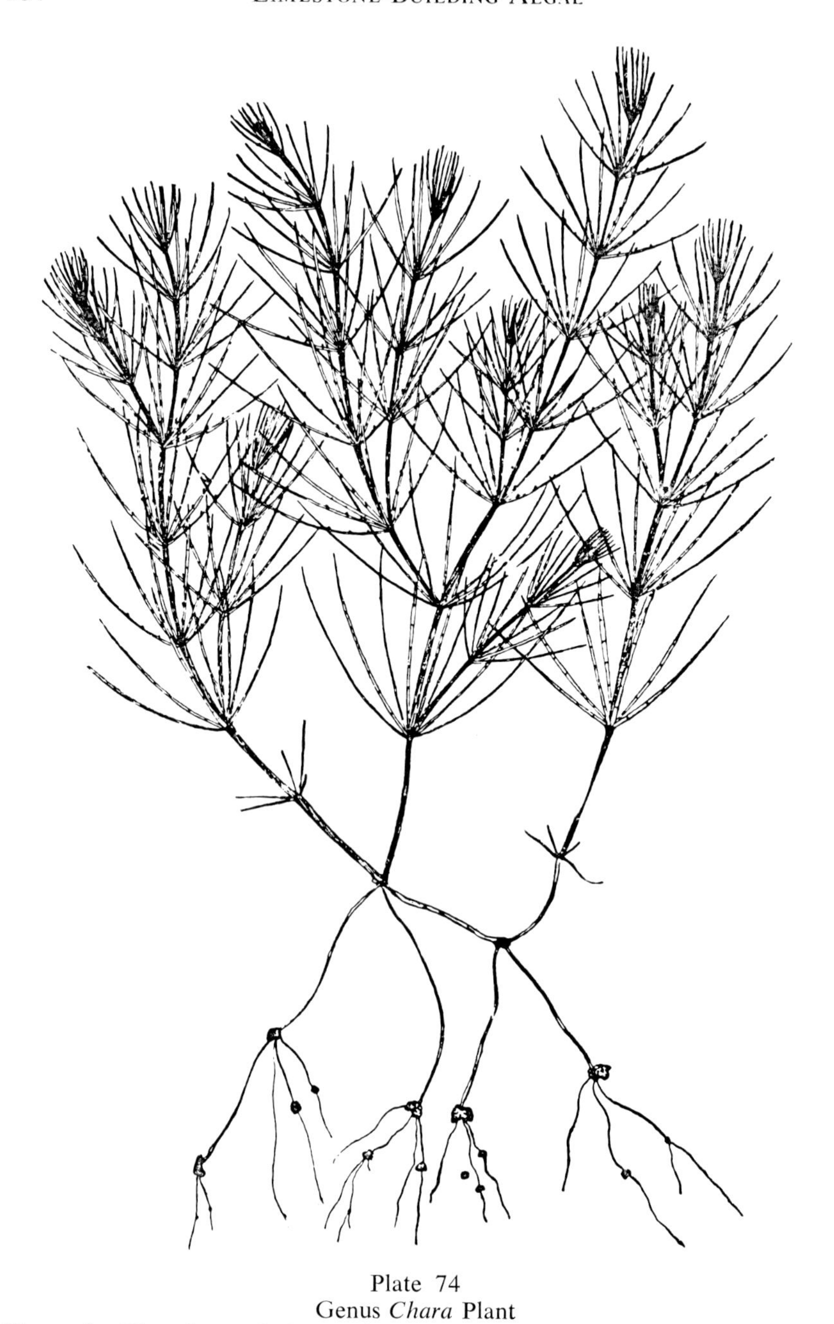

Plate 74
Genus *Chara* Plant
Figure 1. The plant of the modern *Chara fragifera* Durieu (x1). (After Oltmans.)

Plate 75
Genus *Chara*
Figures 1-3. *Chara wrighti* Salter in Forbes (x30).
Figures 4-6. *Chara helicteres* Brongniart (x30).
Figures 7-9. *Chara medicaginula* Brongniart (x30).
Figures 10-12. *Chara oehlerti* Dollfus (x30).

Plate 76
Genera *Tectochara* and *Chara*
Figures 1-4. *Tectochara grambastorum* Peck (x60), Lower Cretaceous, Lakota sandstone, Fall River County, South Dakota. (From Peck, 1957.) Figures 5-13. From uppermost Cretaceous or Paleocene, Rajahmundry, India. (From K. Sripada Rao and S. R. Narayana Rao, Geol. Survey India, 1939.) 5-7. *Chara strobilocarpa* Reid and Groves (x30). 8-10. *Chara caelata* Reid and Groves (x30). 11-13. *Chara vasiformis?* Reid and Groves (x30).

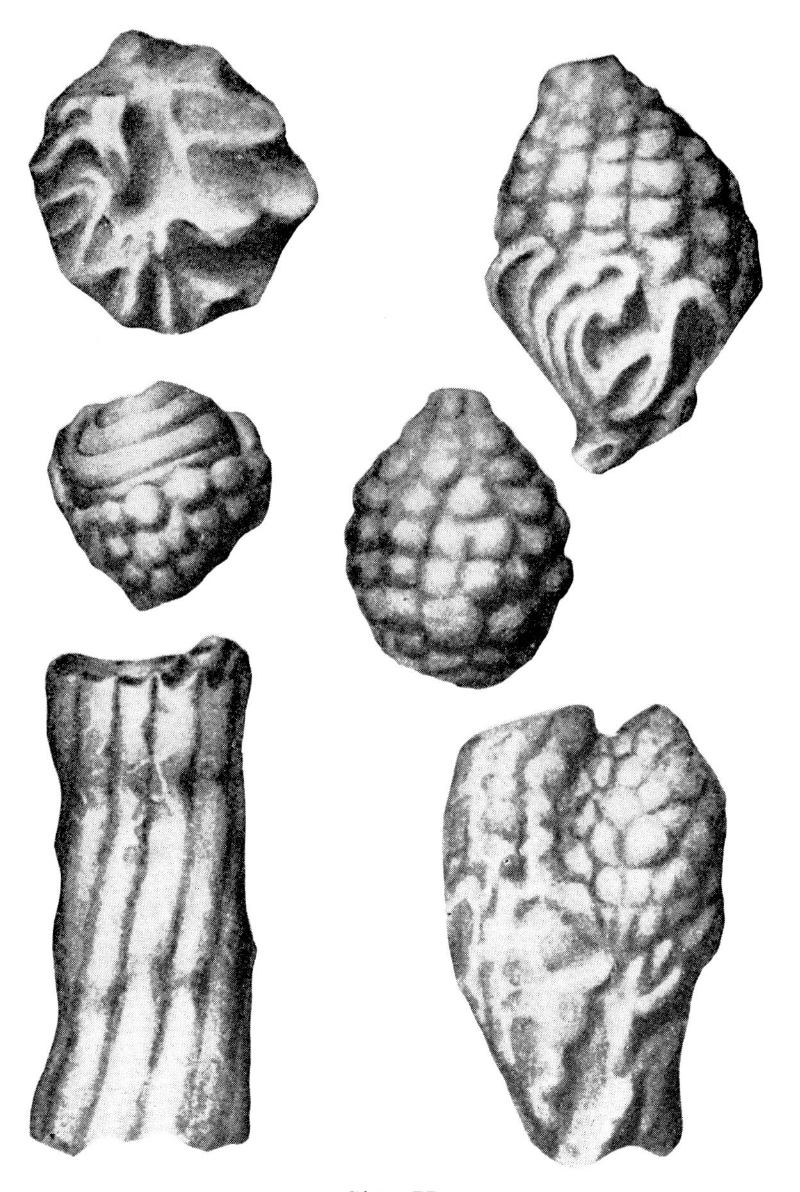

Plate 77
Genus *Clavator*

Figures 1-6. *Clavator nodosus* Peck (x60). Cretaceous, Lakota sandstone, Fall River County, South Dakota. 1. Basal view of utricle; 2. side view partially exfoliated utricle showing gyrogonite; 3. side view internode with spine cell removed; 4-5. utricles, side view; 6. utricle attached to branchlet. (From Peck, 1957.)

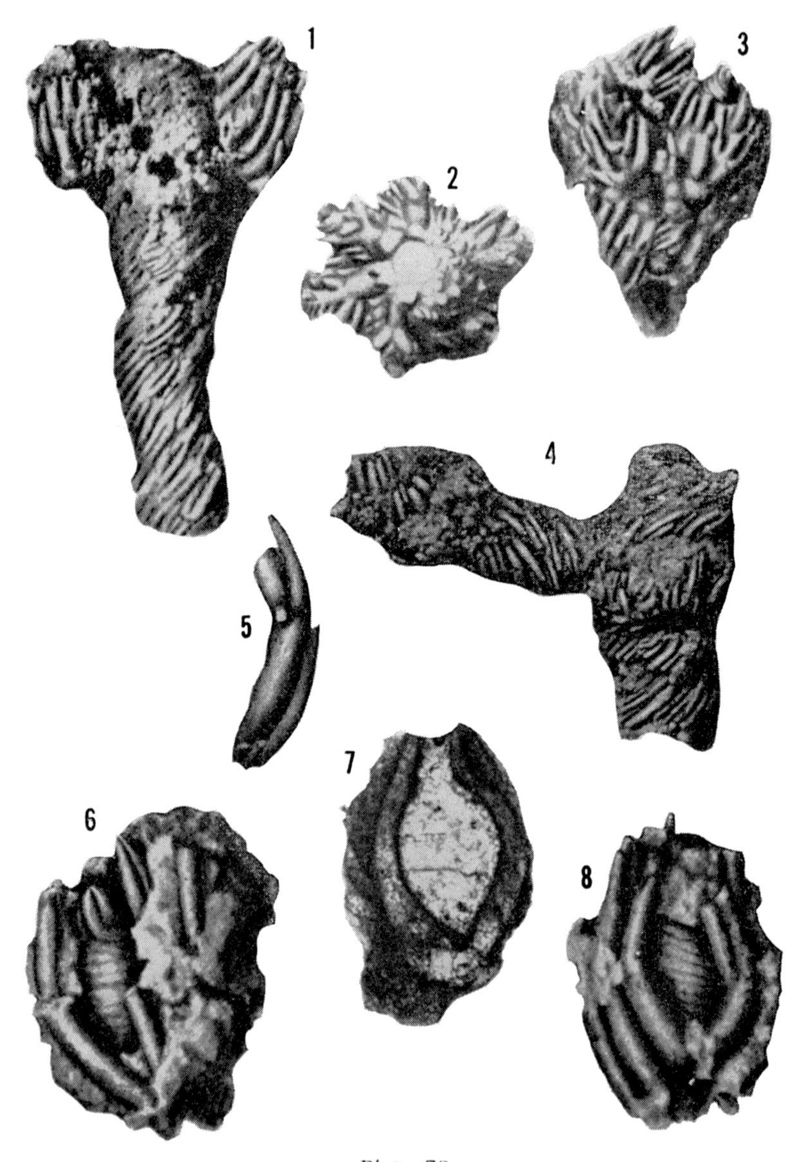

Plate 78
Genus *Echinochara*

Figures 1-8. *Echinochara spinosa* Peck. Upper Jurassic, Morrison formation, Colorado. (From Peck, 1957, U. S. Geol. Survey Prof. Paper 294-A.) 1. Fertile node (x20); 2. fertile node; end view (x20); 3. fertile node (side view) (x20); 4. part of stem with attached branchlet (x15); 6 and 8, side views of gyrogonite and utricle (x40); 7. section through utricle and gyrogonite (x50).

Plate 79
Genus *Latochara*
Figures 1-12. *Latochara latitruncata* (Peck), Jurassic, Morrison formation, Utah. (From Peck, 1957, U. S. Geol. Survey Prof. Paper 294-A).

Plate 80
Genera *Perimneste* and *Praechara*

Figures 1-7. *Praechara voluta* (Peck) (x60), Jurassic, Morrison formation, Crook County, Wyoming.

Figures 8-16. *Perimneste corrugata* Peck (x40). Lower Cretaceous, Peterson limestone, Caribou County, Idaho. (From Peck, 1957.)

Plate 81
Genus *Sphaerochara*
Figures 1-8. From Cretaceous Bear River Formation, Uinta County, Wyoming. 1-3. *Sphaerochara stantoni* (Knowlton) (x60). 4-8. *Sphaerochara latifasciata* Peck (x60). (From Peck, 1957, U. S. Geol. Survey Prof. Paper 294-A.)

Plate 82
Genus *Stellatochara*

Figures 1-6. *Stellatochara arguta* Peck (x60); Jurassic, Morrison formation, Emery County, Utah.

Figures 7-11. *Stellatochara sublaevis* Peck (x60), Middle Jurassic, Piper formation, Fergus County, Montana (from Peck, 1957).

Plate 83
Genera *Trochiliscus* and *Sycidium*
Figures 1-5. *Trochiliscus laticostatus* Peck (x16).
Figures 5a-7. *Sycidium clathratum* Peck (x16).
Figures 8-10a. *Trochiliscus decacostatus* Peck (x16).
Figures 11-13. *Trochiliscus octocostatus* Peck (x16).
Figures 14-24. *Sycidium foveatum* Peck (x16). All from Williamsburg, Missouri. (From Peck, 1934.)

Phylum SCHIZOPHYTA (Falkenberg) Engler, 1892
"Section" POROSTROMATA Pia, 1927

The term "Porostromata" was introduced by Pia in 1927 to include all fossil algae which had a definite microstructure consisting of masses or bundles of well-defined tubes, but which were of unknown systematic position. A considerable number of genera were included in the group at that time, but with the gradual increase in knowledge of fossil algae, some of the genera, such as *Garwoodia* and *Ortonella,* have been taken out of this group and classed with the green algae, mostly under the family Codiaceae, division 1. At the present time, probably the only well known genera remaining in this section are *Girvanella,* some of the forms originally described under the name *Sphaerocodium* from the Lower Paleozoic, and such forms as *Ottonosia* and *Somphospongia* from the Upper Paleozoic.

Genus *Girvanella* Nicholson and Etheridge, 1880
Plate 84, figures 1-3; plate 85; plate 86; plate 87; plate 116; plate 122, figure 2.

Description—

Flexuous tubes of uniform diameter, commonly with thick, well-defined walls. The tubes are simple cylinders without cross partitions or perforations in the side walls. Branching may occur, commonly in some species, rarely in others. Sporangia unknown.

Frequently, the tubes develop around a nucleus and form small rounded or bean-shaped masses.

Remarks—

The *Girvanella* have been separated into species on the basis of diameter of the tubes. The thickness of the wall also appears to be a specific feature. The character and frequency of branching may be specific features or they may be the result of local environmental conditions. About 20 species have been named.

Fremy and Dangeard (1935) present strong evidence that the *Girvanella* were probably green algae.

Girvanella were important limestone builders in many regions during the upper half of the Cambrian, the Lower Ordovician, and the Jurassic, and in restricted areas during the Silurian, Devonian, Mississippian, and Triassic.

Generic range—

Cambrian to Cretaceous.

Geographic distribution—

World wide.

Genus *Ottonosia* Twenhofel, 1919
Plate 88, figures 1-2.

Description—

The algae form rounded, flattened, "biscuits" up to 3 inches in diameter. The surfaces of the "biscuits" are covered with irregular rounded protuberances.

Twenhofel described the plants as follows:

> "Irregularly shaped coenoplases which begin as incrustations around other substances and increase in size through the deposition of material over and around that already deposited. The diameters vary up to about 85 millimeters. Most of the coenoplases are biscuit-shaped; a few are spherical. In small specimens the shapes appear to have been determined by the shapes of the nuclei. One specimen which has the convex valve of a *Derbya* for a nucleus still retains that shape, the shell being covered with about one-eighth inch thickness of algal material. The exteriors are irregular through the presence of little cylindrical-shaped domes and their separating depressions. The elevations arise from one another and there is great irregularity in shapes, sizes and degrees of divergence.
>
> "The interior structure consists of very thin and closely placed concentric laminae and these repeat the irregularities of the exteriors. A depression on the surface is apt to be continued into the interior by small lines or streaks of fine sand and mud. These streaks are interpreted as arising from small quantities of mud and sand becoming lodged in the depressions between the domes. Some tube-like structures which penetrate portions of the interior may have been produced by the boring of annelids or mollusks, or they may be molds of the thalli of algae. These are also filled with fine sand and mud. The laminae vary in thickness; at one place eight were distinguished in a thickness of 5 millimeters; at another place only five laminae are present in the same thickness." (Twenhofel, 1919, p. 348-349).

Specimens studied by Twenhofel show little in the way of microstructure beyond the laminae or growth layers. Johnson (1946) described some specimens formed of laminae composed of a felt of thread-like filaments, 0.009 mm to 0.017 mm in diameter, with an average of 0.0115 mm. The threads are arranged more or less parallel to the surface in the thin laminae, but in the thicker laminae they may also form rounded or dome-shaped masses of radiating threads. Branching of filaments is common in the latter, and tufts of branches may occur.

Remarks—

This genus is described in detail and Twenhofel's original description is quoted because this was one of the first fossil algae to be described as such from Late Paleozoic rocks, and Twenhofel's description and remarks concerning it have been quoted (and misquoted) in numerous publications.

At the type locality near Otto, Kansas, colonies of this form are so abundant as to form beds as much as a foot thick which can be traced for several miles.

Pia (1927, p. 818), following Twenhofel's description, classed this genus in the family Spongiostromata on the basis of its lack of internal structure. Microstructure, however, shows that *Ottonosia* belongs in the "section" Porostromata.

Generic range—

Lower Permian.

Geographic distribution—

Kansas.

Genus *Somphospongia* Beede, 1899
Plate 89, figure 1.

Description—

This is one of the most spectacular of Pennsylvanian algae. Early geologists observed it and considered it to be of uncertain affinities. Beede (1899, p. 128) described it as a sponge, giving the following generic description:

> "A globular to mushroom-shaped calcisponge, attaining a large size, and generally possessing a more or less spherical-shaped cloaca near the base; the canals are all very irregular and crooked, distributed over the entire surface, and moderately large. A rather thick dermal layer is present. They were free, apparently resting with the base in the mud in the adult stage."

During the 1930's, Harold Hawkins, M. K. Elias, W. H. Schoewe, and others of the Kansas Geological Survey staff determined that this organism was a calcareous alga.

Johnson (1946, p. 1104) gave the following revised description:

> "Globular, egg-shaped to mushroom-shaped calcareous algae. Young colony spherical to biscuit-shaped, but as it becomes larger it tends to become inverted conical or mushroom-shaped. Surface irregular, covered with folds, crenulations, or digitate projections. Several small colonies may start separately and then coalesce."

Remarks—

Usually the colony starts as an incrustation around a nucleus consisting of a shell fragment, mud ball, or some other small foreign body. The incrustation is composed of finely laminated layers having a concentric arrangement from which digitate processes develop and grow out in all directions. After the colony attains a thickness of more than 2 cm, growth seemingly is concentrated on the upper surface, and digitate processes develop, become long, and may branch. The spaces between are filled with fine sediment.

Microscopic study shows that the colony is formed of laminae composed of fine branching algal threads enclosing sedimentary particles and some organic debris. Most threads have a diameter of 0.007 mm, but associated with them are some larger ones (diameter 0.024 mm). The laminae are irregular, alternating light and dark; some have crenulated margins. They are composed of molds of fine algal threads of very fine-grained calcite. Some fine clay and silt particles are enclosed. Probably the laminae represent seasonal growth.

Generic range—

Pennsylvanian.

Geographic distribution—

Kansas.

REFERENCES

Banks, M. R., and Johnson, J. H., 1957, *Maclurites* and *Girvanella* in the Gordon River limestone (Ordovician) of Tasmania: Jour. Paleontology, v. 31, no. 3, p. 632-640, pls. 73-74, 2 text-figs.

Beede, J. W., 1899, New fossils from the Kansas coal measures: Kansas Univ. Quart., v. 8, p. 123-130.

Cayeux, L., 1909, Les algues calcaires du groupe des *Girvanella* et la formation des oolithes: Compte Rendu Acad. Sci. Paris, v. 150, p. 359-362.

Fremy, P., and Dangeard, L., 1935, Sur la position systematique des Girvanelles: Bull. Soc. Linn. Norm., ser. 8, v. 8, p. 101-104.

Høeg, O. A., 1932, Ordovician algae from the Trondheim area: Skr. Norske Vidensk.-Akad. i Oslo, Math.-Natur. Kl., v. 1, no. 4, p. 63-96, 7 figs., 11 pls.

Johnson, J. H., 1946, Lime-secreting algae from the Pennsylvanian and Permian of Kansas: Geol. Soc. America Bull., v. 57, p. 1087-1120, 5 figs., 10 pls.

——————, 1952, *Girvanella* in Cambrian stratigraphy and paleontology near Caborca, northwestern Sonora, Mexico: Smithsonian Misc. Coll., v. 119, no. 1, p. 24-26, pl. 6.

Nicholson, H. A., 1888, On certain anomalous organisms which are concerned in the formation of some of the Paleozoic limestones: Geol. Mag., v. 25, p. 15-24.

——————, and Etheridge, R., 1880, A monograph of the Silurian fossils of the Girvan District, Ayrshire, Part 1: Scotland Geol. Survey Mem., v. 23, p. 23, pl. 9.

Pia, Julius, 1937, Die wichtigsten Kalkalgen des Jungpaläozoikums und ihre geologische Bedeutung: 2d Cong. av. etudes stratig. carbon., 1935, Compte Rendu, v. 2, p. 765-856, 2 figs., 13 pls.

Twenhofel, W. H., 1919, Pre-Cambrian and Carboniferous algal deposits: American Jour. Sci., ser. 4, v. 48, p. 339-352.

Wethered, E., 1890, On the occurrence of the genus *Girvanella* in oolitic rocks and remarks on oolitic structure: Geol. Soc. Quart. Jour., v. 46, p. 270-283, pl. 11.

Wood, Alan, 1957, The type-species of the genus *Girvanella* (calcareous algae): Paleontology, v. 1, pt. 1, p. 22-28, pls. 5-6.

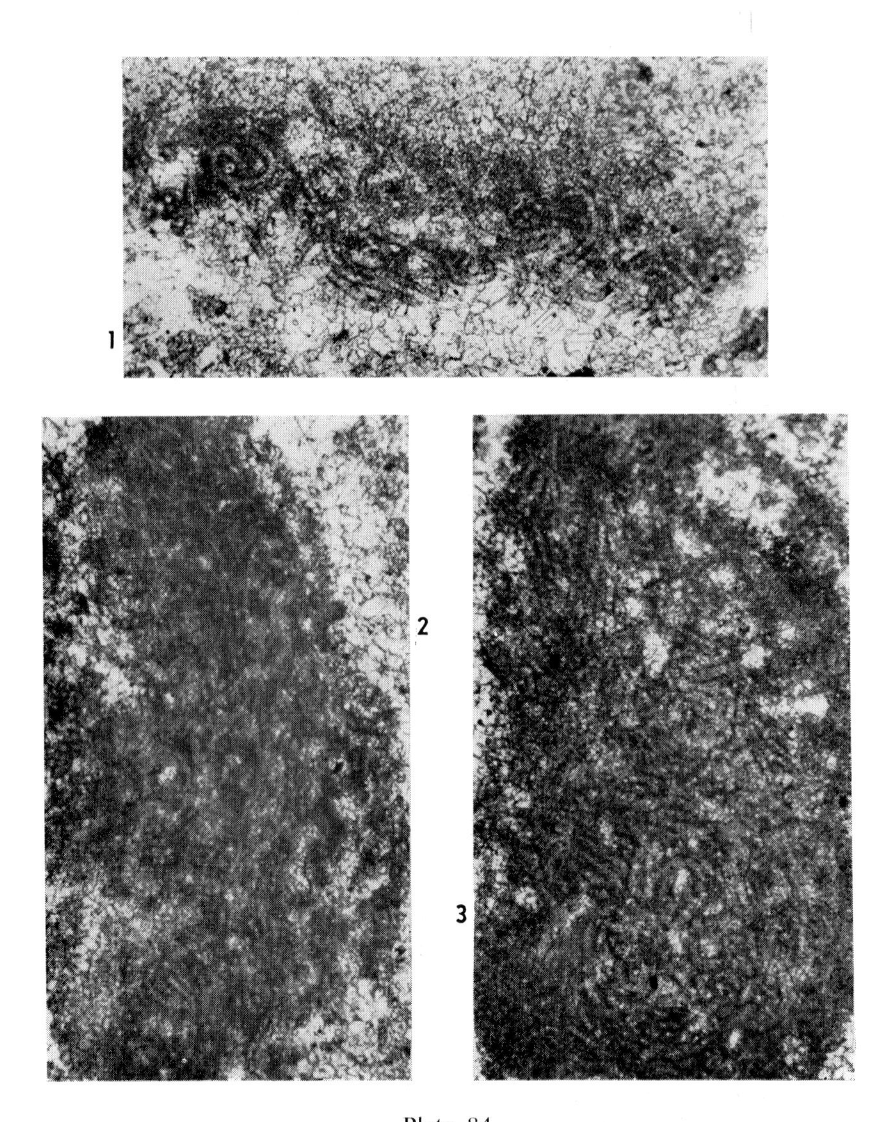

Plate 84
Genus *Girvanella*
Figures 1-2. *Girvanella problematica* Nicholson and Etheridge. (x100), Lower Silurian (?) Gazelle formation, California (Johnson and Konishi, 1959a).
Figure 3. A typical *Girvanlla* from Tasmania (x100) showing the thick-walled tubular filaments.

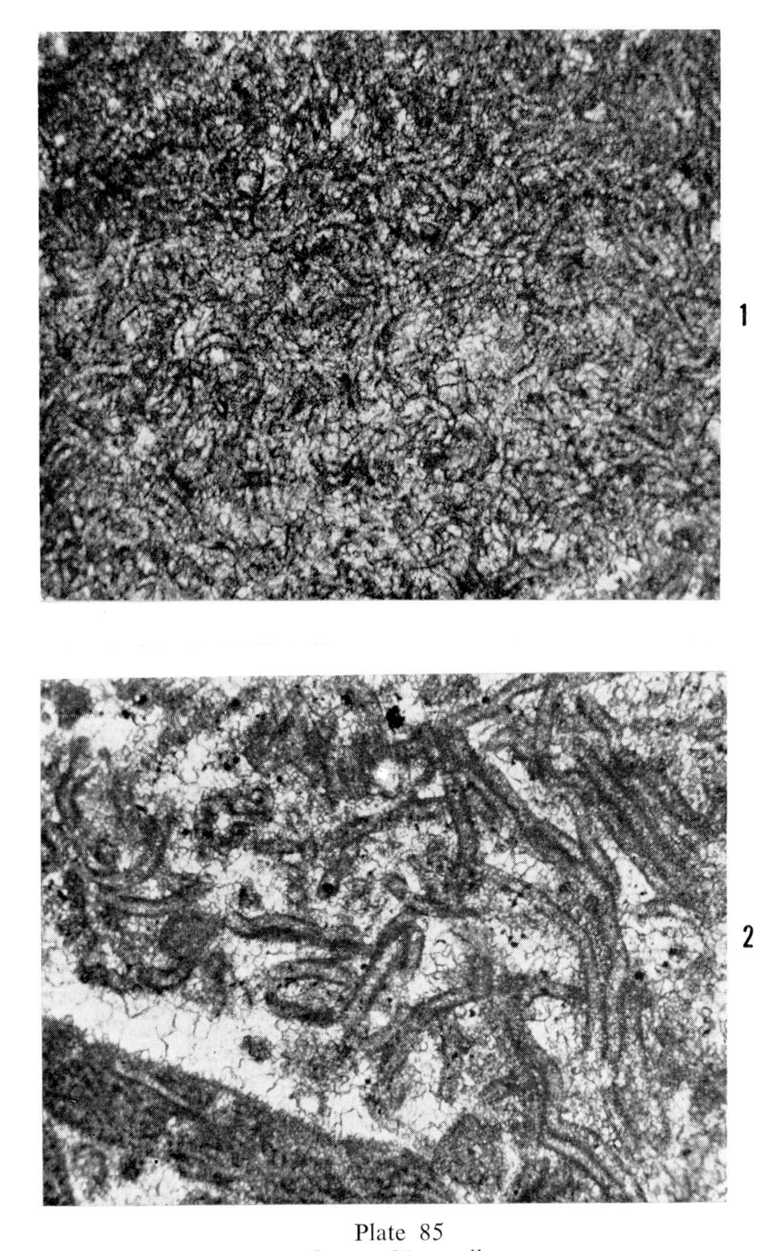

Plate 85
Genus *Girvanella*
Figure 1. *Girvanella problematica* Nicholson and Etheridge (x100).
Figures 2. *Girvanella grandis* Banks and Johnson (x100).
Ordovician of Tasmania.

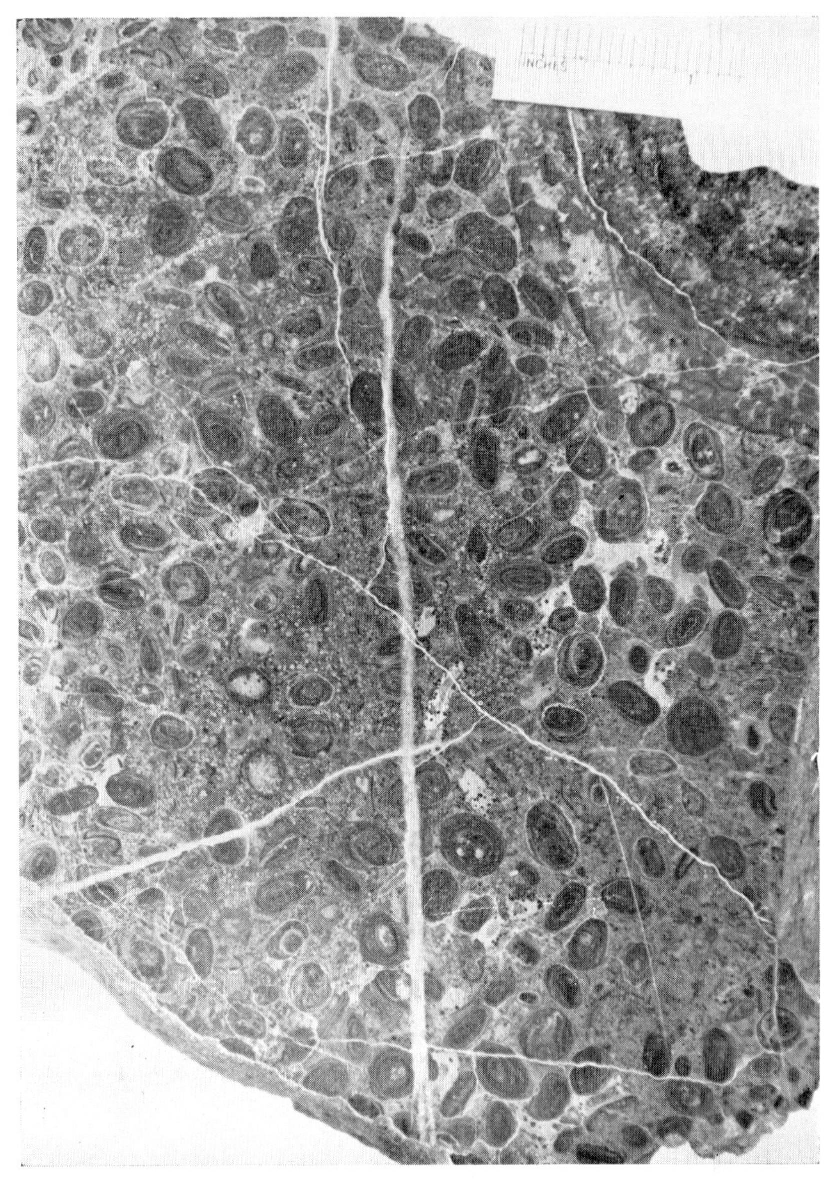

Plate 86
Genus *Girvanella*
Figure 1. Polished slab of *Girvanella* limestone (x2/3). Cambrian of Mexico, showing the characteristic pellets or bean-shaped masses of *Girvanella*. (Johnson, 1952.)

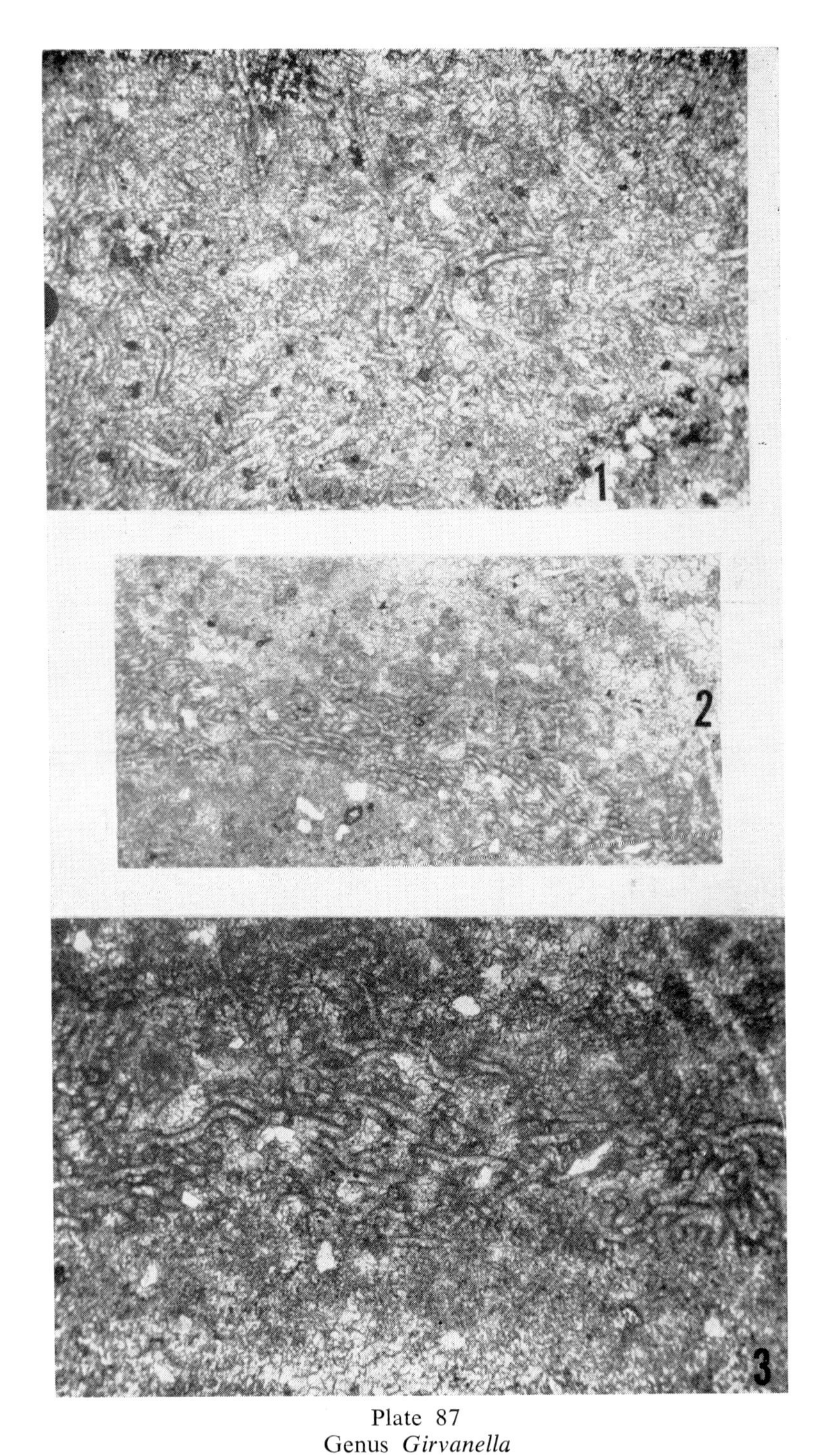

Plate 87

Genus *Girvanella*

Figure 1. *Girvanella mexicana* Johnson, a section of the rock (x75) showing the typical twisted tubes. Cambrian of Sonora, Mexico.

Figures 2-3. *Girvanella* species, Upper Cambrian, Eureka district, Nevada (2-x50, 3-x100).

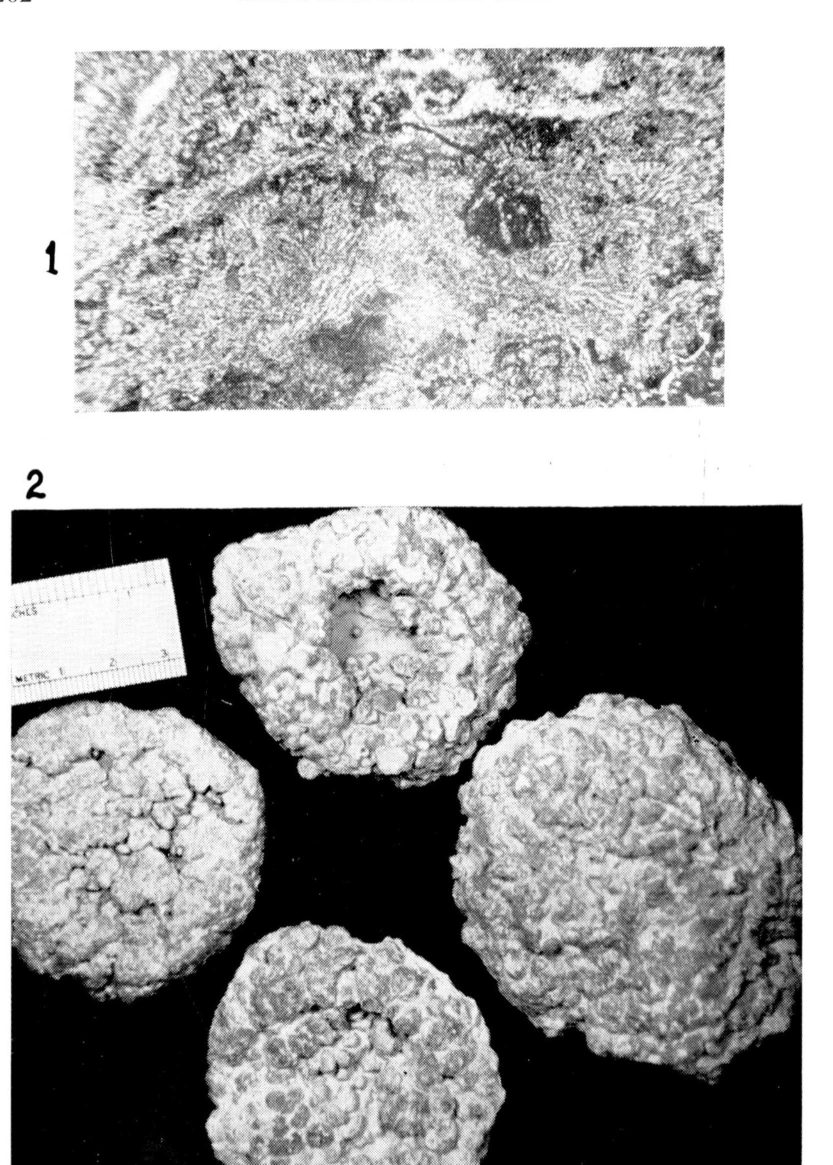

Plate 88

Genus *Ottonosia*

Figures 1-2. *Ottonosia,* Permian of Kansas. 1. A section showing microstructure (x55). 2. Four hand specimens.

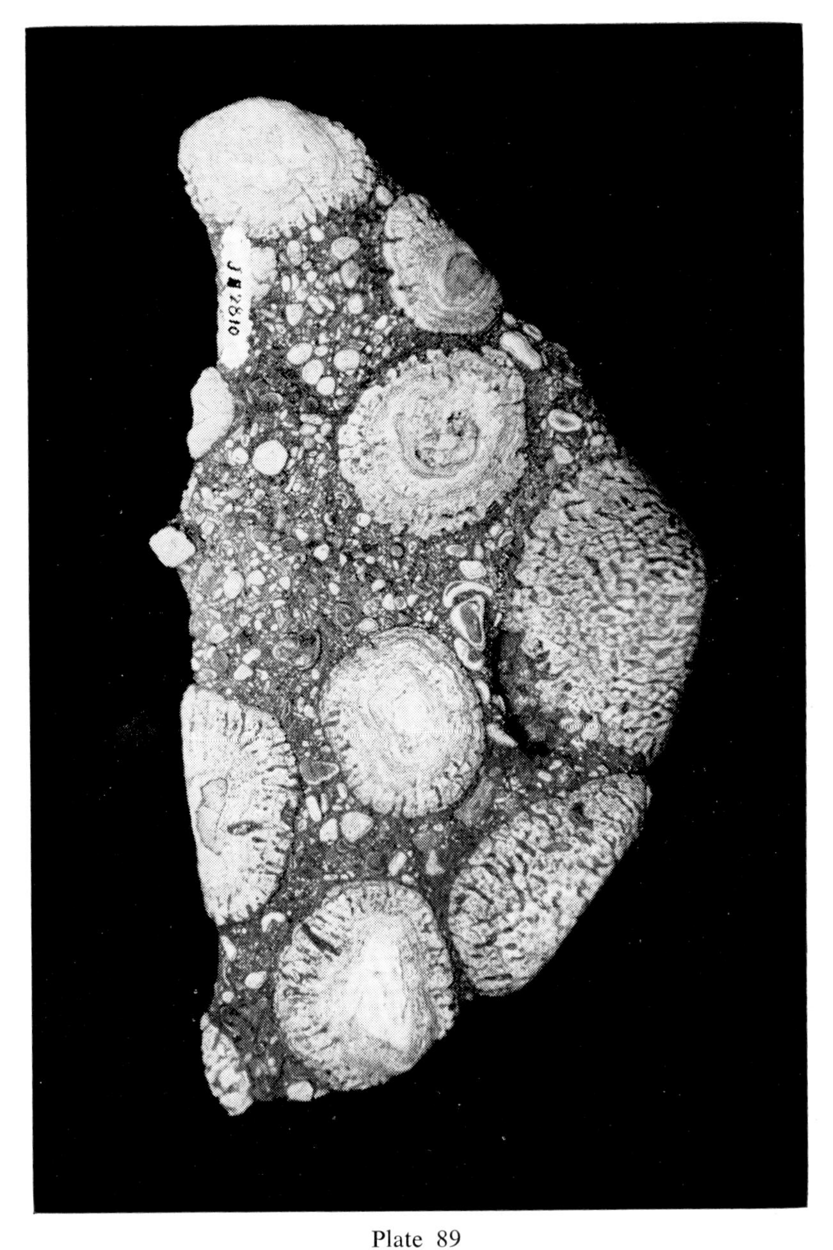

Plate 89

Genus *Somphospongia*

Figure 1. A piece of rock (x2/3) showing a number of specimens. Pennsylvanian Burlingame limestone of Kansas.

"Section" SPONGIOSTROMATA Pia, 1927
(Commonly called Stromatolites)

Pia erected this family to cover a great number of fossil forms built, or presumably built, by algae which show little or no structure but which develop colonies having a constant shape. The name was probably suggested by Gürich's genus *Spongiostroma* and family Spongiostromideae (Gürich, 1906). The term stromatolite was invented by Kalkowski in 1908 for similar fossils generally having a laminated structure, which were attributed, or considered possibly attributable, to the work of blue-green (or green) algae. During recent years, the two terms have become more or less synonymous, the term stromatolite becoming more prevalent in popular usage, especially among American geologists.

The fossils covered by these terms occur in rocks of all ages from the Early Proterozoic to the Recent. Some appear to represent natural genera; others definitely are artificial form genera, which probably were built by several species, or possibly even genera, of algae living in constant intimate association.

These fossils show no clear or usable microstructure. They consisted originally of a mat or felt of fine algal threads which was rather loose in some cases and apparently fairly dense in others. The threads were probably covered by a sticky secretion. Foreign matter such as silt and/or organic debris became entangled in the algal mat in varying quantities. In some cases, the amount of this foreign material was small as compared to the lime of the algal mat; in others, it was very large. Calcium carbonate appears to have been precipitated around the algal threads forming a mold of the algal and enclosed debris. In some cases, especially where large amounts of foreign material were present, the fossils seem to have been formed largely by mats of sediment-binding algae with little or no algal precipitated lime. These types grade imperceptibly from one into the other so that, with our present knowledge, it does not seem possible to separate them.

The fossils included under the family Spongiostromata represent a wide variety of forms from simple laminated plates or tubular masses to cabbage-like heads. As originally formed, these calcareous masses were very porous and easily affected by solution and recrystallization. The fossils commonly show definite shapes and clearly defined surface markings, either on weathered or etched surfaces. However, in petrographic slides, the most than can be seen are vague molds of fine threads and foreign particles. Commonly, the mass is so recrystallized that little or no suggestion of organic structure can be seen.

It, unfortunately, happens that fossils of this type are very common, particularly in the Precambrian, Cambrian, Ordovician, and Pennsylvanian formations. It often happens that, in a given locality, they are the only fossils observed. These fossils may attain large size and be quite spectacular, but, at best, they are very unsatisfactory and should be used only with great caution. They can give some information as to the environment of deposition of the enclosing strata, but they cannot be used in correlation except in very restricted local areas. Actually, stromatolites have been used in mapping as horizon markers in certain restricted areas of sedimentation, particularly in the Precambrian and in the Cambro-Ordovician. Cloud (1942) discussed the utility of this type of fossil and their probable environmental significance. Anderson (1950) made a similar study of the numerous fossils of this type in the Lower Carboniferous beds of northern England and southern Scotland and commented on the probable environments of deposition.

The work started by Black (1933) in the Bahamas and later continued by Newell, Cloud, and Ginsburg suggests that structures of this type are most common and reach their greatest development in very shallow water, from just above low tide level down to depths of 10 or 12 feet, although they may develop considerably deeper.

Ginsburg has been working on structures of this type both in the Bahamas and, more recently, in the Florida region. His work is yielding much information on the origin and ecological significance of fossils of this type. His papers should be considered as "must reading" for anyone interested in stromatolites.

During the last fifteen years, Russian geologists have become very interested in the Spongiostromata. In the course of geological work in Siberia, these geologists found thick Cambrian strata outcropping over large areas. Commonly, the only fossils to be found in these Cambrian beds were algal deposits of stromatolitic type. Such fossils, however, occurred in great numbers, in some places forming bioherms of various sizes, in other places forming thick beds of limestones often of great geographic extent. Naturally, the geologists wished to learn what they could about the fossils, especially their possible uses in correlation and for ecological interpretations.

Korde, in particular, made some very interesting studies. She first studied Recent deposits of Spongiostromata to see how the calcification took place and how the algal structures or traces of them would be preserved. Then studies were made of collections of similar materials from the Cambrian of Siberia. The material was compared, and the structures found in the Cambrian specimens were interpreted on the basis of the data obtained from the Recent material.

It was found that careful microscopic study of well-preserved material revealed indications of structure, and in some cases, definite traces of algal structures, such as filaments and thick-walled cells and chains of cells (Korde, 1950, p. 1109-1112), were observed. Using these data a number of genera and species were described and illustrated.

The results of these studies prove what the Fentons and other authorities have suspected for a long time; that is, that the previously described "form genera" represent the work of blue-green algae and that a single stromatolite may be formed by a number of different species or even several different genera.

This discovery means that much more study should be given to stromatolites. Ultimately, probably most of the artificial "form genera" will have to be abandoned. However, until these genera can be replaced by more solidly based "true genera," they should be retained as convenient niches and labels to be used in sorting and naming fossils of this kind.

Pia, in his 1927 treatise, divided the family Spongiostromata into two divisions. The first, which he called the *Stromatolithi,* included all forms that grew attached to the sub-stratum. The second division he called the *Oncolithi* and used it to include all unattached forms. This classification is used in the present work (see Table XII, showing the more common genera of the Spongiostromata).

TABLE XII
CLASSIFICATION OF SPONGIOSTROMATA

DIVISION	Growth habit	Common genera
Stromatolithi	Attached to the bottom or other objects	*Collenia* *Cryptozoon* (most of the Precambrian, Cambrian and Ordovician forms) *Codonophycus* *Conophyton* *Spongiostroma*
Oncolithi	Unattached forms	Most algal pisolites and small nodular forms. Some Upper Paleozoic forms attributed to "*Cryptozoon*," some "*Osagia*."

Spongiostromata as Rock Builders

The Spongiostromata have been very important as rock builders, particularly during the Late Precambrian and the Early Paleozoic. Their greatest development appears to have been during the Precambrian.

Limestone masses attributed to the work of algae of this type are known in rock as old as the Huronian. They have been found in rocks of this age in a number of localities around the upper Great Lakes, particularly Lake Huron and Lake Superior.

Spongiostromata were extremely important during the Late Proterozoic. The most thoroughly studied and, hence, best known, deposits of this type are in the Belt series of Montana, where stromatolitic algae form thick bioherms and extensive limestone beds. These can easily be seen in and around Glacier National Park. They have been described, discussed, and illustrated by Walcott (1914), the Fentons (1931, 1933, and

1937), and, more recently, by Rezak (1957). Similar deposits in the Grand Canyon region and in a number of areas in southern Arizona are known to exist but have not been adequately described.

In Asia, limestones formed by this type of algal deposit appear to be locally common and very widespread in rocks of Proterozoic age. Such limestones have been mentioned from a number of areas in India. Grabau, in his work on the stratigraphy of China (1924-1928) and in his publication on the Sinian system (1922), repeatedly mentioned the occurrence of algae and algal limestones of this type. He stressed their important development and widespread distribution in the Late Proterozoic rocks of China.

Recent studies have shown a widespread distribution of similar materials in the continent of Africa. Young (1933 and 1940) and others have described numerous examples from the Union of South Africa and other areas in southern Africa. Choubert, in a number of publications (1950 and 1952) mentions their occurrence in Morocco, while the Termiers (1949 and 1950) and others mention occurrences of stromatolitic deposits at various localities in Algeria. The greatest development of such deposits, however, occurs in and around the Congo Basin. These deposits are described by Cahen and other members of the Geological Survey of the Belgian Congo (1946). Studies made by these geologists indicate continuous outcrops of algal limestones, from 20 to 30 feet thick, which may be traced for distances of over 400 miles.

In Australia, algal limestones seem to be fairly common throughout a great thickness of sediments ranging in age from Late Proterozoic to Early Paleozoic. These limestones are mentioned by Fairbridge (1950) and other Australian geologists but so far have not been described.

Algae of this type seem to have been important rock builders during the Late Cambrian and earliest Ordovician in North America. Numerous beds of limestone formed by such algae have been reported from the Llano region of Texas (Cloud and Barnes, 1948). The famous *Cryptozoon* reefs of the Saratoga Springs region of New York are well known, and photographs of them appear in most textbooks of geology. Similar deposits also occur through Vermont and the Maritime Provinces of Canada and extend southward through Pennsylvania, Maryland, and Virginia. They have also been reported from the northern Mississippi Valley, particularly in Wisconsin and Minnesota, and farther south in Missouri.

Subsection STROMATOLITHI

Genus *Aphrostroma* Gürich, 1906

Plate 90, figure 1.

Description—

Colonies nodular or encrusting, about the size of a nut or slightly larger. Colonies formed of irregular layers, alternately loose and more compact. The layers seem to represent molds of dense growths of relatively coarse algal filaments enclosing an appreciable amount of foreign matter, although only faint or vague suggestions of the algal threads are seen in thin sections.

Remarks—

Gürich considered these organisms to be colonial animals somewhat similar to stromatoporoids, and in his original description, he interpreted the visible structure in terms of layers, pillars, and canals.

Generic range—

Mississippian (Visean).

Geographic distribution—

Belgium.

Genus *Collenia* Walcott, 1914

Plate 91, figure 1; plate 92; plate 115, figure 1.

Description—

The original description given by Walcott (1914, p. 110) is as follows:

> "More or less irregular dome-shaped turbinate or massive, laminated bodies that grew with the arched surface uppermost. The growth appears to have been by the addition of external layers of lamellae of varying thickness with interspaces that vary greatly even in the same specimen."

Rezak (1957, p. 133) added the following information:

> "Colones begin as incrustations on a surface of the substratum and grow upward by addition of convex upward laminae. Gross form cylindrical or hemispheroidal."

Generic range—

Precambrian to Silurian (?).

Geographic distribution—

Practically world wide. Abundant in Precambrian of Montana, British Columbia, China, Australia, South Africa, Congo, Morocco, Algeria, and in Cambrian of the Llano uplift of Texas, Colorado, Missouri.

Genus *Codonophycus* Fenton and Fenton, 1939

Plate 93, figures 1-2.

Description—

Bell-shaped conical or columnar algal colonies which are connected with neighboring colonies by convex or undulating areas. In some cases, two or more connected colonies rise from a single convex basal area. Structure consists of laminae and indistinct, horizontally radial pillars.

Remarks—

Locally, these are rock builders. *Codonophycus* biostromes occur in the Mississippian Madison limestone at many exposures along the western slopes of the Big Horn Mountains in Wyoming.

Generic range—
Mississippian (Madison limestone).

Geographic distribution—
Wyoming.

Genus *Cryptozoon* Hall, 1883
Plate 91, figure 2; plate 94; plate 95; plate 115, figure 2.

Description—
The typical *Cryptozoon* forms a flattened spherical (cabbage-shaped) colony with definite growth laminae. Only rarely can any microstructure be observed; the most that can be seen are vague molds of algal threads.*

Remarks—
In general appearance and structure *Cryptozoon* and young *Collenia* are very similar, but the two differ in that *Collenia* develop digitate or turbinate extensions from the upper surface. These extensions do not develop in *Cryptozoon.* In a few cases, *Cryptozoon* appear to bud. The buds develop along the sides and form small rounded forms similar in character to the parent but of much smaller size.
Cryptozoon colonies commonly attain diameters of 12 to 18 inches and may grow much larger. Locally, they are important rock builders, forming layers of limestones and bioherms.

Generic range—
Precambrian to Permian (?).

Geographic distribution—
Practically world wide, particularly in the Late Proterozoic and Cambrian.

Genus *Malacostroma* Gürich, 1906
Plate 98, figure 1; plate 99, figure 1.

Description—
Colonies form laminated masses, commonly hemispherical or turreted. In some cases, the colonies rise from a nearly flat base, and nearly flat areas may connect several turrets or domes. The laminations show an alternation of dense and loose layers. The laminae consist of dark compact granules which are tightly crowded together in the dense layers and are separated and easily distinguished in the loose layers.

Remarks—
The structural features observed in this genus are similar to those seen in Recent shallow-water algal deposits formed by lime-depositing and some sediment-binding algae. Similar structures appear to be rather common in Late Paleozoic deposits the world over. They

*Wieland (1914) reported observations on microstructure including possible sporangia, but in a study of several hundred slides from over twenty localities, the present author has not been so fortunate.

have been described or mentioned under a variety of names by a number of writers.

Generic range—

Mississippian to Permian. Probably started earlier.

Geographic distribution—

Belgium, New Mexico, Siberia.

Genus *Pycnostroma* Gürich, 1906

Plate 100; plate 101, figures 1-2; plate 124, figures 1-2.

Description—

Concentric crusts forming undulating arcs, sometimes nearly flat with parallel layers. Crusts 1/2 to 1 mm thick. Tissue fine, composed of tubular or thread-like elements 1/50 mm thick, arranged in concentric layers, although sometimes showing a vertical arrangement. Often indications of a radial arrangement of the threads suggested inside the layers.

Remarks—

This form also was originally interpreted by Gürich as a colonial animal.

Generic range—

Mississippian to Permian (?).

Geographic distribution—

Belgium, Russia, Siberia.

Genus *Spongiostroma* Gürich, 1906

Plates 103 and 104.

Description—

Incrustations and irregular crusts composed of irregular layers which are alternately compact and loose. In thin section, these layers appear as an aggregate of dark granules cut by numerous short, thick tubes which branch frequently at a fairly wide angle. The tubes are of various sizes, and appear to be circular in section.

Remarks—

This genus served as the basis for Pia's family Spongiostromata. It was originally classed in the animal kingdom by Gürich.

Generic range—

Devonian to Pennsylvanian.

Geographic distribution—

Belgium, Canada, Colorado, New Mexico.

Genus *Tetonophycus* Fenton and Fenton, 1939

Plate 107, figures 1-2.

Description—

The following description was given by the Fentons (1939, p. 99):

"Calcareous algae forming massive biscuit shaped colonies, which consist of closely compacted or confluent columns and subconical mammillae. In the one known species, each conical mass consists of a basal concavo-convex dome formed of radiating columns divided by irregular spaces, and an outer zone composed of irregularly laminated mammillae whose outer surface is channeled and pitted."

Remarks—

These algae are sufficiently abundant locally in Grand Teton National Park to form limestones or biostromes. Individual colonies attain sizes of 2 to 5 feet in diameter and 2 feet in height.

Generic range—

Upper Cambrian (Pilgrim formation).

Geographic distribution—

Grand Teton National Park in Wyoming and adjoining areas in Wyoming and Montana.

Subsection ONCOLITHI

Genus *Gouldina* Johnson, 1940

Plate 96, figure 1.

Description—

Calcareous algal masses, more or less circular in plan. They begin with a gently arched basal dome of undulating laminae and have an outer portion consisting of arched layers of irregular mammilae or short finger-like growths.

Remarks—

Gouldina resembles *Shermanophycus* in the early stages of growth, but differs in that the basal dome is overgrown by layers of small irregular mammillae or short digitate processes. The layers develop a larger and relatively higher colony which commonly is widest just below the top. Colonies may attain a diameter of a foot or slightly more, with a height of 8 to 10 inches. Locally, *Gouldina* is a limestone builder.

Generic range—

Upper half of the Lower Pennsylvanian.

Geographic distribution—

Central Colorado.

Genus *Leptophycus* Johnson, 1940

Plate 97, figures 1-2.

Description—

Calcareous algae forming small, loosely knit, nearly spherical colonies, composed of conical, bell-shaped, or plumose individuals, each built of thin arched laminae.

Remarks—

Differs from *Shermanophycus, Gouldina,* and *Stylophycus* by forming very loosely knit colonies with the individual plants more or less separated from each other. Also, the colonies are smaller and more irregular. Locally, this genus forms beds of limestone.

Generic range—
Pennsylvanian (Belden shale).

Geographic distribution—
Central Colorado.

Genus *Shermanophycus* Johnson, 1940
Plate 102, figure 1.

Description—
Calcareous algae which form rounded biscuit-shaped colonies consisting of a central basal mass of thin irregular concentric layers, surrounded by an irregular zone of laminae, surmounted by finger-like branching growths of arched laminae. The surface is irregularly pitted and looks much like that of a small cauliflower. Colonies commonly 2 to 4 inches in diameter and 1 1/2 to 2 1/2 inches high. Upper surface gently domed.

Remarks—
At the type locality, *Shermanophycus* occur as loose colonies and as as thin limestone masses containing many colonies, surrounded by black shale.

Generic range—
Pennsylvanian (Belden shale). Essentially in the upper half of the Lower Pennsylvanian.

Geographic distribution—
Colorado.

Genus *Stylophycus* Johnson, 1940
Plates 105 and 106.

Description—
Calcareous algae which form wide colonies roughly circular in outline. The upper surface may be gently arched or concave. The colony consists of a gently arched basal portion formed of thin concentric laminae and a larger outer portion of layers of relatively long digitate processes.

Remarks—
Differs from *Gouldina* and *Shermanophycus* in having a relatively larger basal portion (1/3 to 1/2 the diameter and about 1/4 to 1/3 the height of the colony) and in having the upper portion formed of longer finger-like growths which often have some space between them.

Generic range—
Pennsylvanian, mainly the upper half of the Lower Pennsylvanian.

Geographic distribution—
Central Colorado.

REFERENCES

Anderson, F. W., 1950, Some reef-building calcareous algae from the Carboniferous rocks of northern England and southern Scotland: Yorkshire Geol. Soc. Proc., v. 28, pt. 1, p. 5-27, illus.

Bass, N. W., and Northrop, S. A., 1953, Dotsero and Manitou formations, Colorado: Am. Assoc. Petroleum Geologists Bull., v. 37, no. 5, p. 889-904.

Black, Maurice, 1933, The algal sediments of Andros Island, Bahamas: Royal Soc. (London) Philos. Trans., ser. B, v. 222, p. 165-192, pls. 21-22.

Cahen, L., Jamotte, A., Lepersonne, J., and Mortelmans, G., 1946, Note preliminaire sur les algues der series calcaires anciennes du Congo Belge: Service Geol. Congo Belge Bull., no. 2, fasc. II, p. 171-236, 23 figs.

Choubert, G., du Dresnay, R., and Hindermeyer, J., 1950, Sur les calcaires a *Collenia* de la region Safsaf-Ain Chair: Service Geol. du Moroc, Notes et Memoirs, no. 76, p. 93-103, 4 pls.

————, Hindermeyer, J., and Holland, H., 1952, Note preliminaire sur les *Collenia* de l'Anti-Atlas: Service Geol. du Moroc, Notes, v. 6, p. 85-102, pls. 1-12.

Cloud, P. E., Jr., 1942, Notes on stromatolites: Am. Jour. Sci., v. 240, p. 363-379, pls. 1-2.

————, and Barnes, V. E., 1948, The Ellenberger group of central Texas: Texas Univ. Bur. Econ. Geol. Pub. 4621, 473 p., 45 pls.

Cushing, H. P., and Ruedemann, R., 1914, Geology of Saratoga Springs and vicinity: New York State Mus. Bull. 169.

Fairbridge, R. W., 1950, Precambrian algal limestones in western Australia; Geol. Mag., v. 87, no. 5, p. 324-330, pl. 15.

Fenton, C. L., and Fenton, M. A., 1931, Algae and algal beds in the Belt series of Glacier National Park: Jour. Geol., v. 39, p. 670-686.

————, 1933, Algal reefs or bioherms in the Belt series of Montana: Geol. Soc. America Bull., v. 44, p. 1135-1142.

————, 1937, Cambrian calcareous algae from Pennsylvania: Am. Midland Naturalist, v. 18, p. 435-441, 3 pls. 1 fig.

————, 1937, Belt series of the north: stratigraphy, sedimentation, paleontology: Geol. Soc. America Bull., v. 48, p. 1873-1970, 19 pls., 20 figs.

————, 1939, Precambrian and Paleozoic algae: Geol. Soc. America Bull., v. 50, p. 89-126, 11 pls.

Ginsburg, R. N., Isham, L. B., Bein, S. J., and Kuperberg, Joel, 1954, Laminated algal sediments of south Florida and their recognition in the fossil record: Final report to the National Science Foundation, no. 54-20, Univ. of Miami Marine Lab., 33 p., 16 figs.

Goldring, Winifred, 1938, Algal barrier reefs in the lower Ozarkian of New York with a chapter on the importance of coralline algae as reef builders through the ages: New York State Mus. Bull. 315, p. 1-75, 22 figs.

Grabau, A. W., 1922, The Sinian system: Geol. Soc. China Bull., v. 1, p. 44-88.

————, 1924-28, Stratigraphy of China, part 1: p. 1-528, 306 text-figs., 6 pls. (reprinted 1928).

Gürich, Georges, 1906, Les spongiostromides du Viseen de la province de Namur: Mus. Royal Hist. Nat. de Belgique Memoires, v. 3, 55 p., 23 pls.

Johnson, J. Harlan, 1940, Lime-secreting algae and algal limestones from the Pennsylvanian of central Colorado: Geol. Soc. America Bull., v. 51, no. 4, p. 571-596, 10 pls.

————, 1942, Permian lime-secreting algae from the Guadalupe Mountains, New Mexico: Geol. Soc. America Bull., v. 53, p. 195-226, 7 pls., 5 figs.

————, 1943, Limestones formed by plants: Mines Mag., v. 33, p. 527-533, 15 illus.

————, 1946, Late Paleozoic algae of North America: Am. Midland Naturalist, v. 36, no. 2, p. 264-274, 2 pls.

——————, 1946, Lime-secreting algae from the Pennsylvanian and Permian of Kansas: Geol. Soc. America Bull., v. 57, p. 1087-1120, 10 pls., 5 figs.

Kalkowski, Ernst, 1908, Oolith und Stromatolith in norddeutschen Buntsandstein: Deutsche Geol. Gesell. Zeitschr., v. 60, p. 68-125, figs. 1-3, pls. 4-11.

Korde, K. B., 1950, Structure microscopique des strates de Stromatolithes et types de conservation des Cyanophyceae fossiles: (In Russian) Comptes Rendus Acad. Sci. U.S.S.R., v. 71, no. 6, p. 1109-1112, 4 figs.

MacGregor, A. M., 1940, A Precambrian algal limestone in southern Rhodesia: Geol. Soc. South Africa Trans., v. 43, p. 9-15, 2 figs., pls. 2-5.

Newell, N. D., et al, 1951, Shoal-water geology and environments, Eastern Andros Islands, Bahamas: Am. Mus. Nat. Hist. Bull., v. 97, art. 1, p. 5-29.

Pia, Julius, 1927, Thallophyta *in* Hirmer, M., Handbuch der Paläobotanik: Munchen und Berlin, v. 1, p. 1-136.

Rezak, Richard, 1957, Stromatolites of the Belt series in Glacier National Park and vicinity, Montana: U. S. Geol. Survey Prof. Paper 294-D, p. 127-154, pls. 19-24.

Roddy, H. J., 1915, Concretions in streams formed by the agency of blue-green algae and related plants: Am. Philos. Soc. Proc., v. 54, no. 218, p. 246-258, 2 figs.

Stauffer, C. R., 1945, *Cryptozoons* from the Shakopee dolomite (Minn.): Jour. Paleontology, v. 19, no. 4, p. 376-379, pl. 58.

Steinmann, G., 1911, Über *Gymnosolen ramsayi*, eine Coelenterate von der Halb'nsel Kanin: Soc. Geol. Finlande Bull., v. 31, p. 18-23, pl. 3.

Termier, H., and Termier, G., 1949, Les sediments antecambriens et leur pauvrete en fossiles: La Revue Scientifique, Annee 87, fasc. 2, no. 3302, p. 74-84.

——————, 1950, Affinities des faunes nord-africaines au cours de l'ere primaire: La Revue Scientifique, Annee 88, fasc. 4, no. 3308, p. 209-216.

Twenhofel, W. H., 1919, Precambrian and Carboniferous algal deposits: Am. Jour. Sci., ser. 4, v. 48, p. 339-352.

Vologdin, A. G., 1955, Explication de l'origine des Stromatolithes: (In Russian) Priroda, U.S.S.R., no. 9, p. 39-46, 8 figs.

Walcott, C. D., 1884, Genus *Cryptozoon* Hall: Smithsonian Misc. Coll., v. 57, p. 257.

——————, 1914, Pre-Cambrian algal flora: Smithsonian Misc. Coll., v. 64, no. 2, p. 77-156, pls. 4-23.

Young, R. B., 1933, The occurrence of stromatolitic or algal limestones in the Campbell Rand series, Griqualand West: Geol. Soc. South Africa Trans., v. 35, p. 29-36, 1 fig., 2 pls.

——————, 1940, Further notes on algal structures in the Dolomite series: Geol. Soc. South Africa Trans., v. 43, p. 17-22, 1 fig., 3 pls.

——————, and Mendalssohn, E., 1948, Domed algal growths in the Dolomite series of South Africa, with associated fossil remains: Geol. Soc. South Africa Trans., v. 51, p. 53-62, pls. 7-15.

Plate 90
Genus *Aphrostroma*
Figure 1. *Aphrostroma tenerum* Gürich. A section (x5). Mississippian of Belgium. (From Gürich, 1906).

Plate 91

Genera *Collenia* and *Crytozoon*

Figure 1. *Collenia undosa* Walcott from the Precambrian, Missoula group, near Logan Pass, Glacier National Park, Montana.

Figure 2. *Cryptozoon occidentales* Dawson, Missoula group (Proterozoic) near Nyak, Montana.

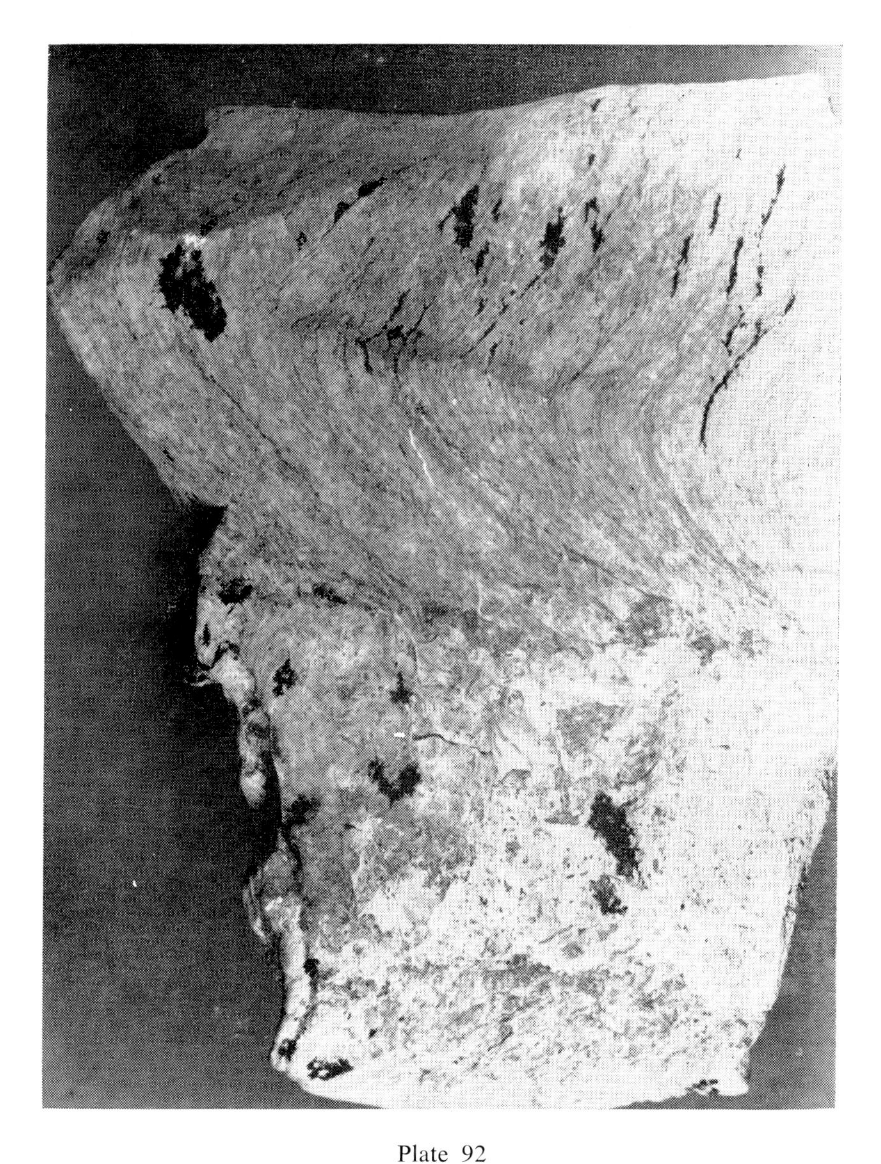

Plate 92
Genus *Collenia*
Figure 1. *Collenia* sp. (x2/3) Upper Cambrian, Dotsero formation, Clinetop algal member, near Trappers Lake, Colorado.

Plate 93
Genus *Codonophycus*
Figures 1-2. *Codonophycus austinii* Fenton and Fenton. Mississippian, Big Horn Mountains, Wyoming. 1. Weathered surface transverse section (x1/2). 2. Vertical weathered section (x1/2). (From Fenton and Fenton, 1939.)

Plate 94
Genus *Cryptozoon*

Figure 1. *Cryptozoon proliferum* Hall. Glaciated outcrops. 1. At Lester Park, 3 miles N.W., Saratoga Springs, New York. (Photo by M. F. Dening.) 2. At Saratoga Springs, New York (from Walcott).

Plate 95

Genus *Cryptozoon*

Figures 1, 2, 4, 5. *Cryptozoon rosemontensis* Stauffer.

1. A weathered surface showing sections of the nesting shallow saucer-like cups that make up the pillars (x1/3). 4. A polished vertical section of the same specimen as above (x1). 5. Top view of the type specimen showing the upper ends of the pillar-like or finger-like stacks of saucer-shaped masses forming the colony (x1/3). Specimens from the Shakopee dolomite at Pine Bend, Minnesota.

Figure 3. A specimen similar to that shown in figure 6 split open to show concentric structure (x1/3). Specimen from the Shakopee dolomite at Pine Bend. Minnesota.

Figure 6. A mushroom-like mass "exploded" from the surface of a large hemispherical specimen. These wart-like protuberances may be new colonies formed by a bud-like process (x1/3). From the Shakopee dolomite at Pine Bend, Minnesota.

Plate 96
Genus *Gouldina*
Figure 1. A colony of *Gouldina magna* Johnson, Pennsylvanian, Belden Shale, South Park, Colorado.

Plate 97
Genus *Leptophycus*

Figure 1. *Leptophycus* limestone. Top view.
Figure 2. Side view. Pennsylvanian of South Park, Colorado. (From Johnson, 1940.)

Plate 98
Genus *Malacostroma*
Figure 1. *Malacostroma concentricum,* (x1). Mississippian of Belgium (from Gürich, 1906).

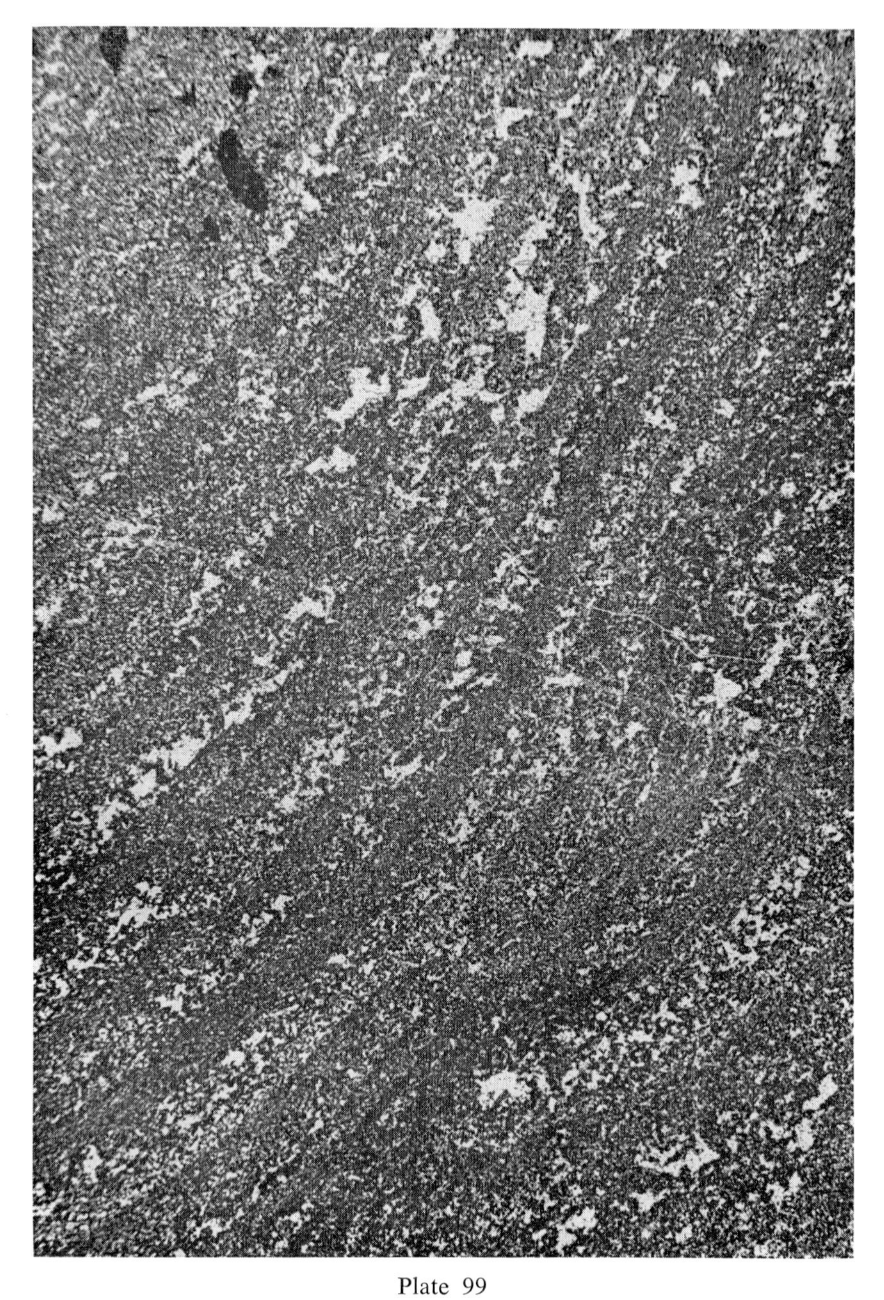

Plate 99
Genus *Malacostroma*
Figure 1. *Malacostroma concentricum* (x20) (from Gürich, 1906).

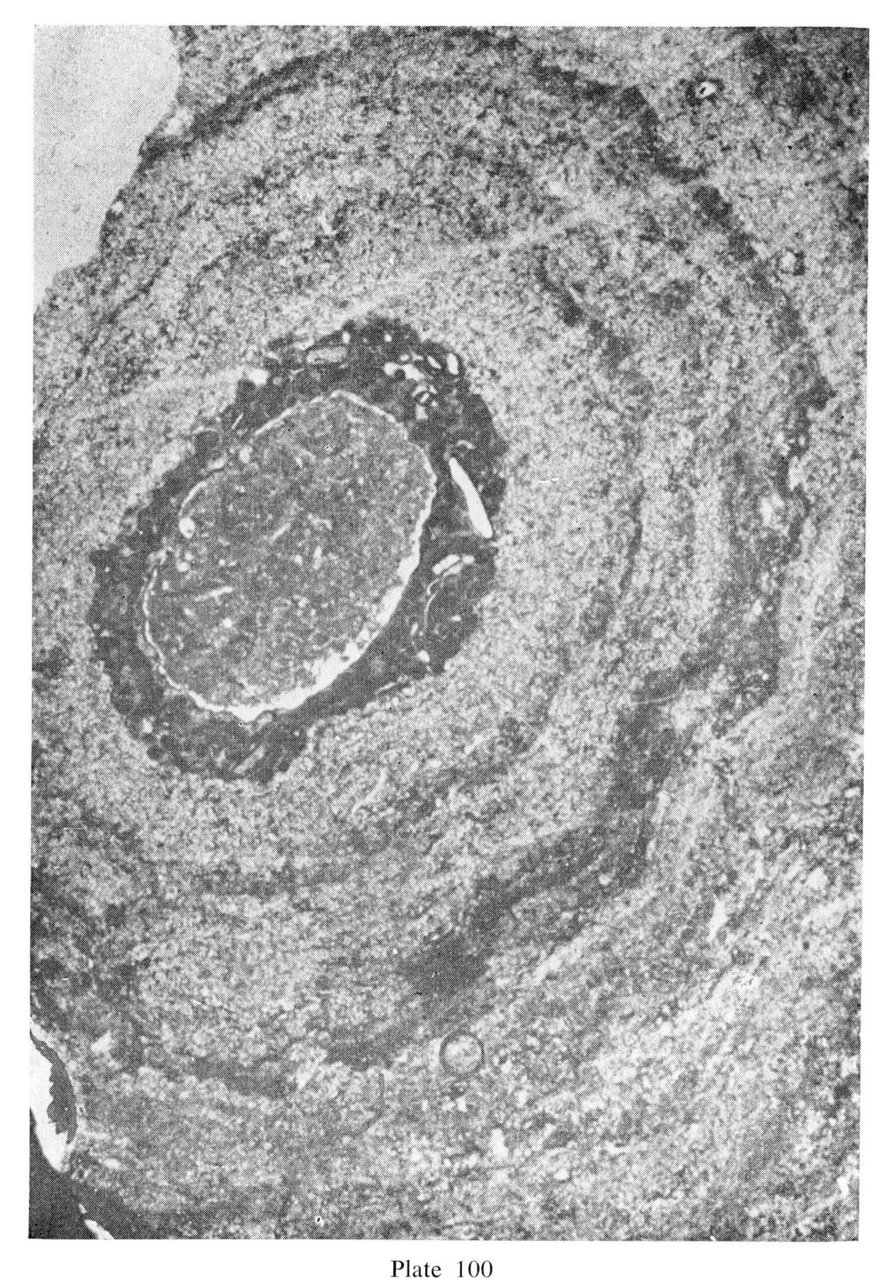

Plate 100
Genus *Pycnostroma*
Figure 1. *Pycnostroma densius* (x20) (from Gürich, 1906).

1

2

Plate 101

Genus *Pycnostroma*

Figures 1-2. *Pycnostroma* sp. Pia, (about natural size). Devonian, south of Leningrad, Russia. (After Pia, 1932).

Plate 102
Genus *Shermanophycus*
Figure 1. Top view of a weathered slab showing a number of specimens (x2/3).

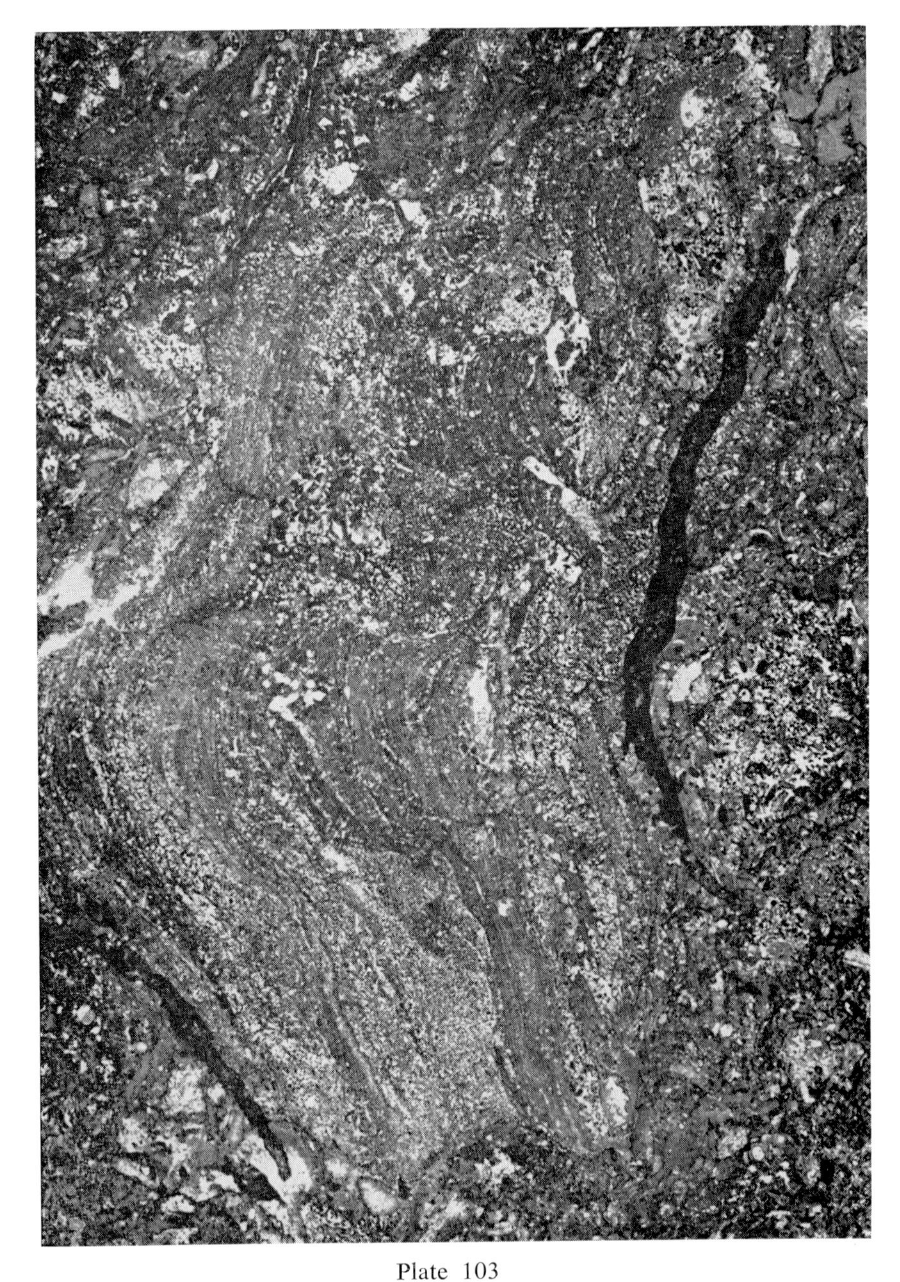

Plate 103
Genus *Spongiostroma*
Figure 1. *Spongiostroma granulosum* (x5), Mississippian of Belgium. (From Gürich, 1906.)

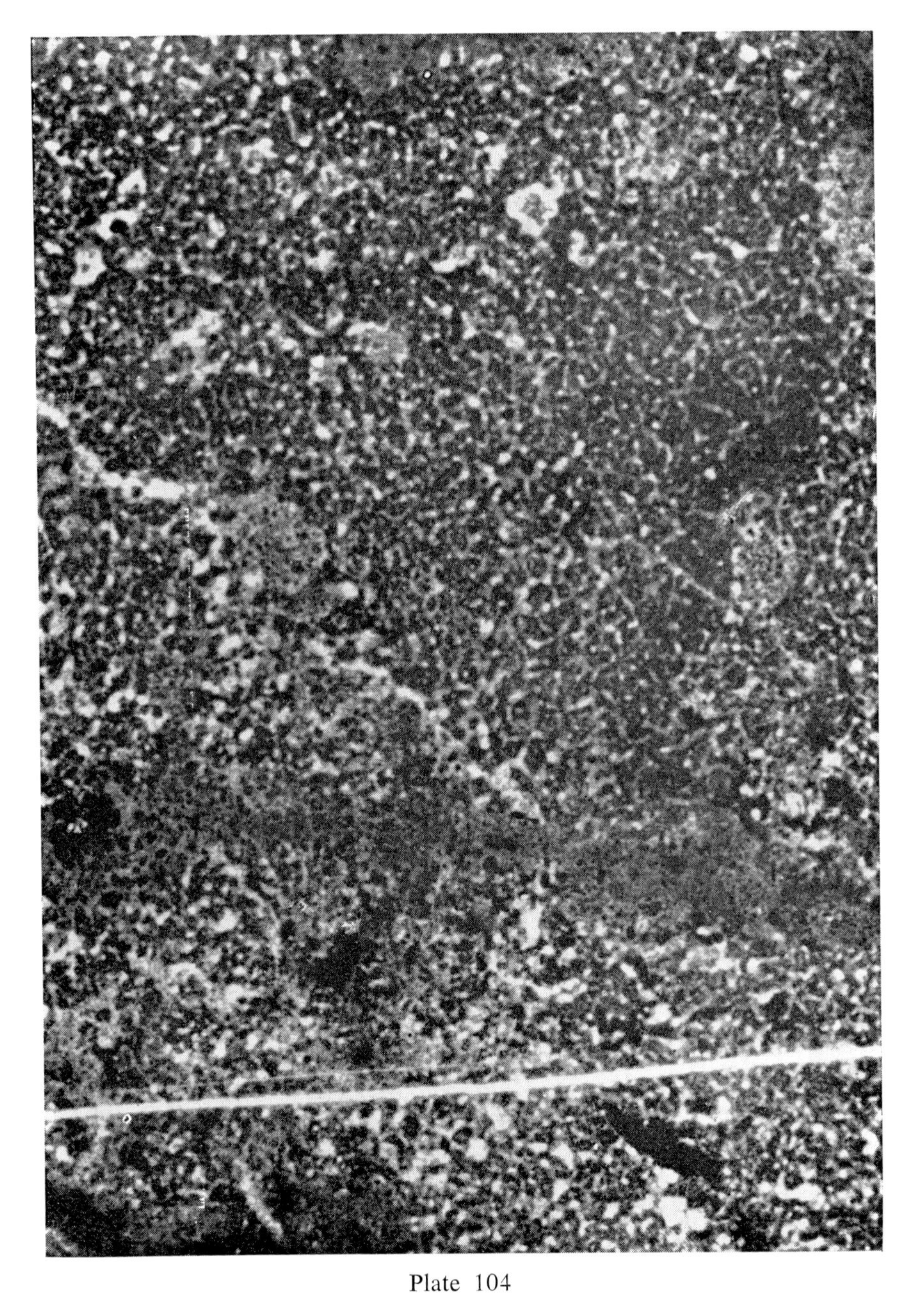

Plate 104
Genus *Spongiostroma*
Figure 1. *Spongiostroma maedrinum* Gürich. Section. (x20). (From Gürich, 1906.)

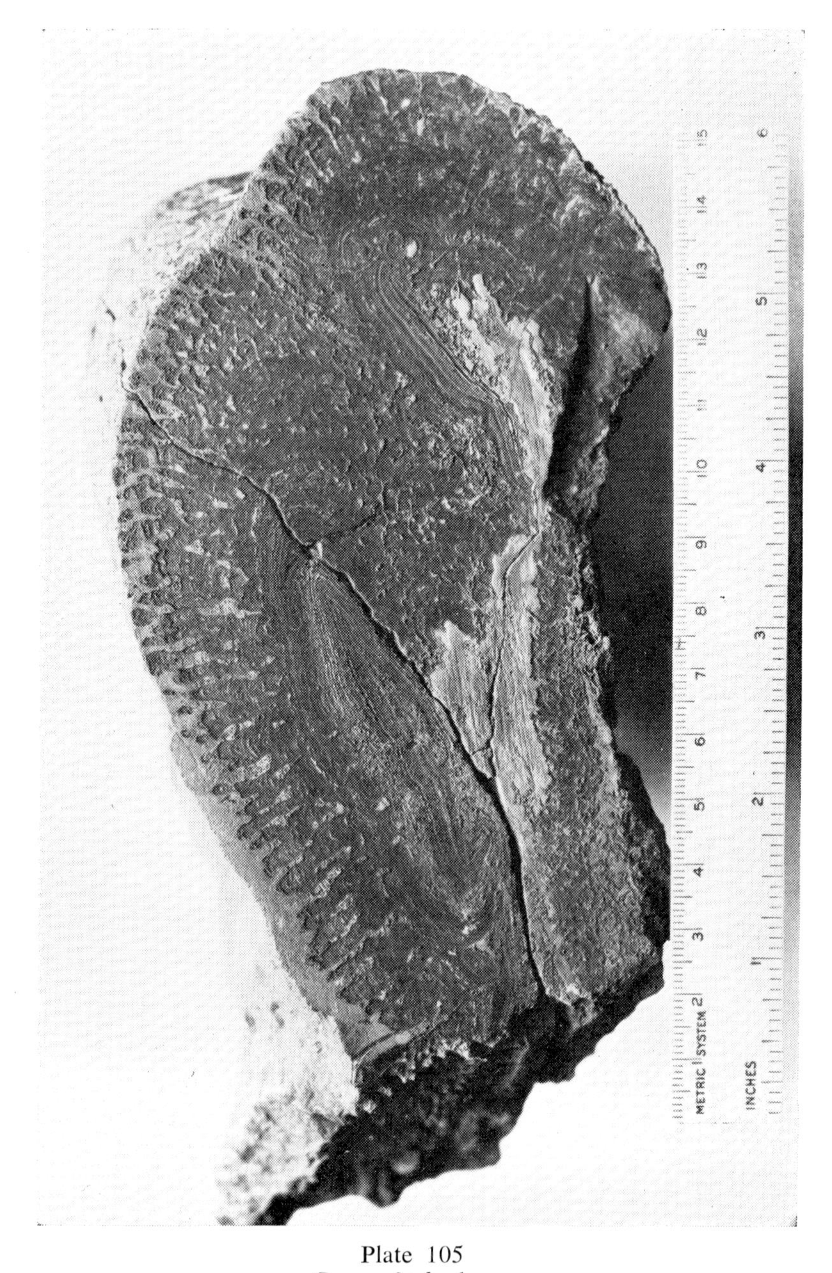

Plate 105
Genus *Stylophycus*

Figure 1. *Stylophycus calcarius* Johnson. Pennsylvanian, South Park, Colorado. (Verticle slice.)

Plate 106
Genus *Stylophycus*
Figure 1. *Stylophycus calcarius* Johnson. (Top view.) Pennsylvanian, South Park, Colorado.

1

Plate 107
Genus *Tetonophycus*
Figures 1-2. Outcrops of the limestone, Upper Cambrian, Grand Teton National Park, Wyoming.

Algae of Uncertain Affinities
Phylum RHODOPHYCOPHYTA ?
Genus *Dasyporella* Stolley, 1893
Plate 108, figures 1-4.

Description—

Thallus large, cylindrical, unbranched, often constricted, with a thick calcareous wall and a central uncalcified hollow. The peripheral calcareous wall consists of a sheaf of septated cell rows which are disposed perpendicular to the central hollow as well as to the surface of the thallus, and are coherent to each other without any space between the adjacent rows. The septation occurs at very irregular intervals. A tangential view of the peripheral cell rows reveals a characteristic hexagonal pattern.

Remarks—

Because of the septated and very compact nature of the peripheral cell rows, Johnson and Konishi (1959, p. 151-154) excluded this genus from the family Dasycladaceae.

Generic range—

Ordovician, Lower Silurian (?).

Geographic distribution—

Norway, Esthonia, Oregon.

Phylum CHLOROPHYCOPHYTA ?
Genus *Ivanovia* Khvorova, 1946
Plates 109 and 110.

Description—

Thallus forms thin irregular, sometimes wavy plates or blades. Surface smooth, or nearly so. Thallus consists of a central portion surrounded by an outer cover. Calcification commonly only affected the outer portion, hence inner structure seldom preserved. The outer portion formed of cylindrical cells arranged perpendicular to the surface.

Remarks—

There is some question as to the systematic position of this genus, but here it is provisionally put among the green algae, tentatively with the Codiaceae. Among the many specimens observed, calcification normally was limited to the outer portions of the tissue so that only the outer structure was preserved. The structure shown and the habit of calcification are similar to those of the Codiaceae. Remains of these algae are very abundant in Pennsylvanian rocks (especially Upper Pennsylvanian). Recent work has shown that they continue into the Lower Permian.

Generic range—

Upper Pennsylvanian to Lower Permian.

Geographic distribution—

West Texas and New Mexico, Moscow Basin in U.S.S.R.

Genus *Lancicula* Maslov, 1956

Plate 111, figures 1-11; plate 112, figures 1-4.

Description—

The thallus appears as a series of saucers or bowl-shaped discs surrounding a thick central stem. The upper surfaces of the discs show numerous "pores" representing the ends of internal tubes. The internal structure consists of a framework of coarse, branching tubes encased in a precipitate of calcium carbonate. In the upper disc, some of the tubes terminate in swollen ovoid bodies which may represent sporangia.

Remarks—

Externally, these plants resemble some dasyclads, but structurally, they lack the whorls of branches and resemble codiaceans much more closely.

Generic range—

Lower Devonian.

Geographic distribution—

Siberia.

Genus *Litanaia* Maslov, 1956

Plate 113, figures 1-5.

Description—

A calcified cylindrical form with irregularly arranged branches growing at an acute angle from the central stem.

Remarks—

Maslov's illustrations show that the central parts of the cylinders consist of a number of coarse, tubular strands, with the branches developing from them.

Generic range—

Lower Devonian.

Geographic distribution—

Central Siberia.

Genus *Orthriosiphon* Johnson and Konishi, 1956

Plate 114, figures 1-7.

Description—

Thallus moderately large, club-shaped, consisting of a thick medulla, with an irregularly calcified cortex. The cortex is penetrated by branching tubes, which branch dichotomously two or three times. The branches are not all in a single plane and thicken toward the ends, terminating after the second or third bifurcation in small funnels. Conceptacles subspherical in shape and subcortical in position.

Remarks—

Many codiacean algae with a dendritic growth habit closely resemble dasycladaceans and are difficult to distinguish in fossil fragments. Superficially, *Orthriosiphon* strongly resemble dasycladaceans but are distinguished from them by the following features: 1) An assymetrical and nonlinear alignment of the branching tubes. 2) The tubes (utricles) bifurcate two or three times in a very irregular pattern. 3) The conceptacles are larger than is normal for dasycladaceans and are irregularly arranged. All these features suggest that *Orthriosiphon* belong among the Codiaceae.

Generic range—

Mississippian.

Geographic distribution—

Saskatchewan, Canada.

REFERENCES

Johnson, J. H., and Konishi, Kenji, 1956, Studies of Mississippian algae: Colorado School of Mines Quarterly, v. 51, no. 4, p. 1-133, illus.

——————, 1959, Studies of Silurian (Gotlandian) algae: Colorado School of Mines Quarterly, v. 54, no. 1, pt. 3, p. 131-158.

Khvorova, L. V., 1946, On a new genus of algae from the Middle Carboniferous deposits of the Moscow basin: Acad Sci. U.R.S.S., Comptes Rendus (Dokl.), v. 53, no. 8, p. 737-739, 2 figs.

Maslov, V. P., 1956, Fossil calcareous algae in U.S.S.R.: U.S.S.R. Acad. Sci. Inst. Geol. Sci., no. 160, 301 p., 86 pls.

Stolley, E., 1893, Über silurische Siphoneen: Neues Jahrb. Mineralogie, v. 2, p. 135.

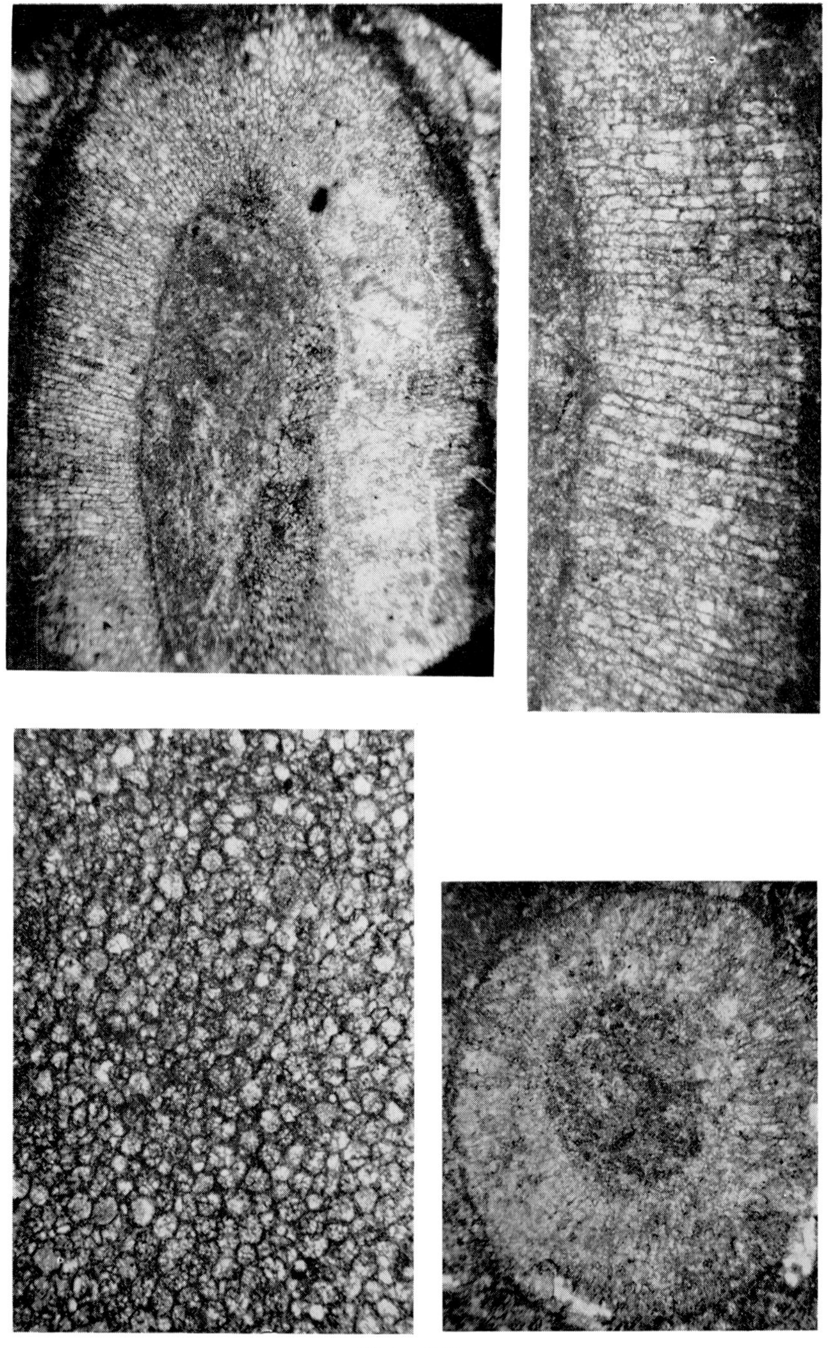

Plate 108
Genus *Dasyporella*

Figures 1-4. *Dasyporella norvegica* Høeg. Upper left: Slightly oblique, longitudinal section, indicating the large thallus, thick central stalk, and thick calcified wall. (x25) Lower Silurian (?), Gazelle formation, McGonahue Gulch, about 10 miles southwest of the Graystone area, Siskiyou County, northern California. (Slide 2367; U.S.G.S. Algae No. a784-b.) Upper right: Portion of the same thallus with the preceding figure. (x50) Notice septated cell structure. Same as the above.

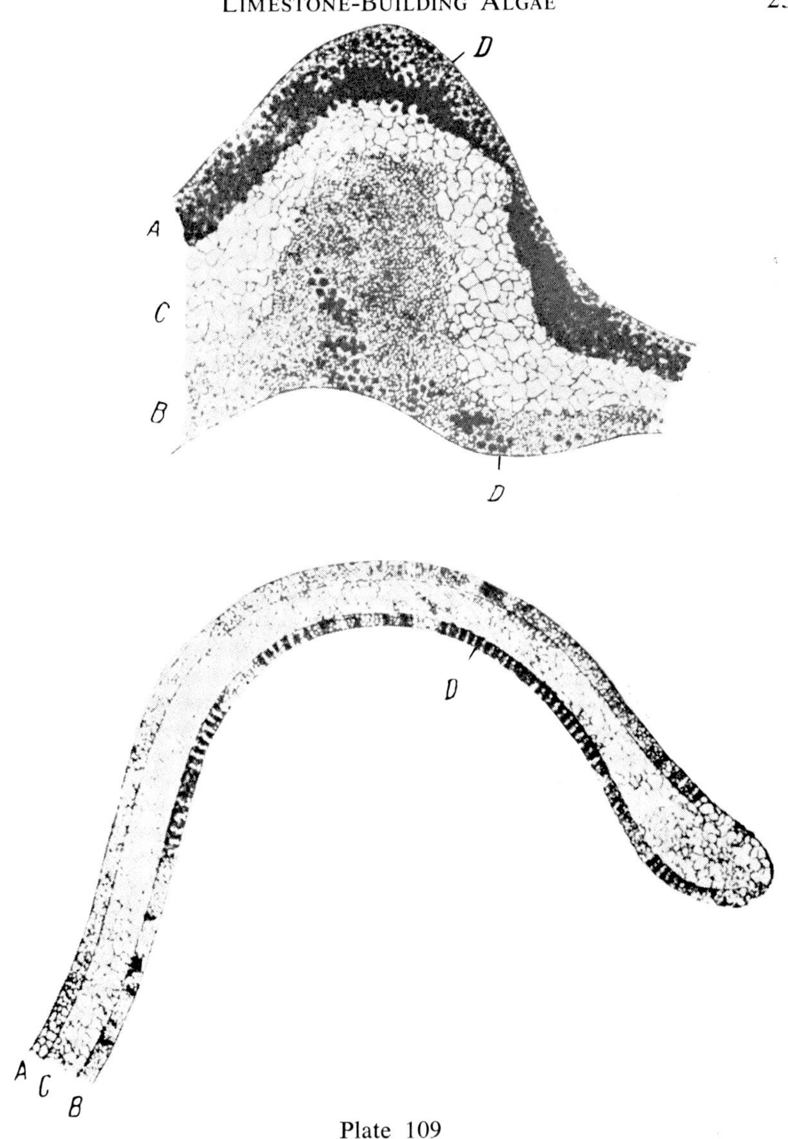

Plate 109
Genus *Ivanovia*

Figures 1-2. *Ivanovia tenuissima,* (x20), Pennsylvanian of Russia. Khvorova's original illustrations.

Plate 108 (Continued)

Lower left: Tangential section (x50). Notice hexagonal pattern. Lower Silurian (?) Gazelle formation, McGonahue Gulch, about 10 miles southwest of the Graystone area, Siskiyou County, northern California. (Slide 2366; U.S.G.S. Algae No. a783.) Lower right: Transverse section, poorly preserved (x25). Lower Silurian (?) Gazelle formation, Graystone area, 4 miles southwest of Gazelle, Siskiyou County, northern California. (Slide 2407; U.S.G.S. Algae No. a785.)

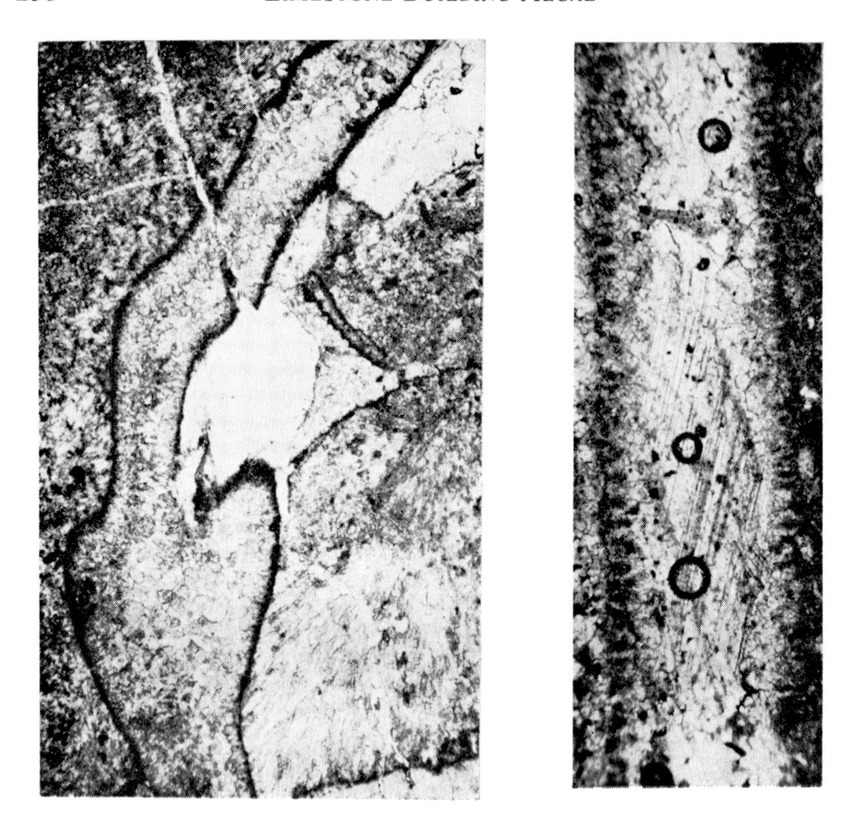

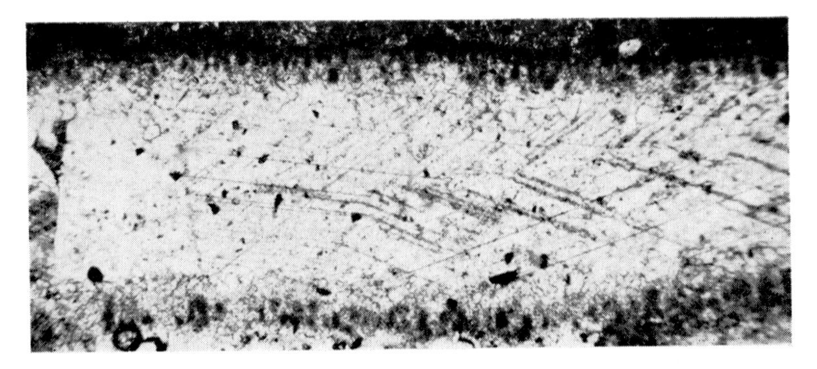

Plate 110
Genus *Ivanovia*

Figure 1. *Ivanovia* sp. (x9). Permian, Laborata formation, Sacramento Mountains, New Mexico.

Figures 2-3. *Ivanovia* sp. (x40). Pennsylvanian, Four Corners area, New Mexico.

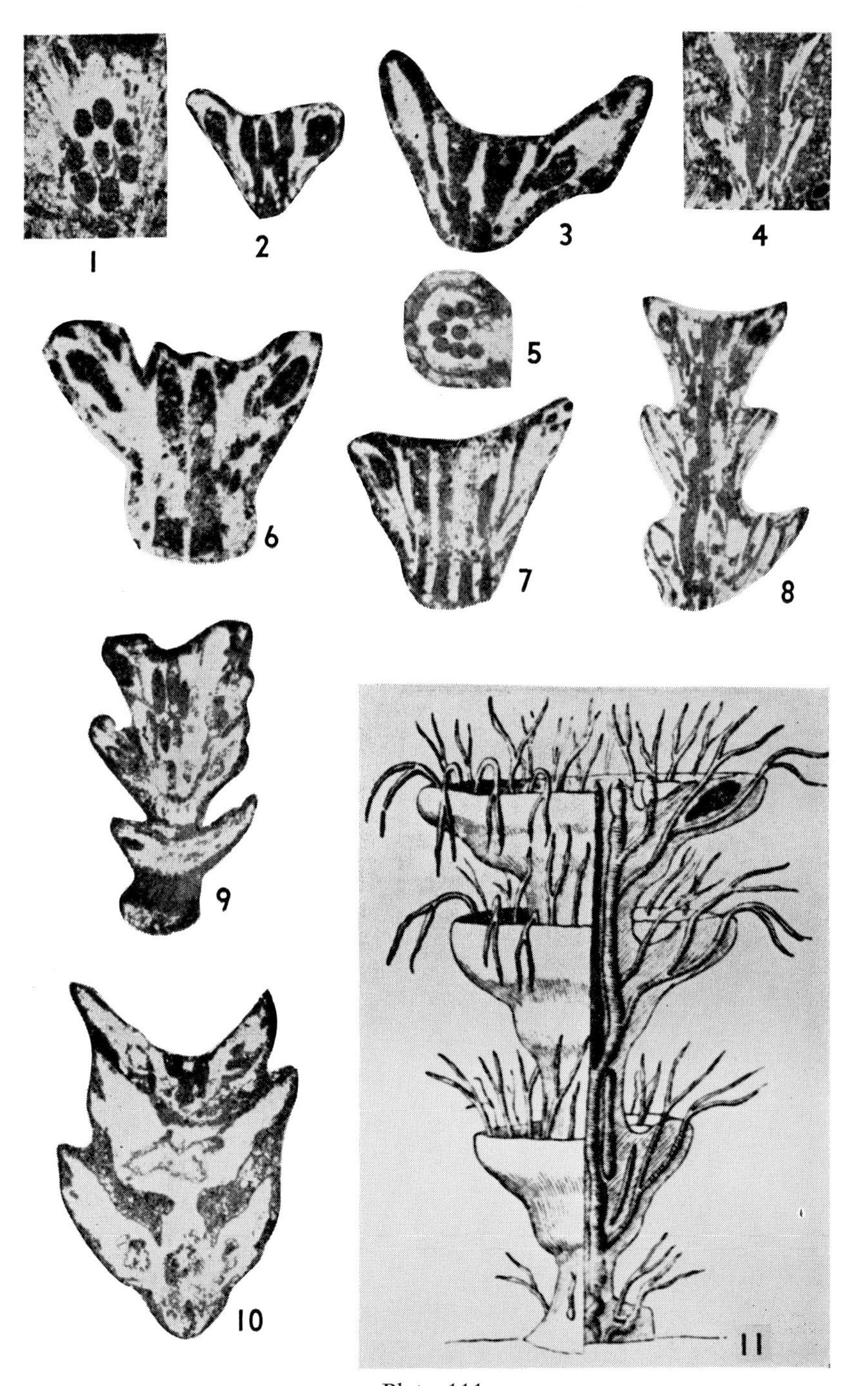

Plate 111
Genus *Lancicula*
Figures 1-11. *Lancicula alta* Maslov. 1. An oblique cross section (x20). 5. A cross section (x20). 2-4, 6-10. Longitudinal (x46). 11. A restoration of the plant—at the left it shows the external appearance, the right shows the internal structure. (From Maslov, 1956.)

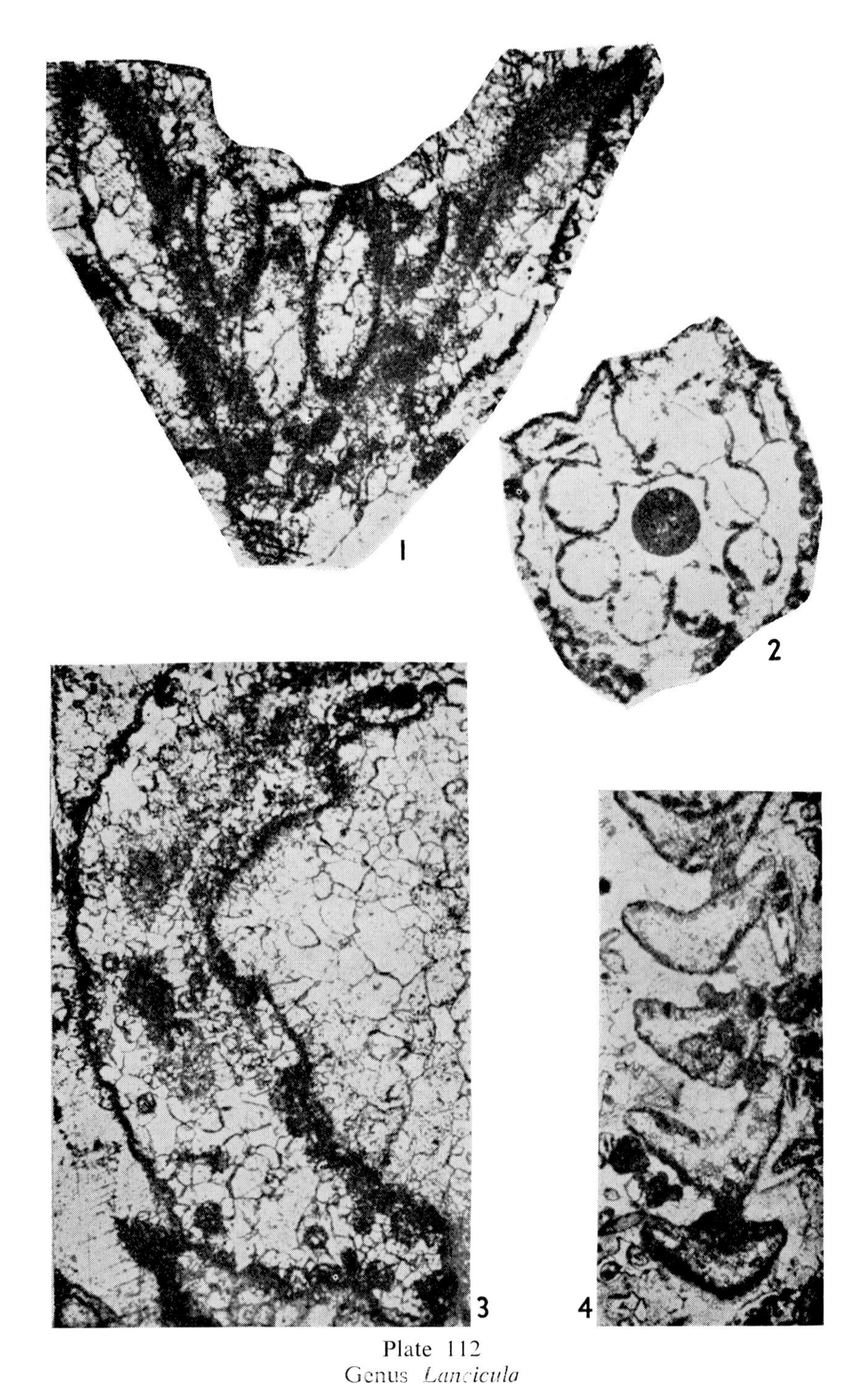

Plate 112

Genus *Lancicula*

Figures 1-4. *Lancicula alta* Maslov. Lower Devonian, Kuznetsk basin, central Siberia. (From Maslov, 1956.) 1. Longitudinal section (x20) showing possible sporangia. 2. Cross section (x46). 3. Cross section of wall (x46). 4. Tangential section cut from central area showing five cups (x46).

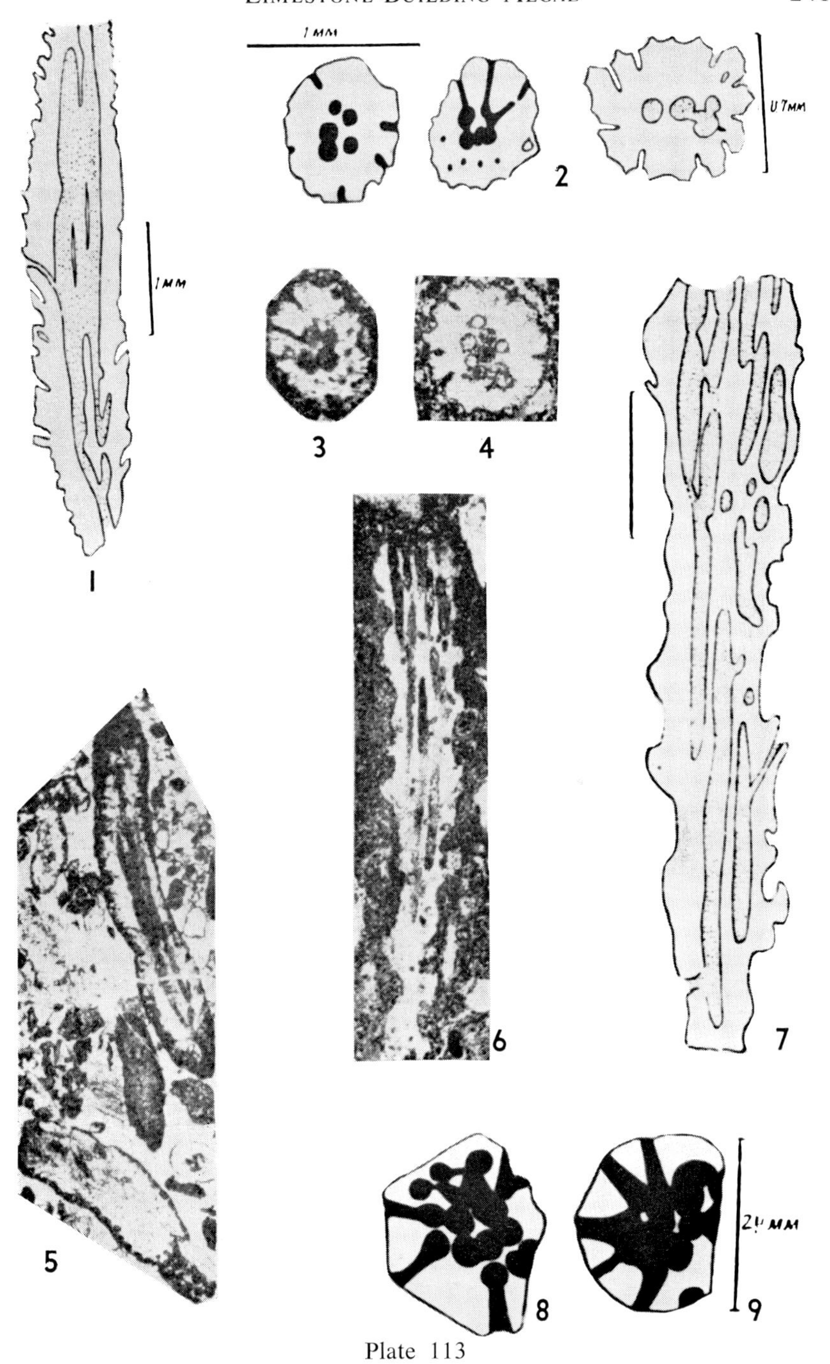

Plate 113
Genus *Litanaia*
Figures 1-5. *Litanaia mira* Maslov; Figures 6-9. *Litanaia anirica* Maslov. Lower Devonian, Kuznetsk basin, U.S.S.R. (After Maslov, 1956.)

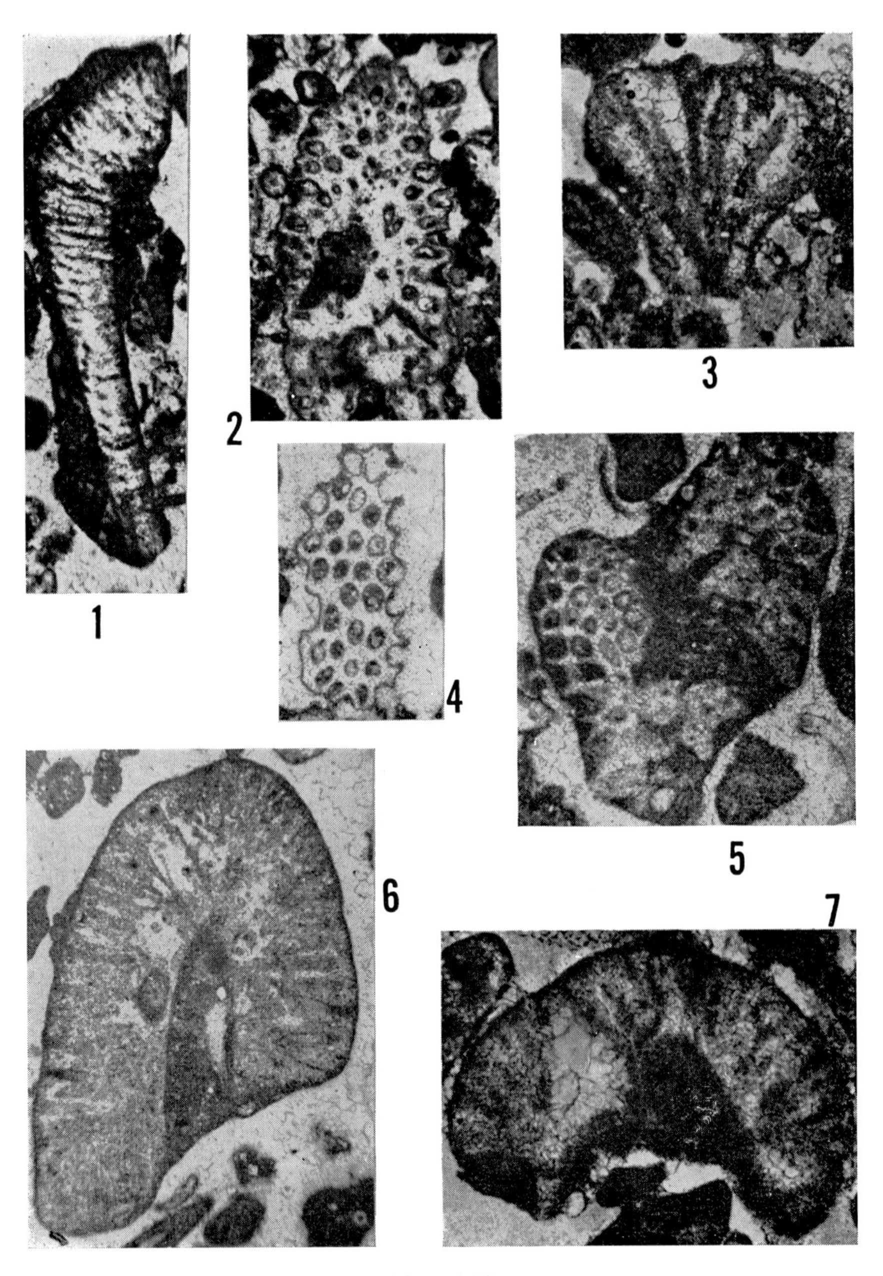

Plate 114

Genus *Orthriosiphon*

Figures 1-7. *Orthriosiphon saskatchewanensis* Johnson and Konishi. 1. Fragment of cortical tissue showing branches (x25). 2. Tangential section cutting near outer margin (x25). 3. A fragment showing branching (x50). 4. A section through cortex near surface (x25). 5. An oblique section (x40). 6. A tangential section almost through central axis (x25). A conceptacle is shown on left side, subcortical in position. 7. A nearly tangential section (x50).

GEOLOGICAL IMPORTANCE OF CALCAREOUS ALGAE

The calcareous algae are of considerable geological importance, probably much more so than is generally realized even by field geologists. The geological significance of such algae may be considered under four categories: 1) as builders of limestone, 2) as builders of bioherms and reefs, 3) as indicators of age, and 4) as indicators of environment.

Calcareous Algae as Builders of Limestone

It has already been pointed out that a few members of the major groups of red, green, and blue-green algae have, during the long course of geologic time, acquired the habit of secreting or depositing calcium carbonate. The relative importance of these groups in this activity has varied from time to time and from place to place.

The Red Algae

Coralline algae. Today these are the red algae which are important as rock builders. Locally, where conditions are particularly favorable, as in clear, shallow, warm waters, rich in lime and containing an abundance of food materials, these algae may grow rapidly in enormous numbers and contribute appreciably to the formation of limestone.

In the Mediterranean region, the West Indies, and around the islands of the tropical Pacific, most of the limestones contain at least some algae, and, in many cases, the algae rank among the principal contributors to the rock. Algal-foraminiferal limestones are especially abundant in the Cenozoic rocks of those regions and in the area of the Tertiary Tethyan Sea. The algae mainly responsible for this work are the crustose corallines, both crustose and branching forms. The articulate coralline algae seldom contribute much to the volume of the rock, although fragments of these delicate plants are to be found in most of the rocks studied.

Rocks of this type are described and illustrated in the section on algal limestones.

Solenoporaceae. During the Paleozoic and most of the Mesozoic, the Solenoporaceae occupied much the same position in the general marine ecology as the coralline algae occupy today, and they accomplished the same geologic work. However, one gets the impression that during the Paleozoic, at least, the Solenoporaceae were not as abundant nor as important as the coralline algae are today. At different times and in rather restricted localities, however, a few species developed in such numbers as to contribute to the formation of limestones. During Ordovician times, Solenoporaceae contributed to the formation of limestones at a number of localities in western Ohio, Indiana, Ontario, Great Britain, and

Norway. During the Silurian, they were locally important in the Baltic region, southeastern Canada, and the northern Mississippi Valley. Solenoporaceae occur widely distributed in Devonian limestones, but in very few areas were they of sufficient abundance to really be of any importance as rock builders. During the Mississippian, Solenoporaceae aided in limestone building at several areas in Great Britain, Belgium, Germany, and the Mississippi Valley. They commonly occur associated with reefs and bioherms in widely distributed areas during the Permian. They do not appear to have been very important during the Triassic, but spread and became very important during the Jurassic. In that period, Solenporaceae were important as builders of limestones, bioherms, and reefs over wide areas in France, Germany, Switzerland, Great Britain, the Near East, India, and Japan. During the Cretaceous, they were rock builders in such widely separated areas as Baja California, Lebanon, southern France, the Near East, and India.

Gymnocodiaceae. Members of this family were particularly abundant during the Permian when they were very widespread. They appear to have been important rock builders in northern Italy, Yugoslavia, Greece, the Near East, and Japan.

The Green Algae

The rock-building green algae all belong to the families Codiaceae and Dasycladaceae. To these forms might be added the genus *Girvanella* which, while of uncertain systematic position, very possibly includes at least some green algae. The Codiaceae and Dasycladaceae are known from the Cambrian to the present, but, while numerous fossils have been observed in the Lower Paleozoic, they rarely occurred in sufficient abundance to be rock builders until Mississippian times. Since the Mississippian, green algae have been rock builders in localized areas on numerous occasions. Algal limestones constructed essentially by green algae are widespread in many areas in the Mississippian of Great Britain, Belgium, France, and Germany, and the Mississippi Valley and Rocky Mountain regions of the United States. During the Pennsylvanian, green algae assisted in limestone building in the Mid-continent region, Texas, several areas in the Rocky Mountain region and the Pacific Northwest, and also in the Near East and in Japan. During the Permian, their work was even more important in the formation of bedded limestones, bioherms, and reefs. Green algae seem to have been important reef builders in most areas where reefs were formed during the Permian, and they seem to have contributed appreciably to Permian limestones in the Mediterranean region, the Near East, India, Japan, and a number of widely scattered areas in the western and southwestern United States and Mexico. Russian geologists comment on the importance of green algae in the Moscow Basin and in a number of areas in Siberia. During the Triassic, these algae were very important as reef builders in the Tyrol region of Austria and adjoining countries. Algal limestones formed by green algae are extremely widespread in Jurassic deposits. They have been recorded from

Great Britain, France, Spain, North Africa, the Near East, Japan, Guatemala, Texas, Arkansas, and the West Indies. They do not appear to have been as important during the Cretaceous, but such limestones have been reported from the Cretaceous of Guatemala, Texas, North Africa, southern Europe, the Near East, and India.

Girvanella are known from rocks of all ages from the Cambrian to the Lower Cretaceous. They were very important as rock builders during the Cambrian, Mississippian, and Jurassic periods. *Girvanella* limestones have a nearly worldwide distribution during the late Middle and Upper Cambrian and continued in fair importance through the Ordovician and Silurian. They are known from the Devonian but were not very important or widespread. However, in the Mississippian, *Girvanella* again became abundant and widespread, and they have been reported from Great Britain, France, Belgium, North Africa, the Mississippi Valley, western Canada, China, Japan, and Siberia. They were very widespread during the Jurassic.

The Blue-Green Algae

These comprise a large and rather varied group. Structurally, blue-green algae are among the most primitive of the algae. They are marvelously adaptable and, as a result, are found in probably a greater range of environments than any other group of algae. Blue-green algae occur in surface marine waters, near-shore marine waters, brackish waters, fresh waters, highly saline waters, and many species develop in the soils and on moist surfaces. Some species have adapted themselves to living in hot springs, thriving in temperatures surprisingly close to the boiling point. As a group, these algae seem to be quite active chemically, and different types carry on a wide range of chemical activities. As a result, they have varied geological significance, although their geologic work has not been carefully studied and is not adequately understood. One group actively promotes the deposition of bituminous materials which range in character from material that resembles coal to highly carbonaceous shales, types of oil shale, and types of oil. Those blue-green algae which live in hot springs appear to assist actively in the deposition of various types of tufas, both calcareous and siliceous. They play an active role in the formation of travertine and similar calcareous deposits. They bring about the deposition of calcareous material in some of the highly saline lakes of Utah, Nevada, and California, and assist, at least, in the formation of certain types of calcareous shales and bituminous sediments in the deposits of many large lakes. Blue-green algae also are active in the sedimentation of many brackish-water environments. One of their most spectacular activities is the formation of stromatolites and stromatolitic limestones.

Stromatolites. Stromatolites and stromatolitic limestones are among the most common of all types of algal deposits or of deposits attributed to the work of algae. The term stromatolite has been applied to a variety of rounded, nodular, or irregular masses which normally show good laminated structure on weathered or etched surfaces, but which, in thin section, show little or no microstructure. The most that may be shown in the best preserved specimens consists of a number of vague molds of branching threads. The presumed algal nature of these fossils rests entirely on analogy with similar Recent structures which it can be demonstrated are formed by algae.

Stromatolites and stromatolitic limestones are discussed at some length and are illustrated in the chapter on algal limestones. Here we will merely mention their geologic importance.

Stromatolites form the oldest known fossils. They have been reported from rocks considered to be Late Archaeozoic in age. In the Huronian beds of the Great Lakes region, numerous examples of stromatolites and stromatolitic limestones have been reported. In some cases, these attain considerable size; deposits 20 feet or more in thickness may extend over a sizable area. Deposits of this type are abundant and widespread in rocks of Late Proterozoic age, practically the world over. In the United States, the best known and probably the most extensive Late Proterozoic deposits of this type are those found in the Belt series of Montana, particularly in and around Glacier National Park. These have been described in some detail by the Fentons and, more recently, by Rezak (1957). In addition, such deposits are also known from the Late Proterozoic of the Great Lakes region and of the Grand Canyon.

The work of Grabau and associated Chinese geologists has shown stromatolites to be very important in beds presumed to be of about Late Proterozoic age in the Sinian system of China. Australian geologists have told the author that in the Late Proterozoic and beds intermediate between the Proterozoic and the Cambrian, there are numerous thick, widespread algal limestones of this type in Australia.

In Africa, the reports of the Geological Surveys of Morocco and Algeria list numerous occurrences of such algal deposits in Proterozoic rocks. Perhaps the greatest and most extensive deposits known are those occurring in the Congo Basin. Cahen and his associates on the Geological Survey of the Belgian Congo report rims of stromatolitic limestone, up to 40 feet thick, which can be traced around the basin for hundreds of miles.

Similar deposits are also known in the Cambrian from practically all continents. In the United States, such deposits seem to be particularly well-developed during the Late Cambrian and in beds transitional between Late Cambrian and Early Ordovician. Probably, the best known examples of these are found in the so-called Ellenberger formation of the Llano uplift of Texas, in the *Cryptozoon* beds and reefs in the region around Saratoga Springs, New York, and in the limestones of the Dotsero formation of north-central to northwestern Colorado.

Calcareous Algae as Builders of Bioherms, Biostromes, and Reefs

Perhaps, to start with we should define these terms. To judge by current usage, particularly among the petroleum geologists, one gets the impression that most writers do not have too clear a picture of the meaning of these terms. During the last ten years, the term reef has been stretched to the point that some writers appear to consider the term reef-limestone as synonymous with any highly organic limestone.

In 1932, Cumings coined the terms *bioherm* and *biostrome*. He defined a bioherm as any dome-like, mound-like, or otherwise circumscribed mass, built exclusively or mainly by sedentary organisms, such as corals, bryozoa, algae, brachiopods, mollusks, crinoids, and so forth, the deposit being unbedded and enclosed in normal rock of different lithologic character. He defined a biostrome as a similar deposit of organic limestone, built by the same types of sedentary organisms, which is bedded and does not form mound-like or lens-like shapes.

In his definitions and discussion, there was no implication as to the size of the deposits. The term bioherm can be used for a lens or mound of unbedded organic limestone of very small size, for example, only a few feet high and a few feet across. On the other hand, some bioherms develop to enormous dimensions. After a bioherm attains a certain size, it will affect sedimentation, with different types of sediments being deposited on different sides or around different parts of it. When this happens, we consider the structure to be a *reef*. In our graduate lecture courses at the Colorado School of Mines on reefs and associated facies, we have found it very helpful to define a reef as a bioherm of sufficient size to develop associated facies. Thus, while all reefs are bioherms, all bioherms are not reefs.

Algae assist in the formation of bioherms, biostromes, and reefs, and apparently they have done so since early geologic times. They may assist in three different ways: first, by their actual contributions to the volume of the rock; second, by acting as binding organisms to cement and hold together colonies of other organisms, such as corals, stromatoporoids, large bryozoans, rudistid mollusks, and other large lime-secreting animals and plants; and third, the algae may form a covering over and around portions of the bioherm or reef mass and protect it from wave erosion. This covering of the bioherm or reef mass permits it to grow to considerable size and, in some cases, partially directs the resultant shape of the mass. In any given deposit, the algae may assist in any one of these three ways or in any combination of them. The Precambrian bioherms and reefs seem to have been built largely by algae. In most of the Paleozoic bioherms, biostromes, and reefs, however, the algae seem to have taken the much less conspicuous role of binding organisms. A large number of the Silurian and Devonian reefs were built with stromatoporoids and corals as the principal contributors, and with algae acting primarily as binders. In some cases, however, the algae contributed a minor amount of material to the reef. This arrangement appears to have been largely the case throughout the Mesozoic, although there are some striking exceptions, particularly among the Jurassic reefs of France and England. In France and Switzerland, we find a number of reefs where solenoporoid algae were the main constituents, contributing much of the bulk of the limestone, while in the famous Jurassic reefs of the Oxford region of England, the algae were quite subordinate, and their main role seems to have been as reef binders, although one case has been pointed out where protective rims of algae may have been developed.

During the Cenozoic, the red coralline algae were the important reef builders, and we find them acting much as they do in the present-day reefs of the tropics, where all three roles are important. As an example, we may consider Bikini Atoll which undoubtedly is the most thoroughly studied and best known atoll in the world (see U.S.G.S. Prof. Paper 260, parts A and M). In different parts of the reef at Bikini, the algae are active in all three roles.

Along the seaward side of Bikini Island, there is a magnificently developed algal ridge forming the algal rim of the reef (plate 138, figures 1-2). Taylor, in his 1950 book, states that this algal ridge is the finest botanical exhibit on the island. It is indeed a spectacular thing, running for several miles, rising appreciably above low tide level, and extending back in some places for as much as 100 yards. The ridge has a strong pink to brownish-rose color. The outer portions are made of large rounded heads of algae belonging to the genus *Porolithon,* while the inner part of the rim is formed of crustose species of the same genus (see plate 139, figures 1 and 2). This algal rim is strongly developed along the windward side of the island but is poorly developed or absent along the lee side.

Studies of hand samples of the reef rock, taken from a number of localities on all sides of the island, show that the content of algal material in the rock ranges from a low of about 5 percent to a high of more than 90 percent, with a probable average in the magnitude of 25 or 30 percent. The area richest in algal material is, of course, the outer algal rim which is practically pure algae. Some of the sediments dredged from the bottom of the lagoon often run very high in fragments of the green algae *Halimeda.*

Algae are also actively at work binding together the various reef-building organisms. One can observe thin crustose coralline algae spreading over and between coral heads and large heads of algae. Algae of this type can also be seen in a number of places where eroded remnants of the raised reef are visible, and they are quite evident in large specimens of the limestone.

Algae as Indicators of Age (Time Fossils) and for Use in Correlation

Correlation

The question as to the value of algae for use in correlation is asked frequently, because in many regions there are thick series of sedimentary beds in which the only fossils are algae.

Locally, algae are commonly used in mapping and for correlation because, in limited areas, algal limestones may be very widespread. Thus, in the Congo Basin of the Belgian Congo, in the thick Upper Proterozoic section, algal limestones which can be traced for hundreds of miles have been used as reference beds in mapping. The same is true in the Belt series in Montana. The author knows of several areas in Montana where the mapping has been done largely on the basis of beds of algal limestone, and similar cases have been reported in the Appalachian region where *Cryptozoon* beds are extensively developed.

When we come to wide range correlation, however, another problem is involved; that is, the actual time range of given genera and species.

Possibilities of Algae as Time Fossils

Under this heading, we will consider the different groups of algae separately.

Coralline algae. The available evidence clearly points out that, while the genera of coralline algae are quite long ranged, the individual species are not. The coralline algae of the western Mediterranean have been studied more carefully than those of any other region, thanks very largely to the work of Madame Lemoine, Mademoiselle Pfender, and several Italian paleontologists. These studies have shown that the individual species are relatively short lived; that is, they seldom last longer than a geologic epoch. For example, if a species starts in the early Eocene, it seldom reaches up into the Oligocene, or if it starts in the middle Eocene, it does not extend above the middle Oligocene. This limited geologic range of species is also shown by the work that has been done in the tropical Pacific. In our studies of the fossil algae of the Marshall Islands and the Marianas, for example, it was observed that none of the Eocene species are found in the Miocene, and, similarly, very few of the Miocene species are found in the late Pliocene or early Pleistocene. On the other hand, the Pleistocene flora contains most of the Recent species plus a number of others. It is believed that once the coralline algae of a given area have been carefully studied and the range of the species noted, these algae can definitely be used for correlation of strata. This use of coralline algae as time fossils would be especially helpful because many of the present-day species have a very wide geographical distribution and the same was true during the past. We find, for example, that the Eocene flora of Guam and Saipan in the Mariana Islands contains many species originally described from the Eocene of France, Spain, and Algeria. Similarly, the Recent flora of the tropical Pacific contains some species known to extend from the Red Sea to Hawaii, and quite a number of species which grow from the Arabian Sea to the Marshall Islands.

Green algae: family Codiaceae. Most of the genera attributed to this family appear to have been rather long ranged. For example, *Ortonella* is recorded from the Devonian to the Late Permian; *Cayeuxia* appears to extend throughout the Jurassic and quite a way into the Lower Cretaceous; and *Halimeda* is known from the Late Cretaceous to the present and is very abundant and widespread in the tropics today. On the other hand, individual species appear to have been rather short ranged. Most of the described species of *Ortonella,* for example, appear to have lasted for only about one-half or one-third of a geologic period. Mr. Konishi, in his recently prepared monograph on the Paleozoic Codiaceae (1961), discussed this aspect of codiacean algae at some length.

A number of both Recent and fossil codiacean algae have wide geographical ranges. *Halimeda opuntia* Lamouroux has been observed at Key West, Florida, numerous localities in the West Indies, the Red Sea, most of the islands in the Indian Ocean and the East Indies, and is widely distributed in the tropical Pacific. Similarly, *Halimeda gracilis* Harvey is common throughout the tropical Pacific and Indian Oceans and is found, though somewhat less abundantly, in the tropical Atlantic. During the Late Jurassic, a species of *Cayeuxia* ranged from Guatemala to Texas, across the Mediterranean region to the Middle East. During the Permian, several genera developed species which ranged from Japan to western Texas. Several Mississippian species, originally described from British material, have been found in Colorado and Alberta.

Green algae: family Dasycladaceae. The indications are that the various species of dasycladacean algae will probably make the best time fossils of any of the algae. It is true that some of the genera are long ranged. Several genera are known that lived through several geological periods. On the average, however, the range of a given genus seems to have been about equal to the length of a geological period or slightly less, and the individual species appear to have been very short ranged. Probably the most detailed studies that have been made on the stratigraphic distribution of dasycladacean algae are the studies made by the Morellets on the Dasycladaceae from the Eocene of the Paris Basin. These studies have shown quite clearly that, in the Paris Basin at least, most of the species had a range about equal to one stage of the Eocene. The work of Pia on the Triassic Dasycladaceae of Austria and surrounding areas demonstrated the same type of specific range. In fact, Pia, in several of his papers, definitely asserts that the Dasycladaceae have great possibilities as time fossils.

Charophyta. The Charophyta have been used for some time in correlating non-marine beds in western Europe and in the western United States. Most of the species are quite short ranged in terms of geological time, although they may attain quite a wide geographical distribution. Relatively few species appear to have a range greater than about half of a geological period, while quite a number of the species are much more restricted. Inasmuch as, in a number of cases, charophytes are practically the only recognizable fossils that can be obtained from thick series of non-marine beds, the use of these fossils is a matter of considerable importance. Peck (1953) summarized the situation as follows:

> "The charophytes are proving to have some stratigraphic value. Where they occur in considerable abundance, as in the Devonian, Jurassic, Cretaceous, and lower Tertiary deposits of North America, it is found that assemblages, at least, are recognizable and can be placed stratigraphically. It is probable that the gyrogonites will never be of great value in determination of detailed stratigraphic units, but they are becoming useful in the determination of geologic periods, series, and sometimes stages."

Algae as Indicators of Environment

Since algae are plants, they need light for their growth and development, and this, in turn, means that they live abundantly only in well-illuminated waters. In the tropics, good light may penetrate the water to a depth of several hundred feet, but in the temperate zones, adequate light does not extend down nearly so far.

The red coralline algae will grow from tide level down to depths of as much as 1200 feet. At such depths, however, only a few, thin, infertile crusts are found. Large, strongly branching forms occur from tide level down to depths of 50 or 60 feet, with their optimum development probably from tide level to about 25 or 30 feet. Thick, irregular or warty crusts occur in about the same depths, while some of the smaller and thinner crusts may live at deeper levels.

The articulated coralline algae also range from tide level, or just above, down to depths of probably 150 feet, with the greatest development taking place in the upper 15 to 30 feet.

The dasyclads also are very shallow-water forms. These algae are known to live at depths of as much as 75 to 100 feet, but abundant growth occurs only in depths from tide level down to 30 or 40 feet.

The codiacean algae have a greater depth range. Around reefs, *Halimeda* have been observed growing vigorously from approximately tide level down to depths of several hundred feet, with a fairly definite zoning of species with depth. In other words, the species growing at and just under tide level are not the same species as those which are found in the bottoms of lagoons and channels.

In general, one can say that algal limestones are indicative of shallow water, usually very shallow water. The great development of stromatolites seems to occur at, or very close to, tide level with the colonies frequently exposed or at least the tops of them uncovered at low tide. Rocks built chiefly of fragments or, particularly, of heads of the highly branching coralline algae may be assumed to have developed close to low tide level, probably at depths of less than 50 feet. Dasycladacean algal limestones usually are formed at depths of not over 35 or 40 feet. Nearly pure *Halimeda* material has been dredged at Bikini and a number of the other atolls in the western Pacific from depths ranging from 150 to over 200 feet. In the same islands, bottom areas covered with *Halimeda* have been observed only a foot or two below tide level.

REFERENCES

Cahen, L., Jamotte, A., Lepersonne, J., and Mortelmans, G., 1946, Note preliminaire sur les algues du series calcaires anciennes du Congo Belge: Service Geol. Congo Belge Bull., no. 2, p. 171-236, 23 figs.

Cumings, E. R., 1932, Reefs or bioherms?: Geol. Soc. America Bull. v. 43, p. 337-352.

Emery, K. O., Tracey, J. I., Jr., and Ladd, H. S., 1954, Geology of Bikini and nearby atolls: U. S. Geol. Survey Prof. Paper 260-A, 265 p., 64 pls.

Fenton, C. L., and Fenton, M. A., 1931, Algae and algal beds in the Belt Series of Glacier National Park: Jour. Geol., v. 39, p. 670-686.

———, 1933, Algal reefs or bioherms in the Belt Series of Montana: Geol. Soc. America Bull., v. 44, p. 1135-1142.

Grabau, A. W., 1922, The Sinian system: Geol. Soc. China Bull., v. 1, p. 44-88.

Howe, M. A., 1912, The building of "coral" reefs: Science, n. ser., v. 35, p. 837-842.

———, 1912B, Reef-building and land-forming seaweeds: Acad. Nat. Sci. Philadelphia Proc., v. 54, p. 137-138.

———, 1932, The geologic importance of the lime-secreting algae, with a description of a new travertine-forming organism: U. S. Geol. Survey Prof. Paper 170-E, p. 57-65, pls. 19-23.

———, 1933, Plants that form reefs and islands: Sci. Monthly, v. 36, p. 549-552.

Johnson, J. H., 1937, Algal limestones, their appearance and superficial characteristics: Mines Mag., v. 27, no. 10, p. 11-13.

———, 1954, An introduction to the study of rock building algae and algal limestones: Colorado School of Mines Quarterly, v. 49, no. 2, 117 p., 62 pls.

———, 1954B, Fossil calcareous algae from Bikini atoll: U. S. Geol. Survey Prof. Paper 260-M, p. 537-545, pls. 188-197.

———, 1957, Calcareous algae of Saipan: U. S. Geol. Survey Prof. Paper 280-E, p. 209-246, 24 pls.

———, 1957B, Algal limestones: Jour. Paleont. Soc. India, v. 2, p. 48-53, pls. 1-6.

Konishi, Kenji, 1961, Paleozoic Codiaceae: scheduled for publication this year in Journal of Kanazawa University.

Lemoine, Mme. Paul, 1941, Les algues calcaires de la zone neritique: Soc. de Biogeographie Mem., v. 7, p. 75-138.

Peck, R. E., 1953, Fossil charophytes: Bot. Rev., v. 19, no. 4, p. 209-227.

Pia, Julius, 1926, Pflanzen als Gesteinsbildner: Berlin, 355 p.

Rezak, Richard, 1957, Stromatolites of the Belt Series in Glacier National Park and vicinity, Montana: U. S. Geol. Survey Prof. Paper 294-C-D, p. 127-154, illus.

Setchell, W. A., 1926, Nullipore versus coral in reef formation: Am. Philos. Soc. Proc., v. 65, no. 2, p. 136-140.

Taylor, W. R., 1950, Plants of Bikini: Univ. of Michigan Press, Ann Arbor, 227 p., 79 pls.

ALGAL LIMESTONES

Almost everyone is somewhat interested in organic limestones. The very thought that thick masses of hard rock can be formed by the activities of tiny organisms is in itself intriguing. Among the many types of organic limestones, the algal limestones have probably received the least study, in spite of the fact that they are, in the opinion of the writer, the most interesting.

Most, but not all, algal limestones have two rather distinctive properties: first, an initial strength which resists compaction, and second, initial pore space. These properties make algal limestones of particular interest to the economic geologists. The first characteristic protects the limestone from compaction and crushing, while the second feature makes the rocks susceptible to alteration and replacement by ground waters and other circulating solutions. Because of these properties, algal limestones make very good host rocks for ore deposits. A number of examples of rich ore deposits which were formed by the replacement of algal limestones are known to the author, and the extent of the ore body in these deposits is definitely limited by the boundaries of the algal limestone. The original porosity, which may be increased by later solution and/or recrystallization, causes the petroleum geologist to be interested in algal limestones because of their possibilities as reservoir rocks for the accumulation of oil.

The question is frequently asked why do algal limestones have these two distinctive properties. The answer is not simple, and there appear to be at least three factors involved: first, the character of the algae responsible for building the particular limestone; second, the way in which the rocks were built; and third, the later history of compaction, consolidation, and alteration of the rocks.

Actually, there are many types of algal limestones, and the various types show a considerable range in properties. Algal limestones may be classified in two ways: according to the algae that helped build them, and according to the structure and other physical properties of the limestones.

Algal Limestones Classified According to the Algae that Built Them

The various types of rock-building algae have already been discussed in previous portions of this manuscript, so we need only to tabulate them here and give a few examples with reference to the plates illustrating them. (Table XIII.)

TABLE XIII
CLASSIFICATION OF ALGAL LIMESTONES
BASED ON THE ALGAE THAT BUILT THEM

Type of Algae	*Illustrated on*
Red algae	
Coralline algae	
Crustose	Plates 131, 134, 135, and 137
Articulated	Plate 14
Solenoporaceae	Plates 119 and 129
Green algae	
Codiaceae	
Ortonella-Garwoodia	Plate 36
Halimeda	Plate 136
Palaeoporella	Plate 117, figure 2
Dasycladaceae	
Coelosphaeridium	Plate 117, figure 1
Mizzia	Plate 128
Diplopora	Plate 52
Charophyta	Plate 130
Green and/or blue-green algae	
Girvanella	Plate 116
Spongiostromata	Plates 115, 124, and 126

Algal Limestones Classified According to their Structure and Other Physical Properties

The second classification of algal limestones is illustrated on the accompanying table (Table XIV). We will briefly discuss these various types.

Porcelaineous

This term has been applied to algal limestones which are dense and finely textured and which break like a piece of porcelain, showing a luster on the broken surface similar to that shown by a piece of broken porcelain.

All the algal limestones of this type known to the writer were formed by red algae, either coralline algae or members of the Solenoporaceae. Commonly, where the algae grew into large solid heads, it happens that broken portions of the fossilized material show this porcelain-like breakage and luster. Probably, the best known and most publicized examples of porcelaineous algal limestones are to be found in the Jurassic Solenoporaceae limestones where the algae grew in large banks or knolls, with the individual algal heads sometimes attaining a diameter of as much as a foot. Numerous examples of limestones of this type from the Jurassic beds of England, France, and Switzerland have been cited in the literature.

Reef Limestones

As mentioned in the preceding chapter, a reef may be defined as a bioherm of sufficient size to develop associated facies. Hence, in a reef complex we find both the reef proper, usually spoken of as the reef core, and associated facies. The reef core is a mass of nonstratified limestone formed of the remains of sedentary, lime-secreting organisms, in or approximately in the position of growth. This is the biohermal mass. In most of the Recent reefs of the tropical seas, the reef mass is built largely by corals and coralline algae, and the interstices are filled with foraminiferal tests and occasionally with finely ground coral or algal material.

TABLE XIV
CLASSIFICATION OF ALGAL LIMESTONES
BASED ON THEIR STRUCTURE AND PHYSICAL PROPERTIES

Type	
Porcelaineous	Stromatolitic limestones
Reef limestones	Biostromatic
The reef core	Nodular (algal biscuit)
Associated facies	Algal pisolites
Algal-foraminiferal limestones	Algal felts and sponges
	Leached or partially leached algal limestones

Embedded in the rock mass one may also observe occasional shells or pieces of shells of mollusks, fragments of the tests and spines of echinoids, occasional plates from starfish and holothurians, as well as an occasional fragment of other organic debris. During the Paleozoic, the main reef builders were stromatoporoids and corals, together with algae, and, in many cases, considerable amounts of crinoidal debris, bryozoans, and foraminifera. Mesozoic reef builders were largely corals and algae, although in a few localities small reefs and numerous bioherms have been observed which were built largely by bryozoans and hydrozoans.

The reef core usually can easily be recognized in the field by its lack of bedding and by the numerous reef-building organisms in the positions of growth. In hand specimens, however, a piece of the rock core can seldom be differentiated from biostromatic limestones or from some of the associated reef facies since the same organisms will be present in all these rocks.

The limestone facies developed in and around reef cores show considerable variety. Commonly, on the ocean side there will be around the base of the reef a sizeable fan or cone of coarse material broken from the reef. This fan will be formed of fragments of the common reef-building organisms, sometimes showing evidence of breakage and wear and sometimes not, but with the coral and algal heads not in the position of growth and commonly with considerable broken material in the interstices. In the large lagoons of atolls or those developed behind fringing reefs, a great variety of stratified associated limestones may develop. These will show quite a range in character. Some may be largely algal as the *Halimeda* limestones of many of the Recent lagoons. Others will be built essentially of fragments of algae in the midst of great numbers of foraminifera forming rather typical foraminiferal-algal limestones. Locally, there may be a considerable development of well-developed coral limestones or coral-algal limestones. In some cases, the coral and algae may be in the original position of growth; in other cases, the rocks will be composed largely of broken and somewhat worn fragments. In the lagoons of some atolls and, commonly, in the lagoons behind large fringing or barrier reefs, there may be quite a development of marly limestone which may or may not contain considerable algal material. The algae found in these beds may be fragments of coralline algae, both crustose and articulated, Dasycladaceae, and Codiaceae, such as *Halimeda*. In Late Paleozoic materials, one may find considerable amounts of crinoidal material and, in some cases, an abundance of fusulinids. In Mesozoic deposits of this type, Dasycladaceae and Codiaceae of a variety of types may occur.

Stromatolitic Limestones

These limestones are built largely by the activity of algae forming stromatolites and oncolithi. These include several types, the most important of which are the biostromatic limestones, the nodular or algal biscuit limestones, and algal pisolites.

The biostromatic limestones are formed by abundant growths of stromatolites. The stromatolites may be either layered, nodular (that is, large cabbage-like forms), or large branching digitate forms. Rocks of this type are very abundant and widespread in the latest Proterozoic sediments the world over, and locally attain considerable development throughout the Cambrian and into the Early Ordovician. The Precambrian limestones are built largely by the genus *Collenia,* while those of the Cambrian include forms of both *Collenia* and, particularly during the Late Cambrian and earliest Ordovician, the genus *Cryptozoon.* Biostromatic limestones of this type also occur in local areas throughout the geologic column up to the present day.

Nodular limestones develop most commonly in lakes or, more rarely, in streams and estuaries. A number of cases, however, are on record where nodular limestones occur associated with typical marine fossils and apparently represent very shallow-water, near-shore marine deposits. Normally, these limestones are built of algal biscuits which may range in size from pisolitic material up to biscuits that are 3 to 5 inches in diameter. Commonly, the biscuits are rather flattened.

In general appearance, algal pisolites closely resemble typical pisolites. Algal pisolites are composed of more or less spherical masses ranging in size from that of a large oolite to spherules having a diameter as much as an inch across. However, most such forms are about the size of a marble, that is, from a quarter- to a half-inch in diameter.

Commonly, but not always, algal pisolites can be distinguished in superficial examination from true pisolites by the fact that the rounded masses are not as perfectly spherical as is usually the case with the true chemically precipitated pisolites. The two types of pisolites can easily be distinguished from each other by microscopic examination of thin sections. In section, the algal pisolites are seen to consist of layers, more or less concentric, composed of molds of an algal felt, which is quite a contrast to the needle-like structure of the true pisolites. Rocks composed of such forms are not rare, but they almost always have a very restricted local development. Probably the most spectacular examples known to the writer are those found in some of the Permian limestones associated with the reef facies in the Carlsbad, New Mexico, region. Such rocks have developed in a number of the Tertiary lakes scattered across what is now Colorado, Utah, Nevada, northern Arizona, and northwestern New Mexico. Small patches of them have been observed in some of the oolitic Mississippian limestones of eastern Missouri and southwestern Illinois.

Algal Felts and Sponges

Certain green, blue-green, and mixtures of green and blue-green algae may, under favorable conditions, deposit considerable amounts of algal felt which gradually becomes more or less thoroughly calcified. Such deposits have been observed forming in a number of fresh-water and brackish-water environments, particularly along the shores of certain lakes, and on and close to the tidal flats of numerous estuaries and lagoons. Similar deposits are also formed in and around certain hot springs and in some highly saline lakes and lagoons. The rock formed would commonly be classed as a calcareous tufa. Normally, it is light and quite porous. A thin-section examination would show it to consist largely of calcareous molds of algal threads or filaments. Such a rock recrystallizes easily and seldom shows much in the way of real microstructure. The deposits so formed may vary greatly in shape, ranging from rounded or irregular masses to thin, laminated beds.

Leached or Partially Leached Algal Limestones

Under favorable conditions, limestones containing appreciable amounts of algal debris may undergo a selective leaching in which the algal fragments are leached out. The appearance of the resulting rock may vary considerably depending on the type of algae and the size and shape of the pieces. In the Pleistocene deposits of many of the limestone islands of the tropical Pacific, one finds large amounts of *Halimeda* limestone in which the *Halimeda* segments have been leached away, leaving a peculiar, very porous limestone which is riddled with holes or cavities about the size and shape of pieces of dried oatmeal. Such rocks were observed in a number of places capping the limestone Pleistocene terraces on Saipan, Guam, Tinian, and on Peleliu and Angaur in the southern Palau group. Somewhat similar limestones have also been observed in the Pennsylvanian and Permian of the Mid-Continent region and Texas where dasycladacean and codiacean algae have been leached from the rock. Similar occurrences have been observed in dasycladacean limestones of Triassic and Jurassic age at a number of widely separated geographic localities.

REFERENCES

Johnson, J. H., 1937, Algal limestones, their appearance and superficial characteristics: Mines Mag., v. 27, no. 10, p. 11-13.

————, 1954, An introduction to the study of rock building algae and algal limestones: Colorado School of Mines Quarterly, v. 49, no. 2, 117, p. 62 pls.

————, 1957, Algal limestones: Paleont. Soc. India Jour., v. 2 (D. N. Wadia Jubilee Number), p. 48-53, pls. 1-6.

Pia, Julius, 1926, Pflanzen als Gesteinsbildner: Berlin, 355 p.

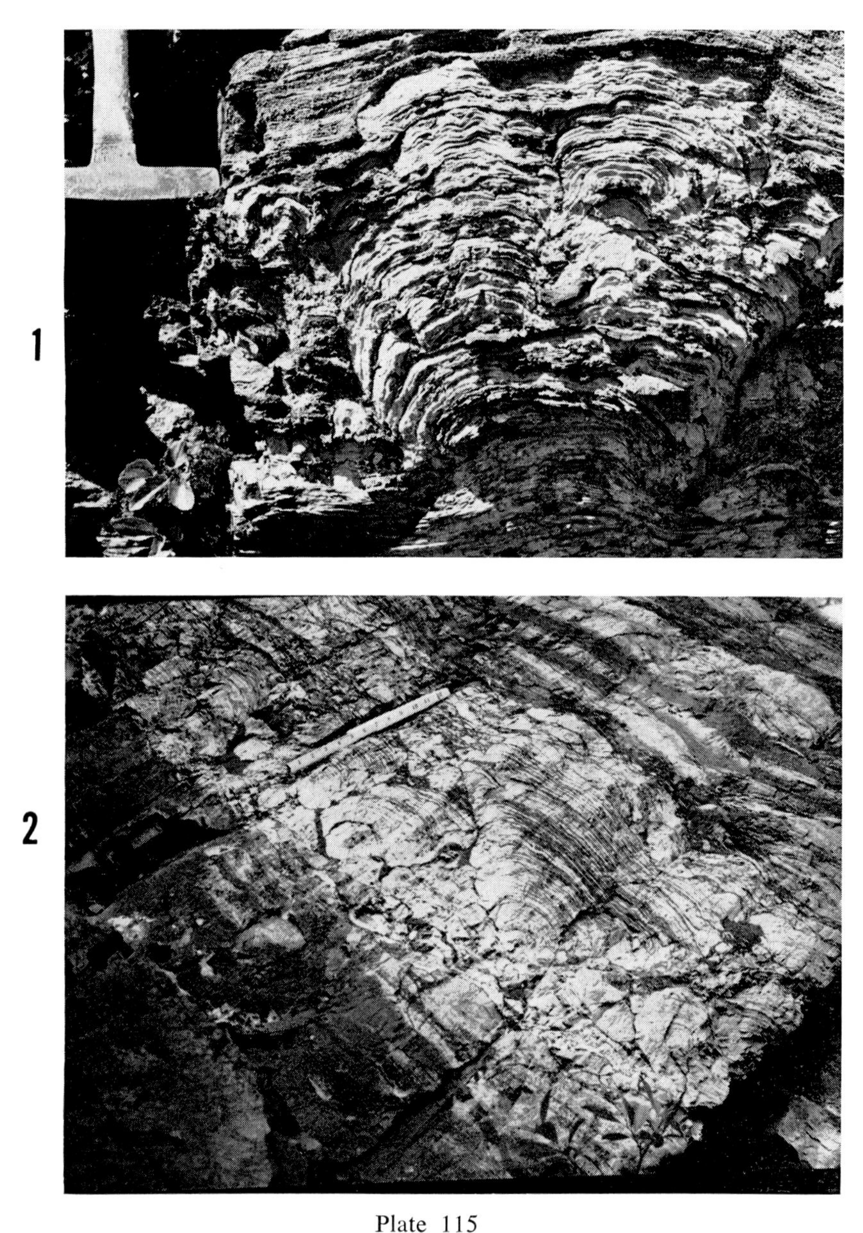

Plate 115
Algal Limestone — Precambrian
(Photos by R. Rezak, U. S. Geological Survey.)

Figure 1. *Collenia undosa* Walcott, Proterozoic, Missoula group, near Logan Pass, Glacier National Park, Montana.

Figure 2. *Cryptozoon occidentales* Dawson, Missoula group, near Nyak, Montana.

Plate 116
Algal Limestones — Cambrian
Figure 1. Limestone composed largely of colonies of *Girvanella manchurica* Yabe and Ozaki (x1). Lower Cambrian, Manchuria.

Plate 117

Algal Limestones — Ordovician

Figure 1. Dasycladacean *(Coelosphaeridium)* limestone (x1). Middle Ordovician, Helgφya, Norway.

Figure 2. Codiacean *(Palaeoporella)* limestone (x1). Upper Ordovician, Frognφya, Ringerike, Norway.

Plate 118
Algal Limestones — Silurian
Figure 1. A slab composed largely of *Rothpletzella gotlandica* (Rothpletz) (x3/4). Silurian of Gotland, Sweden. (In U. S. National Museum.)

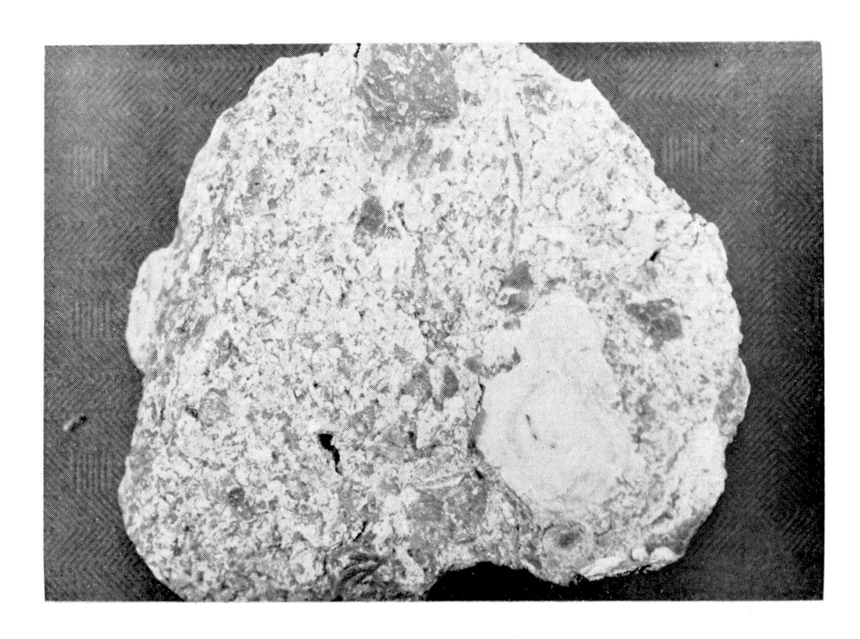

Plate 119
Algal Limestones — Silurian
Figures 1-2. Specimens of Silurian limestone (x1) near reef facies, composed largely of fragments of *Solenopora*. Galgberget, Visby, Gotland, Sweden.

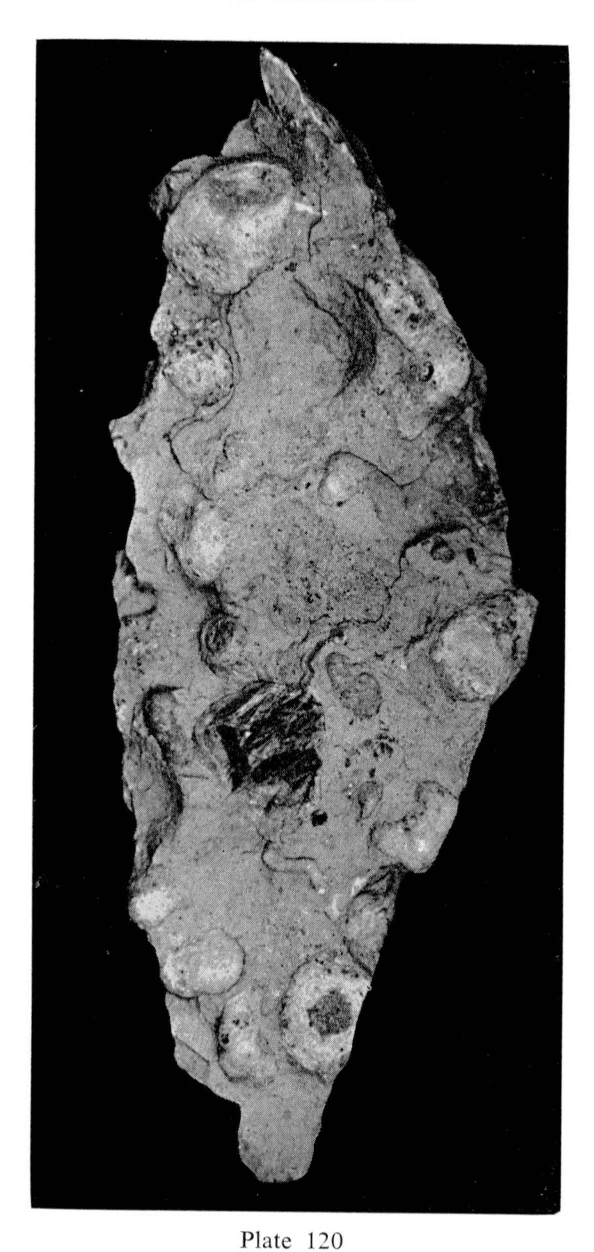

Plate 120

Algal Limestone — Mississippian

Figure 1. An "algal nodule" limestone (x1), Mississippian (Ste. Genevieve formation), St. Louis, Missouri.

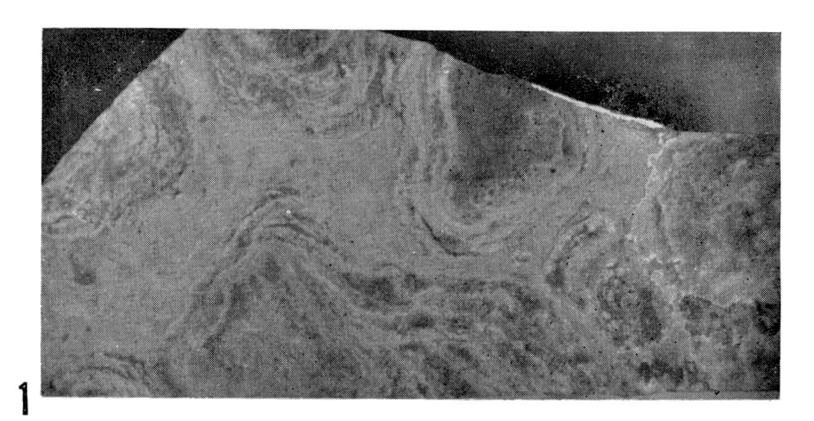

Plate 121
Algal Limestones — Mississippian
Figure 1. "Algal nodule" limestone (x1), Ste. Genevieve formation, St. Louis, Missouri. A slab of the rock shown in Plate 120.
Figure 2. Another type of algal limestone (x1) from the Ste. Genevieve formation, St. Louis, Missouri.

Plate 122

Algal Limestone — Mississippian

Figure 1. A colony of Spongiostromatoid algae from the Mississippian upper Leadville limestone, near Ashcroft, Pitkin County, Colorado.

Figure 2. An algal limestone including codiacean algae, *Girvanella,* and algal pellets. Upper Leadville limestone near Glenwood Springs, Colorado.

Plate 123

Algal Limestone — Mississippian

Figure 1. An algal pellet limestone (x1). Mississippian, upper Leadville formation, Lime Creek, Eagle County, Colorado.

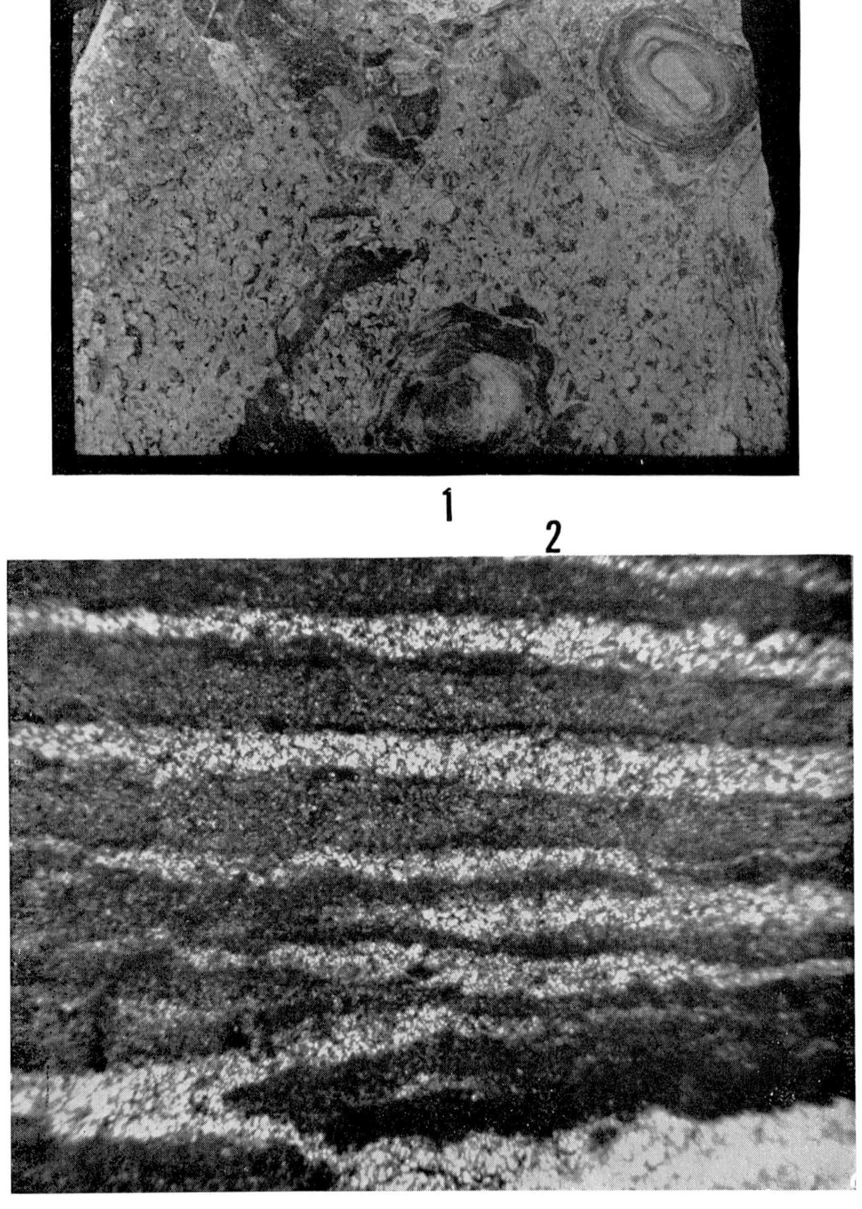

Plate 124

Algal Limestones — Mississippian

Figures 1-2. *Pycnostroma* limestone. 1. A section of the core cut and etched, showing several specimens (x1). 2. A detail of a specimen showing structure of several laminae (x100).

Plate 125
Algal Limestones — Pennsylvanian
Figure 1. Algal "Pisolite," Lower Pennsylvanian, Chaffee County, Colorado.

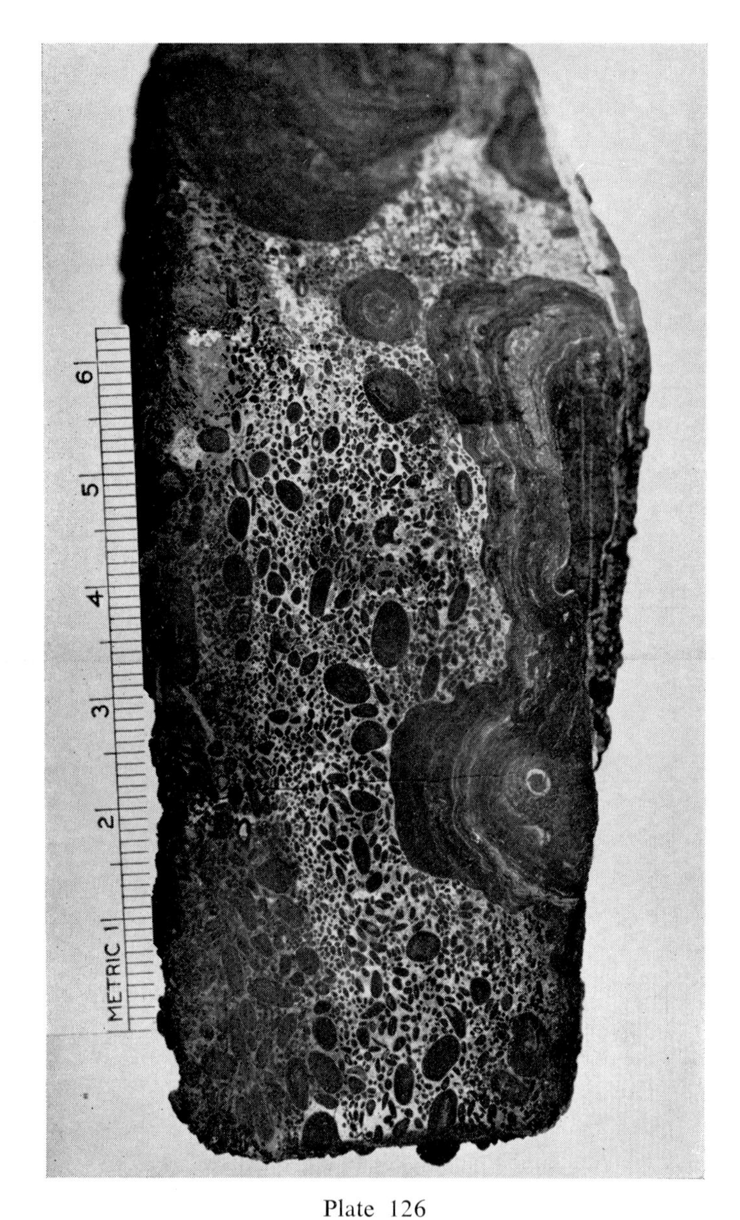

Plate 126

Algal Limestone — Pennsylvanian

Figure 1. Pennsylvanian algal limestone, Chaffee County, Colorado.

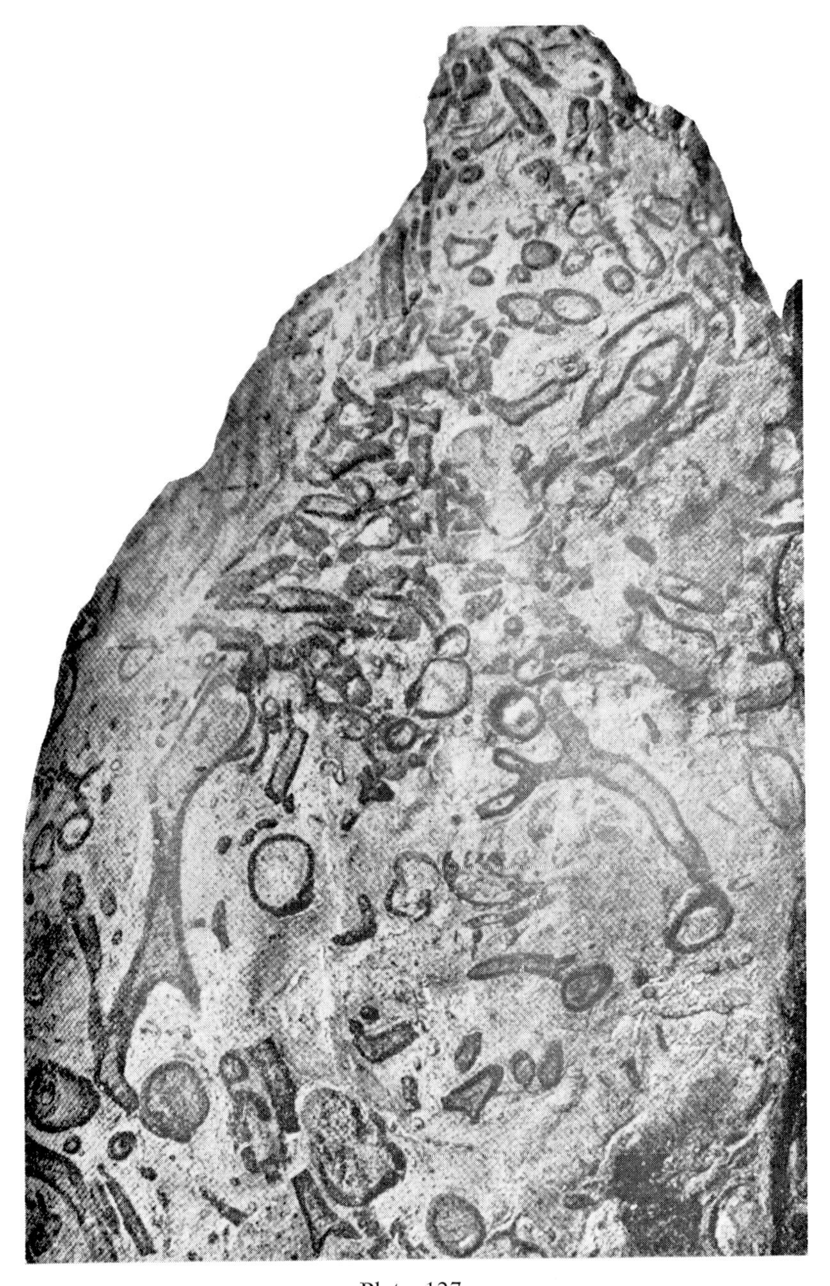

Plate 127
Algal Limestone — Pennsylvanian
Figure 1. *Anthracoporella* limestone (x1-1/2), Pennsylvanian, Carnic Alps, Austria. (From Pia, 1928.)

Plate 128
Algal Limestones — Permian
Figures 1-2. *Mizzia* limestones. Permian, Apache Mountains, Texas. 1. *Mizzia* and *Solenopora* (x4). 2. *Mizzia* limestone (x1).

Plate 129
Algal Limestone — Jurassic
Figure 1. Jurassic *Solenopora* limestone (x1). Switzerland. (From Peterhans, 1930.)

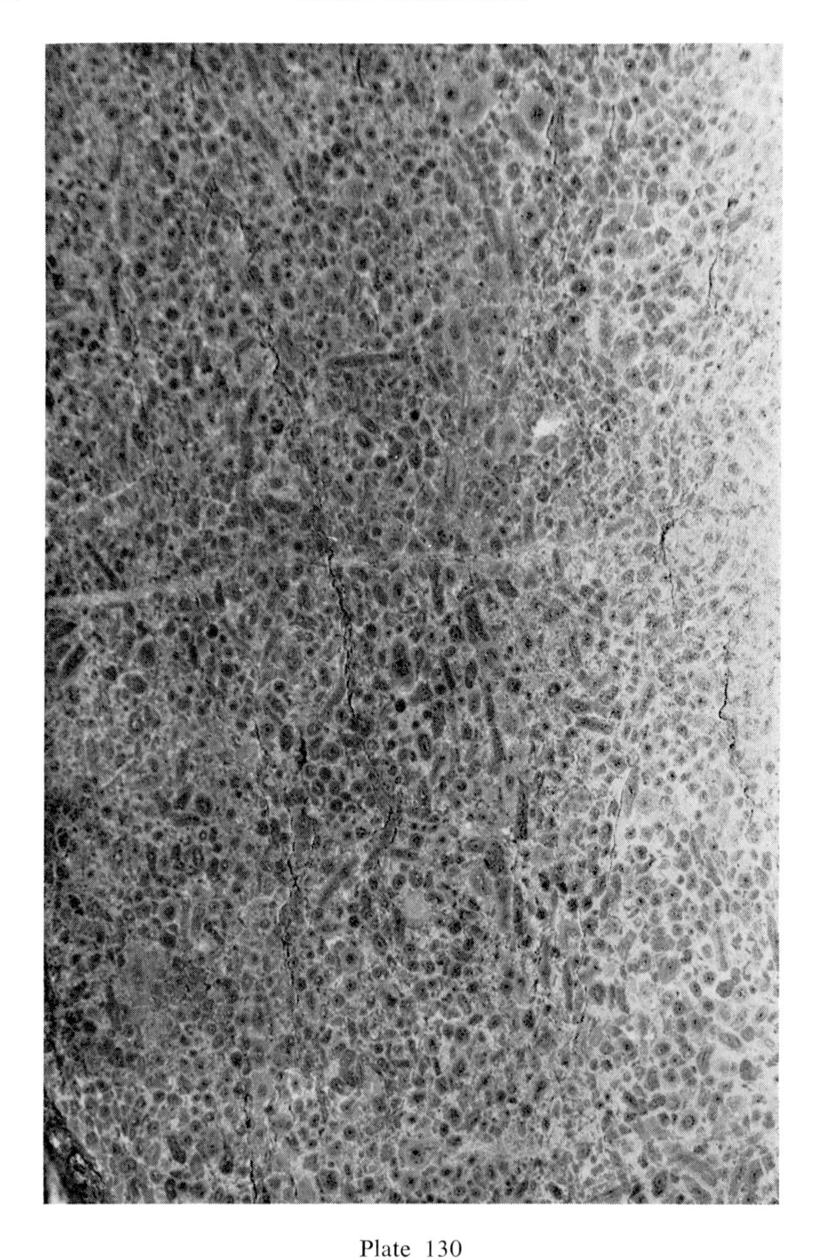

Plate 130
Algal Limestone — Jurassic
Figure 1. *Chara* limestone (polished surface x2), Jurassic, Morrison formation, Perry Park, Colorado.

Plate 131
Algal Limestone — Cretaceous
Figure 1. Cretaceous limestone (x1), with abundant debris of a branching *Archaeolithothamnium*. Southern France. (From Pfender, 1926.)

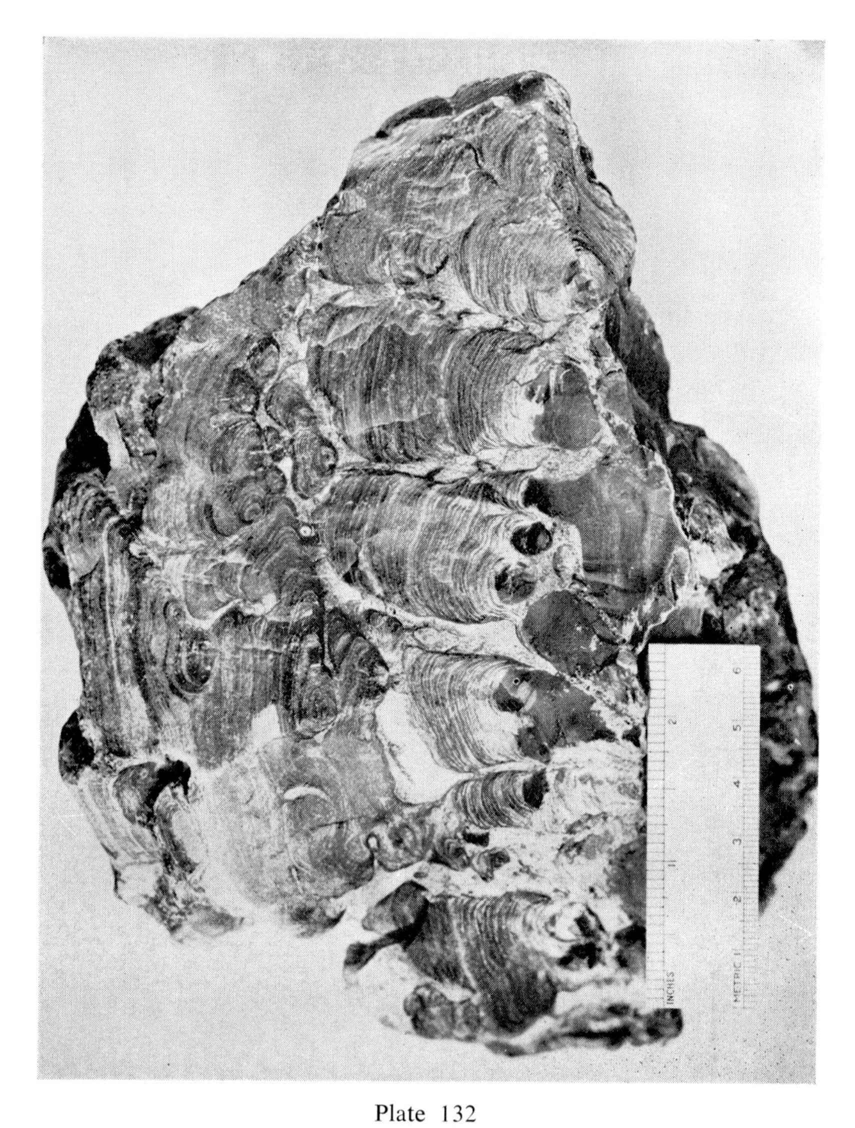

Plate 132

Algal Limestone — Eocene

Figure 1. Fresh water algal limestone. Eocene, Green River beds (Wasatch), Sweetwater County, Wyoming.

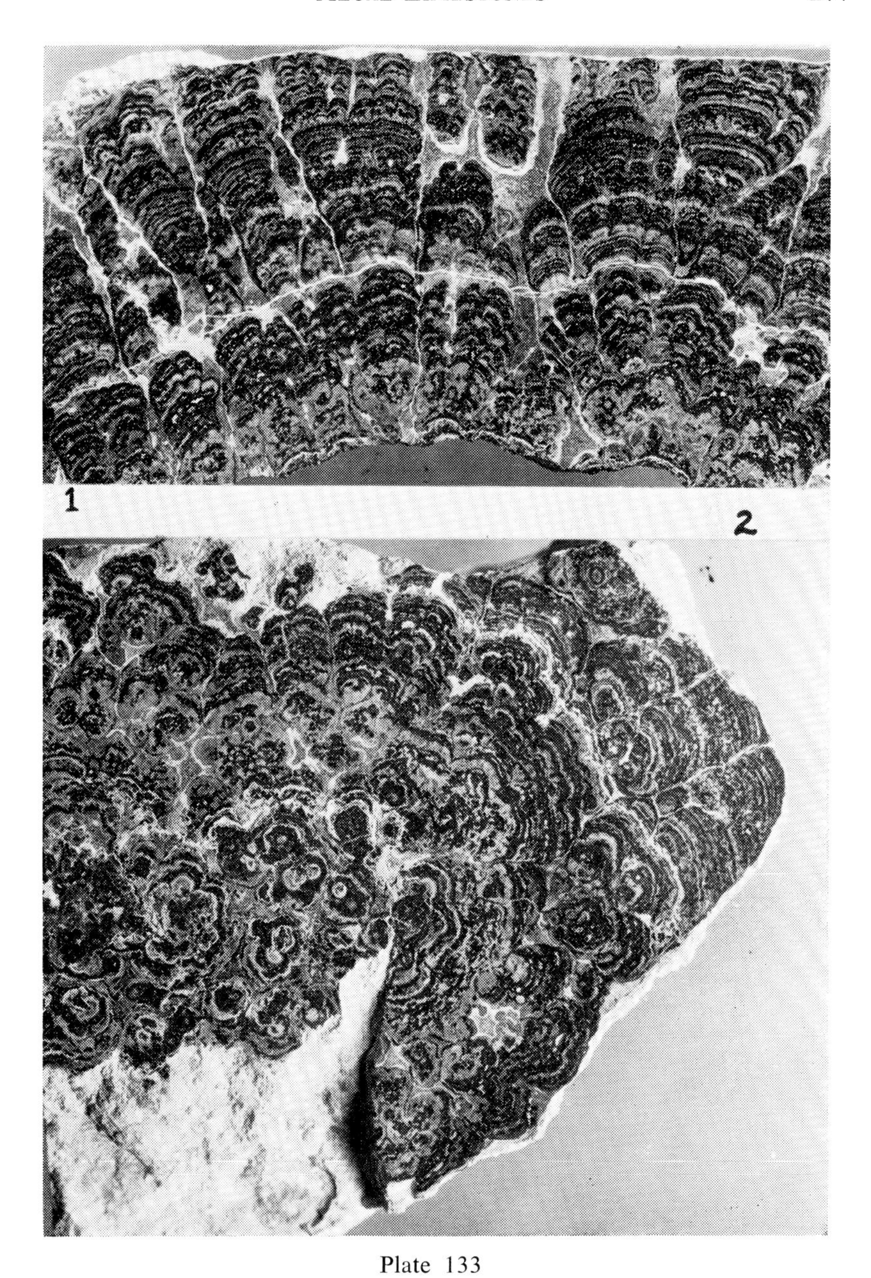

Plate 133
Algal Limestones — Oligocene
Figures 1-2. Oligocene fresh water algal limestone, South Park, Colorado. 1. Side view. 2. Top view.

Plate 134
Algal Limestone — Eocene
Figure 1. Eocene limestone with numerous crusts and fragments of branches of coralline algae, Ishigaki, Ryukyu Islands. (Published by permission, U. S. Geological Survey.)

Plate 135
Algal Limestone — Eocene
Figure 1. Eocene limestone composed largely of fragments of crustose coralline algae (x1), Ishigaki, Ryukyu Islands. (Published by permission, U. S. Geological Survey.)

Plate 136
Algal Limestone — Miocene
Figure 1. *Halimeda* limestone (x1), Miocene, Saipan, Mariana Islands. (Published by permission, U. S. Geological Survey.)

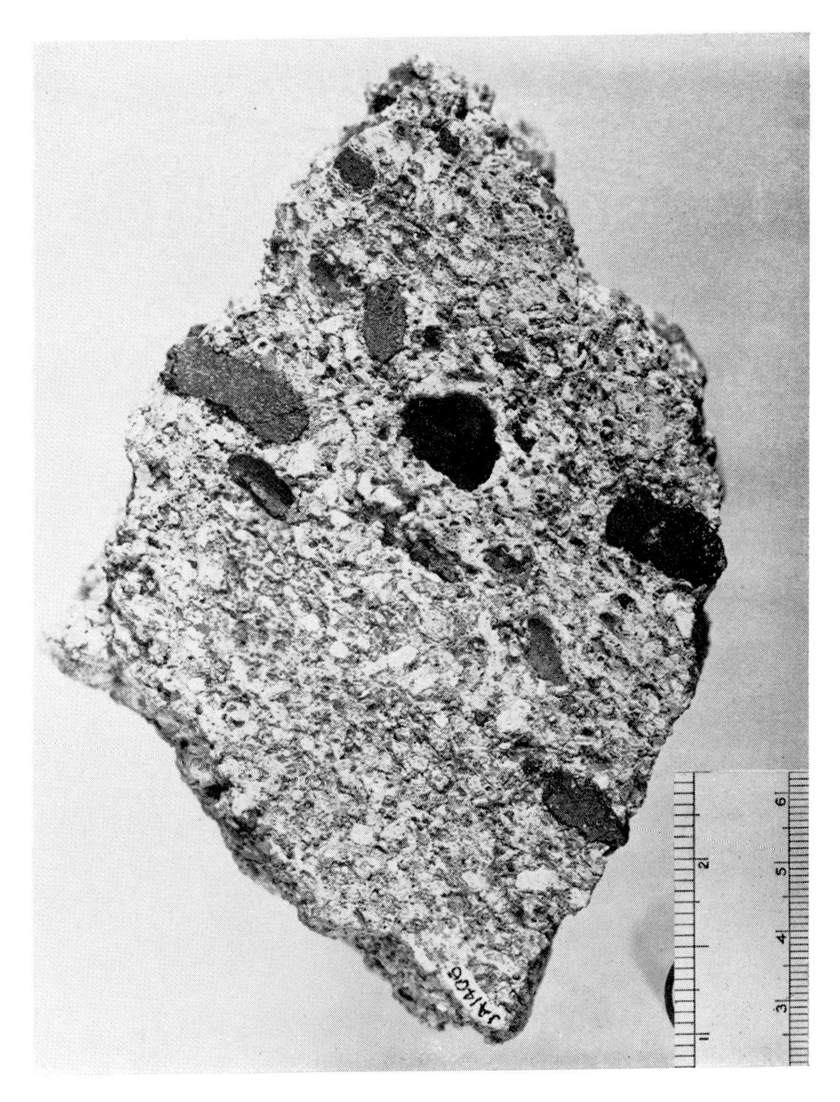

Plate 137
Algal Limestone — Pleistocene
Figure 1. Pleistocene limestone composed largely of coralline algae. Alameda Quarry, California.

Plate 138
Algal Ridge
Figures 1-2. The "Algal Ridge" at Bikini Atoll, at unusually low tide. (Photo by J. H. Johnson.)

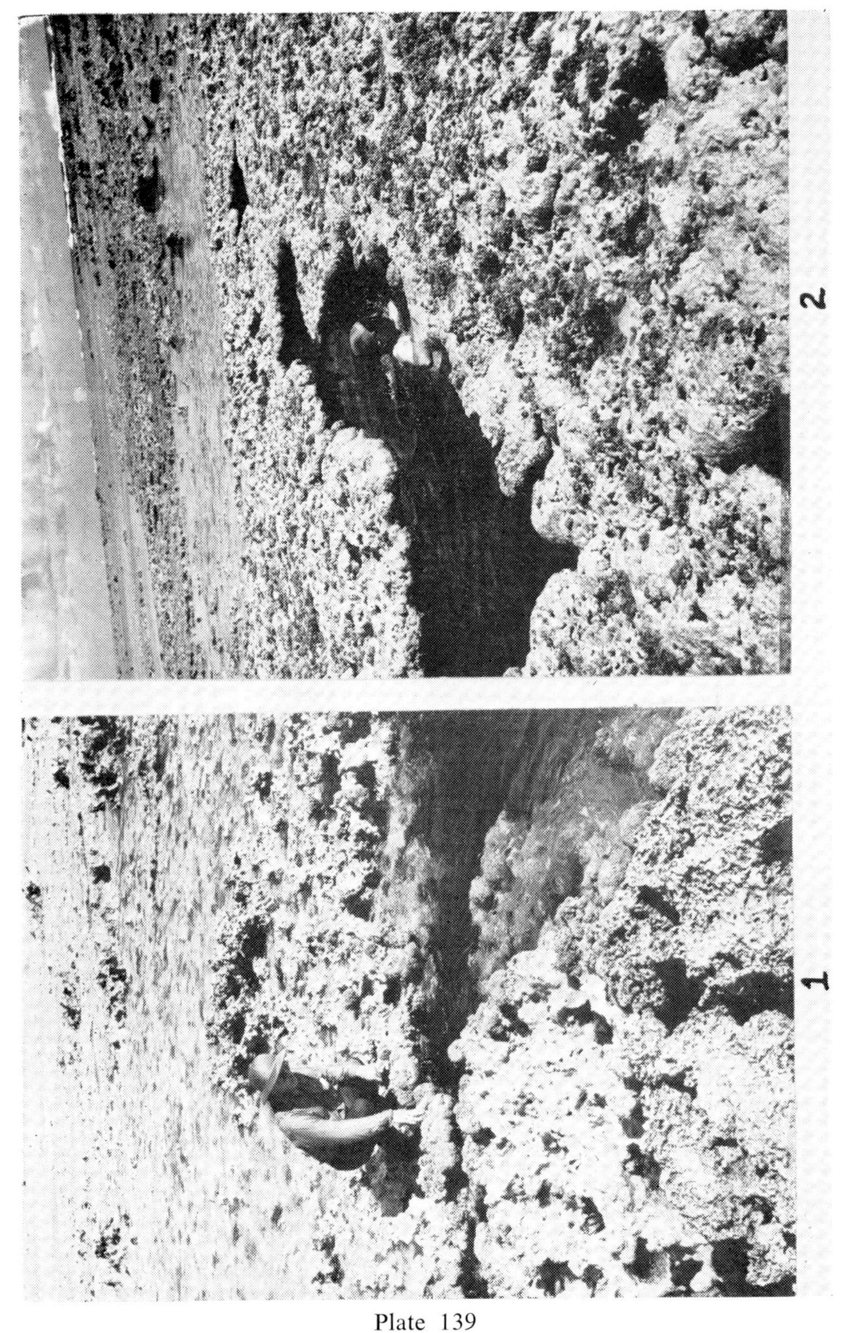

Plate 139
The Algal Ridge

Figure 1. The author on the inner part of the algal ridge on a crust being built over a surge channel by coralline algae. Some rounded heads visible in the water and around the margins.

Figure 2. The algal growth around the end of a surge channel. Reef along Bikini Island at unusually low tide. (Official photographs, U.S. Navy.)

APPENDIX

List of Genera by Geological Periods

Precambrian

Antholithina Choubert and Termier
Archaeozoon Matthew
Collenia Walcott
Cooperia Choubert and Termier
Cryptozoon Hall
Hadrophycus Fenton and Fenton
Pustularia Vologdin
Tazenakhtia Choubert and Termier
Tubiphyton Choubert and Termier

Cambrian

Actinophycus Korde
Amgaella Korde
Anomalophycus Fenton and Fenton
Aulophycus Fenton and Fenton
Bogutschanophycus Korde
Boroldaiphycus Vologdin
Bosworthia Walcott
Botomaella Korde
Cambroporella Korde
Chabakovia Vologdin
Coelenterella Korde
Collenia Walcott
Conophyton Maslov
Copperia Walcott-Maslov
Cryptozoon Hall
Dalyia Walcott
Dictophycus Korde
Dolatophycus Fenton and Fenton
Epiphyton Bornemann
Girvanella Nicholson and Etheridge
Globuloella Korde
Kordeophycus Johnson = **Leptophycus** Korde
Lenaella Korde
Marpolia Walcott
Mejerella Korde
Morania Walcott
Nemaphycus Korde
Nephelostroma Dangeard and Dore
Nubecularities Maslov
Palaeomicrocystus Korde
Palaeoporella Stolley
Palaeorivularia Korde
Poecilophycus Korde
Razumovskia Vologdin
Renalcis Vologdin
Schodackia Ruedemann
Seletonella Korde
Siberiella Korde
Spongiophycus Korde
Stereophycus Korde
Tetonophycus Fenton and Fenton
Thaumatophycus Korde
Vologdinella Korde
Wahpia Walcott
Waputikia Walcott
Yuknessia Walcott

Ordovician

Ajakmalajsoria Korde
Anomalophycus Fenton and Fenton
Apidium Stolley
Buthotrephis Hall
Callithamniopsis Whitfield
Chaetocladus Whitfield
Coelosphaeridium Roemer
Collenia Walcott
Cryptozoon Hall
Cyclocrinus Eichwald (emend.)
Dasyporella Stolley
Dimorphosiphon Høeg
Girvanella Nicholson and Etheridge
Hedstroemia Rothpletz
Katangasia Maslov
Kazakhstanelia Korde
Mastopora Eichwald
Microweedia Toots
Palaeoporella Stolley
Parachaetetes Deninger
Petrophyton Yabe
Primicorallina Whitfield
***Pseudochaetetes** Haug = **Solenopora** Dybowski

Rhabdoporella Stolley
Schodackia Ruedemann
Solenopora Dybowski
Uralella Korde
Vermiporella Stolley

Silurian

Buthotrephis Hall
Callisphenus Høeg
Collenia Walcott
Cryptozoon Hall
Cyclocrinus Eichwald (emend.)
Dasyporella Stolley
Girvanella Nicholson and Etheridge
Goldsonia Shrock and Twenhofel
Hedstroemia Rothpletz
Katangasia Maslov
Mastopora Eichwald
Pachytheca (Hooker) Salter
Palaeodictyota Whitfield
Palaeoporella Stolley
Parachaetetes Deninger
Plectenchymella Kraicz
Prototaxites Dawson
Primicorallina Whitfield
***Pseudochaetetes** Haug = **Solenopora** Dybowski
Rhabdoporella Stolley
Rothpletzella Wood
Solenopora Dybowski
Spongiostroma Gürich
Vermiporella Stolley
Wetheredella Wood

Devonian

Abacella Maslov
Amicus Maslov
Bicorium Maslov
Catena Maslov
Coactilum Maslov
Coelotrochium Schlüter
Drydenia Fry and Banks
Enfieldia Fry and Banks
Girvanella Nicholson and Etheridge
Hedstroemia Rothpletz
Hungerfordia Fry and Banks
Lancicula Maslov
Litanaia Maslov
Nematothallus Lang
Nostocites Maslov
Pachytheca (Hooker) Salter
Parachaetetes Deninger
Paradella Maslov
Parka Fleming
Protosalvinia Dawson
Prototaxites Dawson
***Pseudochaetetes** Haug = **Solenopora** Dybowski
Robertia Choubert
Solenopora Dybowski
Spongiostroma Gürich
Stenophycus C. L. Fenton
Sycidium G. Sandberger
Tasmanites Newton
Thamnocladus White
Trochiliscus Karpinsky
Uva Maslov

Mississippian

Aphanocapsites Maslov
Aphralysia Garwood
Aphrostroma Gürich
Atractyliopsis Pia
Bevocastria Garwood
Calcifolium Shvetzov and Birina
Chrondrostroma Gürich
Codonophycus Fenton and Fenton
Coelosporella Wood
Donezella Maslov
Dvinella Khvorova
Garwoodella Paul
Garwoodia Garwood
Girvanella Nicholson and Etheridge
Globulinea Ulke
Hedstroemia Rothpletz
Hikorocodium Endo
Koninckopora Lee
Malacostroma Gürich
***Mitcheldeania** Wethered = **Garwoodia** Wood
Nostocites Maslov
Orthriosiphon Johnson and Konishi
Ortonella Garwood
Palaeocodium Chiarugi
Parachaetetes Deninger
Paradella Maslov
Penicilloides Paul
Polymorphocodium Derville
***Pseudochaetetes** Haug = **Solenopora** Dybowski
Pycnostroma Gürich
Retephycus Johnson and Konishi
Solenopora Dybowski
Spongiostroma Gürich
Stipulella Maslov
Sycidium G. Sandberger
Tasmanites Newton
Trochiliscus Karpinsky

Pennsylvanian

Anchicodium Johnson
Anthracoporella Pia
Archaeolithophyllum Johnson
Artophycus Johnson
Atractyliopsis Pia
Beresella Maslov
Bevosolen Pia
Buzgulella Korde
Calcifolium Shvetzov and Birina
Calyptophycus Johnson
Chara Valliant
Cryptozoon ? Hall - Johnson
Cuneiphycus Johnson
Donezella Maslov
Dvinella Khvorova
Epimastopora Pia
Garwoodia Wood
Girvanella Nicholson and Etheridge
Gouldina Johnson
Hikorocodium Endo
Ivanovia Khvorova
Komia Korde = **Kordeophyton** Rezak
Leptophycus Johnson
Lithostroma Mamay
Macroporella Pia
Malacostroma Gürich
Nostocites Maslov
Oligoporella Pia
Ortonella Garwood
Osagia Twenhofel
Paleochara Bell
Parachaetetes Deninger
Petschoria Korde
Physoporella Steinmann
Pycnostroma Gürich
Samarella Maslov and Koulik
Shermanophycus Johnson
Solenophyllum Maslov
Somphospongia Beede
Spongiostroma Gürich
Stylophycus Johnson
Trinodella Maslov and Koulik
Tubiphytes Maslov (probably not algal)
Ungdarella Maslov
Unjäella Korde
Uraloporella Korde

Permian

Anchicodium Johnson
Anthracoporella Pia
Archaeocladus Endo
Archaeolithoporella Endo
Atractyliopsis Pia
Bevosolen Pia
Clavaphysoporella Endo
Clavaporella Kochansky and Herak
Collenella Johnson
Diplopora Schafhäutl
Eogoniolina Endo
Epimastopora Pia
Garwoodia Wood
Girvanella Nicholson and Etheridge
Gymnocodium Pia
Gyroporella Gümbel (emend.)
Hikorocodium Endo
Macroporella Pia
Malacostroma Gürich
Mizzia Schubert
Neoanchicodium Endo
Nipponophysoporella Endo
Oligoporella Pia
Ortonella Garwood
Osagia Twenhofel
Ottonosia Twenhofel
Parachaetetes Deninger
Permocalculus Elliott
Permopora Elias
Physoporella Steinmann
***Pseudochaetetes** Haug = **Solenopora** Dybowski
Pseudogyroporella Endo
Pseudovermiporella Elliott
Solenopora Dybowski
Succodium Konishi
Teutloporella Pia
Tubiphytes Maslov (probably a hydrozoan)

Triassic

Aciculella Pia
Diplopora Schafhäutl
Dobrogeites Simionescu
Girvanella Nicholson and Etheridge
Griphoporella Pia
Gyroporella Gümbel (emend.)
Holosporella Pia
Kantia Pia
Macroporella Pia
Oligoporella Pia
Parachaetetes Deninger
Permocalculus Elliott
Physoporella Steinmann
***Pseudochaetetes** Haug = **Solenopora** Dybowski
Sestrosphaera Pia
Solenopora Dybowski
Stellatochara Horn af Rantzien
Teutloporella Pia

Jurassic

Acicularia d'Archiac
Aclistochara Peck
Actinoporella Gümbel
Archaeolithothamnium Rothpletz
Archamphiroa Steinmann
Boueina Toula
Cayeuxia Frollo
Clavator Reid and Groves
Clypeina (Michelin)
Conipora d'Archiac
Cylindroporella Johnson
Echinochara Peck
Eurysolenopora Dietrich
Girvanella Nicolson and Etheridge
Goniolina d'Orbigny
Griphoporella Pia
Gyroporella Gümbel (emend.)
Latochara Mädler
Linoporella Steinmann
Lithoporella Foslie
Macroporella Pia
Marinella Pfender
Munieria Deecke
Neogyroporella Yabe and Toyama
Nipponophycus Yabe and Toyama
Nitellites Horn af Rantzien
Palaeocladus Pia
Parachaetetes Deninger
Permineste Harris
Permocalculus Elliott
Petrascula Gümbel
Polygonella Elliott
Praechara Horn af Rantzien
***Pseudochaetetes** Haug = **Solenopora** Dybowski
Pycnoporidium Yabe and Toyama
Salpingoporella Pia
Sestrosphaera Pia
Solenopora Dybowski
Sphaerochara Mädler
Stellatochara Horn af Rantzien
Stenogrammites Kretschetovitsch
Symploca (Kuetzing) Fremy and Dangeard
Teutloporella Pia
Thyrsoporella Gümbel
Triploporella Steinmann

Cretaceous

Acicularia d'Archiac
Aclistochara Peck
Actinoporella Gümbel
Algites Seward
Amphiroa Lamouroux (emend.)
Arabicodium Elliott
Archaeolithothamnium Rothpletz
Arthrocardia Decaisne (emend.)
Atopochara Peck
Boueina Toula
Broeckella L. and J. Morellet
Cayeuxia Frollo
Chara Valliant
Charaxis Harris
Chondrites Sternberg
Clavator Reid and Groves
Clypeina (Michelin)
Corallina Linnaeus
Cordilites Reuss
Cylindroporella Johnson
Cymopolia Lamouroux
Dissocladella Pia
Girvanella Nicholson and Etheridge
Griphoporella Pia
Gyroporella Gümbel (emend.)
Halimeda Lamouroux
Indopolia Pia
Jodotella L. and J. Morellet
Latochara Mädler
Lithocodium Elliott
Lithophyllum Philippi
Lithoporella Foslie
Lithothamnium Philippi
Mesolithon Maslov
Mesophyllum Lemoine
Microcodium Gluck
Munieria Deecke
Neomeris Lamouroux
Nipponophycus Yabe and Toyama
Orioporella Munier-Chalmas
Parachaetetes Deninger
Permineste Harris
Permocalculus Elliott
Petrophyton Yabe
Phormidioidea Wieland
Praechara Horn af Rantzien
Pseudolithothamnium Pfender
Pycnoporidium Yabe and Toyama
Salpingoporella Pia
Solenopora Dybowski
Sphaerochara Mädler
Stellatochara Horn af Rantzien
Stenoporidium Yabe and Toyama
Symploca (Kuetzing) Fremy and Dangeard
Tectochara Grambast and Grambast
Thyrsoporella Gümbel
Trinocladus Raineri
Triploporella Steinmann

Eocene

Acicularia d'Archiac
Amphiroa Lamouroux (emend.)
Archaeolithothamnium Rothpletz
Arthrocardia Descaine (emend.)
Avrainvilleopsis Forti
Belzungia L. Morellet
Broeckella L. and J. Morellet
Carpenterella Munier-Chalmas
Catellaria Maslov
Chara Valliant
Charaxis Harris
Clypeina (Michelin)
Corallina Linnaeus
Cymopolia Lamouroux
Dactylopora Lamarck
Dermatolithon Foslie
Digitella L. and J. Morellet
Distichoplax Pia (a coelenterate ?)
Ferganella Maslov
Furcoporella Pia
Griphoporella Pia
Halimeda Lamouroux
Jania Lamouroux
Jodotella L. and J. Morellet
Larvaria Defrance
Lemoinella L. and J. Morellet
Lithophyllum Philippi
Lithoporella Foslie
Lithothamnium Philippi
Maupasia Munier-Chalmas
Melobesia Lamouroux
Meminella L and J. Morellet
Mesophyllum Lemoine
Microcodium Glück
Montiella L. and J. Morellet
Morelletpora Varma
Neomeris Lamouroux
Ollaria Maslov
Orioporella Munier-Chalmas
Ovulites Lamarck
Pagodaporella Elliott
Parkerella Munier-Chalmas
Pediastrum Wilson and Hoffmeister
Pseudolithothamnium Pfender
Solenomeris Douville
Sphaerochara Mädler
Subterraniphyllum Elliott
Tectochara Grambast and Grambast
Terquemella Munier-Chalmas
Thyrsoporella Gümbel
Trinocladus Raineri
Uteria Michelin
Zittelina Munier-Chalmas

Oligocene

Acicularia d'Archiac
Amphiroa Lamouroux
Archaeolithothamnium Rothpletz
Arthrocardia Decaisne (emend.)
Chara Valliant
Charaxis Harris
Corallina Linnaeus
Cymopolia Lamouroux
Dermatolithon Foslie
Halimeda Lamouroux
Jania Lamouroux
Leptolithophyllum Airoldi
Lithophyllum Philippi
Lithoporella Foslie
Lithothamnium Philippi
Melobesia Lamouroux
Mesophyllum Lemoine
Microcodium Glück
Neomeris Lamouroux
Sphaerochara Mädler
Subterraniphyllum Elliott

Miocene

Acicularia d'Archiac
Aethesolithon Johnson
Amphiroa Lamouroux
Archaeolithothamnium Rothpletz
Arthrocardia Decaisne (emend.)
Calliarthron Manza
Chara Valliant
Corallina Linneaus
Cymopolia Lamouroux
Dermatolithon Foslie
Eulithothamnium Trabucco
Goniolithon Foslie
Halimeda Lamouroux
Jania Lamouroux
Lithocaulon Menegheni
Lithophyllum Philippi
Lithoporella Foslie
Lithothamnium Philippi
Melobesia Lamouroux
Mesophyllum Lemoine
Microcodium Gluck
Neomeris Lamouroux
Paraporolithon Johnson
Pomatophyllum Conti
Rivularialithus Maslov
Subterraniphyllum Elliott
Tenarea Bory

Pliocene

Acicularia d'Archiac
Amphiroa Lamouroux (emend.)
Archaeolithothamnium Rothpletz
Arthrocardia Descaisne (emend.)
Calliarthron Manza
Chara Valliant
Corallina Linnaeus
Cymopolia Lamouroux
Dermatolithon Foslie
Goniolithon Foslie
Halimeda Lamouroux
Jania Lamouroux
Lithophyllum Philippi
Lithoporella Foslie
Lithothamnium Philippi
Melobesia Lamouroux
Mesophyllum Lemoine
Neomeris Lamouroux
Porolithon Foslie

Pleistocene

Acicularia d'Archiac
Amphiroa Lamouroux
Archaeolithothamnium Rothpletz
Arthrocardia Descaisne (emend.)
Calliarthron Manza
Chara Valliant
Corallina Linnaeus
Cymopolia Lamouroux
Dermatolithon Foslie
Fosliella Howe
Goniolithon Foslie
Halimeda Lamouroux
Jania Lamouroux
Lithophyllum Philippi
Lithoporella Foslie
Lithothamnium Philippi
Melobesia Lamouroux
Mesophyllum Lemoine
Neomeris Lamouroux
Porolithon Foslie
Sclerothamnium Airoldi

GLOSSARY

The following terms are used in the descriptions and discussion of the algae.

Antheridia—the male sex organs.

Apical—the apex, or at the apex.

Arcuate—bent or curved like a bow.

Articulate—jointed.

Articulated corallines—those coralline algae having segmented thalli.

Aspondyl—irregular arrangement of primary branches (rays) of dasycladacean algae.

Attenuate—narrow and gradually tapering.

Axis—the stem-like portion bearing branches, or a primary filament of a branched filamentous thallus.

Cellular furrows—if the enveloping cells are only partially filled with calcium carbonate they are represented on the fossilized oogonia as furrows. (Among Charophyta.)

Central stem (of Dasycladaceae)—the main or central stem from which the primary branches arise.

Choristosporate—a highly specialized type of dasycladaceaen algae in which the reproductive cysts (sporangia) are formed in specialized gametangia, which may be specialized rays (branches) of the second or third order.

Circinate—curled downward from the apex.

Cladosporate—reproductive cysts (sporangia) grow in unmodified primary branches (rays) in dasycladacean algae.

Clavate—club-shaped.

Conceptacle—a cavity containing reproductive organs. In this paper usually a conceptacle of sporangia.

Coralline—referring to algae of the family Corallinaceae.

Cordate—heart-shaped, the point upwards or outwards.

Coronate—crowned, furnished with a crown.

Coronula—one or two tiers of small cells resting on the apical ends of the enveloping cells of Charophyta and forming a more or less erect elevated ring around the summit.

Coronulate—a term used to express the presence or direct indication of the former presence of coronula cells on the fossil oogonia of Charophyta.

Cortex—the thallus tissue between the epidermis and medulla; if no epidermis is present, the region surrounding the medulla or the axial filament.

Crustose—crust-like; in case of algae, the term restricted to thin thalli growing flattened against the substratum.

Cuneate—wedge-shaped.

Dentate—toothed.

Dextral—of, or to the right.

Dextrally spiraled—spiraled to the right.

Dichotomous—a two-fold branching.

Dichotomy—a forking into two similar parts.

Distal—terminal.

Distromatic—two cells in thickness (with refernce to blades) composed of two layers of cells (with reference to thalli).

Endosporate—Reproductive cysts (sporangia) are located in the central stem (stipe) among the Dasycladaceae.

Enveloping cells—the spirally twisted (meridional in Sycidiaceae) cells arising from the node-cell of Charophyta elongating to form a cover or envelope for the oospore.

Epidermis—the outermost cell layer of a thallus.

Epiphytic—growing upon another plant but attached to the surface only.

Epithallus—the outer or dermal portion of the tissue of coralline algae. It is very thin and commonly uncalcified. Seldom preserved in fossil material.
Epizoic—growing upon the surface of an animal.
Euspondyl—the regular arrangement of the primary branches (rays) into whorls (among the Dasycladaceae.)
Falcate—sickle-shaped.
Filament—a branched or unbranched row of cells joined end to end.
Fimbriate—fringed.
Flabellate—fan-shaped.
Foliaceous—leaf-like.
Fruiting—bearing sporangia or gametangia.
Gametangium—a sex organ containing a gamete or gametes.
Gamete—a sexual cell capable of uniting with another sexual cell.
Gametophyte—the sexual or gamete-bearing generation of an alga.
Genicula—a portion between two successive segments of erect, jointed coralline algae.
Gyrogonites—the small spiral calcified parts of the oogonia (Charophyta).
Hypothallus—the tissue of a polystromatic thallus is commonly differentiated into several zones or layers. The most important of these are the *hypothallus* and *perithallus.*
Hypothallus, basal—forms the basal portion of the plant and is below the perithallus in crustose forms.
Hypothallus, medullary—among some genera of the crustose corallines (especially *Lithophyllum*), and among the articulated corallines, the central portion of the branches and stems is composed of larger cells than the outer portion, and has a different arrangement of cells. This is the medullary hypothallus.
Intercellular furrows—if the calcification of the enveloping cells builds up higher than the lateral contacts of adjoining cells the area represented by the lateral sutures becomes furrows. (Charophyta.)
Intergenicula—a segment of an erect, jointed, coralline algae.
Internode—a segment of a jointed thallus.
Isodiametric—oblong or elongated.
Lanceolate—lance-shaped, narrow and tapering.
Ligulate—strap-like and short.
Linear—long and narrow, with sides parallel.
Medulla—the central tissue of an internally differentiated thallus.
Medullary—occurring in or belonging to the medulla.
Megacells—in some genera certain cells or groups of cells grow much larger than the surrounding ones. These are megacells.
Meridional units—the enveloping cells of the Sycidiaceae—twisted neither dextrally nor sinistrally. (Charophyta.)
Metaspondyl—primary branches (rays) occur in clusters of three or six; these clusters being regularly arranged in whorls (among the Dasycladaceae).
Monostromatic—one cell in thickness (with reference to blades). One layer of cells thick (with reference to thalli).
Node—the region between two successive joints of a jointed thallus.
Noncoronulate—a term used to describe the fossil oogonia of Charophyta on which there is no evidence of coronula cells. It does not mean that coronula cells were not present on the oogonia of the living plants, but only that they are not preserved.
Oligostromatic—composed of only a few layers of cells.
Oogonia—the female sex organs.
Oogonium—the female reproductive organ of the Charophyta—the only part of the plant represented in the trochiliscids.
Oosphere—the nonfertilized egg cell of the living Charophyta.
Oospore—the fertilized egg cell of the living Charophyta.
Oospore membrane—the hardened outer covering of the oospore of Charophyta, the only part of the oospore preserved in fossils.
Ovate—in the form of a longitudinal section of an egg and having the broader end toward the base.

Perithallus—the tissue above the basal hypothallus in crustose forms or outside the medullary hypothallus in branching, or articulated forms of coralline algae.
Pinnate—with leaflets or filaments on opposite sides of an axis; featherlike.
Pinnule—a secondary branch of a pinnately divided thallus.
Primary branches (of Dasycladaceae)—branches arising from the main (central) stem.
Polystromatic—composed of many layers of cells.
Rays—the branches of dasycladacean algae.
Rhizoid—a unicellular or multicellular filament functioning as an organ of attachment.
Rugose—wrinkled.
Saccate—sac-like or pouch-like.
Saxicolous—growing on rock.
Secondary branches (of Dasycladaceae)—branches or clusters of branches, arising from the end of the primary branches.
Secund—with branches restricted to one side of the structure bearing them.
Segment—a joint or node among the articulated corallines. A unit of the thallus among certain of the Dasycladaceae.
Septate—with transverse partitions.
Serrate—toothed like a saw and with the teeth pointing toward the apex.
Sinistral—of, or to the left.
Sinistrally spiraled—spiraling to the left.
Sporangia—spore cases in which the spores develop.
Sporangium—singular of sporangia.
Spore—a specialized motile or nonmotile cell which eventually becomes free from the parent plant and is capable of developing into a new plant.
Stipe—the central stem (central axis) of a dasycladacean alga.
Sympodially—a type of branching, where branches grow one upon another to produce a stem.
Sympodium—a stem made up of a series of branches growing on each other; giving the effect of a simple stem.
Tertiary branches (of Dasycladaceae)—branches or clusters of branches developing from the ends of the secondary branches.
Tetrasporangium—a sporangium whose contents divide to form four spores.
Thallus—the plant body of an alga. (Plural *thalli.*)
Utricle—a hull or envelope (commonly calcified) around the spiraled oogonium (among the Charophyte family Clavatoraceae).
Verticil—one whorl in a verticillate system of branching.
Verticillate—with successive whorls of branches along an axis that are arranged like the spokes of a wheel.
Whorl—a circle of branches, commonly about equally spaced, arising around stem, arranged like the spokes of a wheel (a verticil).

INDEX